South China Sea Seeps

Duofu Chen · Dong Feng
Editors

South China Sea Seeps

 Springer

Editors
Duofu Chen
College of Marine Sciences
Shanghai Ocean University
Shanghai, China

Dong Feng
College of Marine Sciences
Shanghai Ocean University
Shanghai, China

ISBN 978-981-99-1496-8   ISBN 978-981-99-1494-4   (eBook)
https://doi.org/10.1007/978-981-99-1494-4

This Springer imprint is published by the registered company Springer Nature Singapore Pte Ltd.
The registered company address is: 152 Beach Road, #21-01/04 Gateway East, Singapore 189721, Singapore

Preface

Following the creation of the term "冷泉" ("cold seeps" in Chinese) in 2002 by the lead editor and his colleagues of this book, such hydrocarbon seeps were found in the South China Sea. Since their discovery, the hydrocarbon seep systems of the South China Sea have become the most intensively studied of any seep communities in the world. The *South China Sea Seeps* is the first monograph that presents an overview of hydrocarbon seep systems in the South China Sea. Hydrocarbon seeps represent the expulsion of near-bottom water temperature, methane-rich gas, and fluids from subsurface reservoirs to the seafloor on continental margins. The subsurface seafloor of continental margins contains large reservoirs of methane, mainly as gas hydrate. This reservoir is dynamic and sensitive to climatic and tectonic perturbations. Marine hydrocarbon seeps are often found above deposits of gas hydrates. Characteristic manifestations of seepage at the seafloor include the release of gas bubbles through subsurface sediments, the development of chemosynthetic-based ecosystems, the precipitation of carbonate crusts, and the formation of mud volcanoes and pockmarks. As such, hydrocarbon seeps are thought to be windows into the deep geosphere and are a fundamental process of Earth's material cycling.

The *South China Sea Seeps* is aimed at students, researchers, governmental organizations, and professionals from the hydrocarbon industry. This book consists of a total of 14 articles written by esteemed researchers. The articles provide both new data and reviews of previously published data. This book opens with an overview and history on marine hydrocarbon seep studies in the South China Sea in Chap. 1 by one of the editors of the book, Dong Feng. The distribution and variability of hydrocarbon seeps are discussed in depth in Chap. 2 by Wei Zhang. Seismic studies of subsurface gas and fluid pathways that feed seafloor hydrocarbon seeps are discussed in Chap. 3 by Kun Zhang and his colleagues. Min Luo and Yuncheng Cao introduce gas hydrates in hydrocarbon seeps in Chap. 4 which, provides insight into the subsurface methane seepage dynamics and the fluctuation of gas hydrates reservoir. The fauna and symbioses of hydrocarbon seeps are discussed in Chaps. 5 and 6 by Yi-Xuan Li, Chaolun Li, and their colleagues. The elemental and isotopic compositions of seep bivalves are introduced by Xudong Wang and his colleagues in Chap. 7. In the next four Chaps. (8–11), Yu Hu, Shanggui Gong, Meng Jin, Hongxiang Guan, and their colleagues highlight our current understanding of the biogeochemical processes at hydrocarbon seeps and their fingerprints in the sedimentary record. The timing of hydrocarbon seep activities and their potential driving forces are discussed in Chap. 12 by Dong Feng. Chapter 13 by Niu Li and Junxi Feng introduces methane seeps in the southern South China Sea. The book concludes with Chap. 14 by Xin Zhang and his colleagues. They introduce seafloor observations at Site F, with the aim of bringing the laboratory to the field and taking only data back from the field.

Through these chapters, all authors have integrated multidisciplinary research results that span the interface between the geology, geophysics, geochemistry, and ecology of marine hydrocarbon seep systems in the South China Sea. They have summarized a large body of data on different seafloor phenomena, authigenic carbonates, and seep-dwelling megafauna by providing numerous illustrations and photographs of marine seep systems that many readers have not yet seen. In all chapters, particular attention has been given to the similarities and differences in the hydrocarbon seep systems between the South China Sea and other regions in the world. Likewise, this book provides a readily accessible background and context for rapidly evolving research foci, for example, the ongoing discoveries of the relationships between carbon cycling in seep systems and climate change; the identification of methanotrophic and thiotrophic endosymbionts in the geological past; and new insights into the interactions between the geosphere, hydrosphere, and biosphere in a highly dynamic environment. This volume provides a useful introduction, with the aim of stimulating further research interest in hydrocarbon seep geology, geochemistry, and biogeochemistry of the South China Sea and elsewhere.

We are grateful to all our colleagues who contributed to this volume by documenting the characteristics of hydrocarbon seep systems in the South China Sea.

Shanghai, China   Duofu Chen
                                                                                   Dong Feng

Acknowledgments

The many graduate students in our groups formerly at Guangzhou and now at Shanghai are thanked for continuous support and a source of new scientific results. We acknowledge the many incredible colleagues we have worked with over the years.

We are grateful to all our colleagues who contributed to this volume documenting the cold seep systems of the South China Sea continental slope. We specifically thank Dai Li and Mingzhu Lin for their editorial support.

We thank the many agencies that have funded our research during the past 20 years, in particular, the National Natural Science Foundation of China (42225603, 41730528), the National Key R&D Program of China, and Chinese Academy of Sciences.

Duofu Chen
Dong Feng

About This Book

This volume details the distribution and variability of the South China Sea hydrocarbon seeps; the geochemical processes involved; and their geological record, macroecology, and symbioses. While several publications exist that cover seeps around the world, the *South China Sea Seeps* constitute only a small component of the literature. As such, many geologists, geochemists, and biologists, as well as undergraduates and graduate students, are not very familiar with the hydrocarbon seeps in the South China Sea. This text is the first to comprehensively discuss the nature of the South China Sea hydrocarbon seeps and emphasize the similarities and differences in the seep systems between the South China Sea and those in other regions of the world. In addition to summarizing available knowledge on these topics for specialists in the field, this book offers the background needed to be of use to students as well as the wider community of geologists, geochemists, and biologists.

Contents

Editors and Contributors

About the Editors

Duofu Chen is a full Professor of Marine Geology in the College of Marine Sciences at Shanghai Ocean University. He received his Ph.D. in Geochemistry from Guangzhou Institute of Geochemistry, Chinese Academy of Sciences, where he was hired as full professor in 1998. He moved to Shanghai Ocean University in 2016 and established the research group of submarine cold seeps and gas hydrates. He created the term "冷泉" ("cold seeps" in Chinese) in 2002 and published the first paper on South China Sea seep in international journal *Marine and Petroleum Geology* in 2005. He has published over 100 peer-reviewed papers during last 20 years that mainly focused on submarine cold seeps and gas hydrates. Duofu was awarded National Science Fund for Distinguished Young Scholars in 2007. His studies have combined geochemical analyses and modeling approacher to reveal the dynamics of methane hydrates in marine sediments and the formation of mineral precipitates at seeps.

 **Dong Feng** is a Full Professor for Marine Geology and Geochemistry at Shanghai Ocean University. His research focuses on understanding the biogeochemical processes and geological record of submarine hydrocarbon seeps. He received his Ph.D. in Geochemistry from Guangzhou Institute of Geochemistry, Chinese Academy of Sciences, China (2008). He was a Post-Doctoral Fellow at Louisiana State University, USA (2008–2012). He became a Full Professor at South China Sea Institute of Oceanology, Chinese Academy of Sciences in 2012. He moved to Shanghai Ocean University in 2019. He has published over 100 peer-reviewed articles, with most focused on submarine hydrocarbon seeps, especially seeps in the South China Sea. He was awarded National Science Fund for Distinguished Young Scholars in 2022. His studies have combined geochemical analyses to establish the link between biogeochemical processes at present and their diagenetic imprint in the geological record.

Contributors

Yuncheng Cao College of Marine Sciences, Shanghai Ocean University, Shanghai, China

Hao Chen Center of Deep-Sea Research, and Key Laboratory of Marine Ecology and Environmental Sciences, Institute of Oceanology, Chinese Academy of Sciences, Qingdao, China;
Laboratory for Marine Ecology and Environmental Science, Qingdao National Laboratory for Marine Science and Technology, Qingdao, China

Jiangxin Chen Key Laboratory of Gas Hydrate, Ministry of Natural Resources, Qingdao Institute of Marine Geology, Qingdao, China

Zengfeng Du Key Lab of Marine Geology and Environment and Center of Deep Sea Research, Institute of Oceanology, Center for Ocean Mega-Science, Chinese Academy of Sciences, Qingdao, PR China

Dong Feng College of Marine Sciences, Shanghai Ocean University, Shanghai, China;
Laboratory for Marine Mineral Resources, Qingdao National Laboratory for Marine Science and Technology, Qingdao, China

Junxi Feng MLR Key Laboratory of Marine Mineral Resources, Guangzhou Marine Geological Survey, Guangzhou, China;

National Engineering Research Center of Gas Hydrate Exploration and Development, Guangzhou, China

Minghui Geng Key Laboratory of Marine Mineral Resources, Ministry of Natural Resources, Guangzhou Marine Geological Survey, Guangzhou, China

Shanggui Gong College of Marine Sciences, Shanghai Ocean University, Shanghai, China

Hongxiang Guan Frontiers Science Center for Deep Ocean Multispheres and Earth System, Key Lab of Submarine Geosciences and Prospecting Techniques, MOE and College of Marine Geosciences, Ocean University of China, Qingdao, China; Laboratory for Marine Mineral Resources, Qingdao National Laboratory for Marine Science and Technology, Qingdao, China

Yu Hu College of Marine Sciences, Shanghai Ocean University, Shanghai, China

Jack Chi Ho Ip Department of Biology, Hong Kong Baptist University, Hong Kong, China

Meng Jin MNR Key Laboratory of Marine Mineral Resources, Guangzhou Marine Geological Survey, Guangzhou, China; National Engineering Research Center for Gas Hydrate Exploration and Development, Guangzhou, China

Steffen Kiel Department of Palaeobiology, Swedish Museum of Natural History, Stockholm, Sweden

Chaolun Li Center of Deep-Sea Research, and Key Laboratory of Marine Ecology and Environmental Sciences, Institute of Oceanology, Chinese Academy of Sciences, Qingdao, China; South China Sea Institute of Oceanology, Chinese Academy of Sciences, Guangzhou, China; University of Chinese Academy of Sciences, Beijing, China

Niu Li Key Laboratory of Ocean and Marginal Sea Geology, South China Sea Institute of Oceanology, Innovation Academy of South China Sea Ecology and Environmental Engineering, Chinese Academy of Sciences, Guangzhou, China

Sanzhong Li Frontiers Science Center for Deep Ocean Multispheres and Earth System, Key Lab of Submarine Geosciences and Prospecting Techniques, MOE and College of Marine Geosciences, Ocean University of China, Qingdao, China

Yi-Xuan Li Department of Biology, Hong Kong Baptist University, Hong Kong, China; Southern Marine Science and Engineering Guangdong Laboratory (Guangzhou), Guangzhou, China

Yi-Tao Lin Department of Biology, Hong Kong Baptist University, Hong Kong, China;

Southern Marine Science and Engineering Guangdong Laboratory (Guangzhou), Guangzhou, China

Boran Liu Laboratory of Coastal and Marine Geology, Ministry of Natural Resources, Third Institute of Oceanography, Xiamen, China

Lei Liu Frontiers Science Center for Deep Ocean Multispheres and Earth System, Key Lab of Submarine Geosciences and Prospecting Techniques, MOE and College of Marine Geosciences, Ocean University of China, Qingdao, China;
Laboratory for Marine Mineral Resources, Qingdao National Laboratory for Marine Science and Technology, Qingdao, China;
Key Laboratory of Gas Hydrate, Qingdao Institute of Marine Geology, Ministry of Natural Resources, Qingdao, China

Zhendong Luan Key Lab of Marine Geology and Environment and Center of Deep Sea Research, Institute of Oceanology, Center for Ocean Mega-Science, Chinese Academy of Sciences, Qingdao, PR China

Min Luo College of Marine Sciences, Shanghai Ocean University, Shanghai, China;
Laboratory for Marine Geology, Qingdao National Laboratory for Marine Science and Technology, Qingdao, China

Jörn Peckmann Institute for Geology, Center for Earth System Research and Sustainability, Universität Hamburg, Hamburg, Germany

Jian-Wen Qiu Department of Biology, Hong Kong Baptist University, Hong Kong, China;
Southern Marine Science and Engineering Guangdong Laboratory (Guangzhou), Guangzhou, China

Haibin Song State Key Laboratory of Marine Geology, School of Ocean and Earth Science, Tongji University, Shanghai, China

Yan Sun Center of Deep-Sea Research, and Key Laboratory of Marine Ecology and Environmental Sciences, Institute of Oceanology, Chinese Academy of Sciences, Qingdao, China;
Laboratory for Marine Ecology and Environmental Science, Qingdao National Laboratory for Marine Science and Technology, Qingdao, China

Yanan Sun Southern Marine Science and Engineering Guangdong Laboratory (Guangzhou), Guangzhou, China;
Department of Ocean Science, The Hong Kong University of Science and Technology, Hong Kong, China

Hao Wang Center of Deep-Sea Research, and Key Laboratory of Marine Ecology and Environmental Sciences, Institute of Oceanology, Chinese Academy of Sciences, Qingdao, China;
Laboratory for Marine Ecology and Environmental Science, Qingdao National Laboratory for Marine Science and Technology, Qingdao, China

Minxiao Wang Center of Deep-Sea Research, and Key Laboratory of Marine Ecology and Environmental Sciences, Institute of Oceanology, Chinese Academy of Sciences, Qingdao, China;
Laboratory for Marine Ecology and Environmental Science, Qingdao National Laboratory for Marine Science and Technology, Qingdao, China

Xudong Wang College of Marine Sciences, Shanghai Ocean University, Shanghai, China

Nengyou Wu Laboratory for Marine Mineral Resources, Qingdao National Laboratory for Marine Science and Technology, Qingdao, China;
Key Laboratory of Gas Hydrate, Qingdao Institute of Marine Geology, Ministry of Natural Resources, Qingdao, China

Ting Xu Southern Marine Science and Engineering Guangdong Laboratory (Guangzhou), Guangzhou, China;
Department of Ocean Science, The Hong Kong University of Science and Technology, Hong Kong, China

Kun Zhang State Key Laboratory of Marine Geology, School of Ocean and Earth Science, Tongji University, Shanghai, China

Wei Zhang Guangzhou Marine Geological Survey, Ministry of Natural Resources, Guangzhou, China

Xin Zhang Key Lab of Marine Geology and Environment and Center of Deep Sea Research, Institute of Oceanology, Center for Ocean Mega-Science, Chinese Academy of Sciences, Qingdao, PR China;
Laboratory for Marine Geology, Pilot Laboratory for Marine Science and Technology (Qingdao), Qingdao, China;
University of Chinese Academy of Sciences, Beijing, China

Zhaoshan Zhong Center of Deep-Sea Research, and Key Laboratory of Marine Ecology and Environmental Sciences, Institute of Oceanology, Chinese Academy of Sciences, Qingdao, China;
Laboratory for Marine Ecology and Environmental Science, Qingdao National Laboratory for Marine Science and Technology, Qingdao, China

Li Zhou Center of Deep-Sea Research, and Key Laboratory of Marine Ecology and Environmental Sciences, Institute of Oceanology, Chinese Academy of Sciences, Qingdao, China;
Laboratory for Marine Ecology and Environmental Science, Qingdao National Laboratory for Marine Science and Technology, Qingdao, China

Chapter 1
A History of South China Sea Hydrocarbon Seep Research

Dong Feng

Abstract As of approximately two decades after the first discovery of marine hydrocarbon seep systems in the 1980s, a number of hydrocarbon seep sites have been found in the South China Sea (SCS). During the past two decades, the SCS has become one of the areas in the world with the most intensive studies on hydrocarbon seep systems. The first major breakthrough was made in 2004, when the "Jiulong methane reef", a large chemoherm carbonate build-up, was discovered during the Chinese–German research cruise in the NE Dongsha area. Continuous exploration in the following ten years has significantly enhanced the understanding of the SCS hydrocarbon seeps, e.g., their distribution, magnitudes, fluid sources, and ages. The second major breakthroughs were achieved during 2013–2015, with the discovery of active cold seeps from Site F to Yam to Haima by submersible vehicles. These active cold seeps have been revisited by remotely operated vehicles, *Faxian*, *Haima*, and *ROPOS*, and a manned submersible, *Deep Sea Warrior*. Submarine vehicles and robots are now essential for scientists to conduct multidisciplinary studies of seeps. South China Sea hydrocarbon seeps have received increasing scientific attention and are now among the best-studied seep areas globally. This chapter introduces the history of the study of SCS hydrocarbon seeps.

1.1 Introduction

Cold seeps are seafloor manifestations of methane-rich fluid migration from the sedimentary subsurface to the seabed and into the water column, and ultimately, some of the methane may even reach the atmosphere (Boetius and Wenzhöfer 2013). Marine hydrocarbon seeps are common features of continental margins worldwide (Suess 2020). Because of their relevance for oceanic emissions of the greenhouse gas

D. Feng (✉)
College of Marine Sciences, Shanghai Ocean University, Shanghai 201306, China
e-mail: dfeng@shou.edu.cn

Laboratory for Marine Mineral Resources, Qingdao National Laboratory for Marine Science and Technology, Qingdao 266061, China

D. Chen and D. Feng (eds.), *South China Sea Seeps*,
https://doi.org/10.1007/978-981-99-1494-4_1

1

methane, widespread chemosynthesis-based ecosystems, and their spatial overlap with areas containing gas hydrate, hydrocarbon seeps have been considered ideal natural laboratories for studying Earth's mass and energy cycling from the geosphere to the exosphere (Boetius and Wenzhöfer 2013; Suess 2020). The term "冷泉" (the term "cold seep" in Chinese) was first introduced by Duofu Chen and his colleagues in 2002, as published in the first issue of the journal Acta Sedimentologica Sinica (Fig. 1.1). Since then, great progress has been made in regard to the hydrocarbon seeps in the South China Sea (SCS). To date, various types of samples, including authigenic minerals, seep-impacted sediments, and seep fauna, have been recovered from more than 40 seep sites, which cover a wide range of water depths in both the northern and southern continental margins of the SCS (Fig. 1.2).

1.2   South China Sea Hydrocarbon Seep Studies

Following their discovery in the 1980s on the Florida Escarpment in the Gulf of Mexico at a water depth of 3200 m (Paull et al. 1984), seeps have been widely found in the world's oceans (Ceramicola et al. 2018; Suess 2020). Seafloor reflectivity or amplitude analysis using 2D/3D seismic datasets has been used for identifying and characterizing cold seeps since the late 1990s (Sun et al. 2011, 2012, 2013, 2017; Chen et al. 2015, 2018; Kunath et al. 2022). Studies on hydrocarbon seeps in the SCS have been driven by growing energy demands, specifically for gas hydrate resources, since the late 1990s. The presence of cold seeps in the SCS was first confirmed in June 2004 during the joint Chinese–German RV *SONNE* Cruise 177 (SO 177; Fig. 1.3; Suess et al. 2005). The cooperative research project between the Guangzhou Marine Geological Survey (GMGS) and Leibniz-Institute für Meereswissenschaften Kiel (IFM-GEOMAR) led to the discovery of a large area of methane-derived carbonate buildups, which are byproducts of hydrocarbon seepage. Carbonate buildups were found in three ridge crest segments (Site 1, Site 2, and Site 3) of the NE Dongsha area, which collectively cover approximately 400 km^2 (Han et al. 2008). Moreover, additional methane-derived carbonates were dredged from the NE Dongsha area and SW Taiwan area of the SCS (Chen et al. 2005, 2006; Lu et al. 2005). Analyses of the carbonate samples, seep-impacted sediments (including pore waters), and animals have produced abundant data that have led to the accumulation of background knowledge on the SCS hydrocarbon seeps, e.g., the sources of seep fluids (Chen et al. 2005, 2006; Han et al. 2008; Yu et al. 2008; Feng and Chen 2015; Feng et al. 2015; Hu et al. 2017, 2018), hydrocarbon–mineral–element–microbe associations (Birgel et al. 2008; Ge et al. 2010, 2011, 2015; Guan et al. 2013, 2016, 2018; Han et al. 2013; Wang et al. 2014; Chen et al. 2016; Li et al. 2016; Lin et al. 2016a, b, c, d; 2017; Lu et al. 2017, 2018; Feng et al. 2018a; Gong et al. 2018, 2022; Wang et al. 2018, 2020; Yang et al. 2018), the timing of seepage, and driving forces (Tong et al. 2013; Han et al. 2014; Luo et al. 2014, 2015; Feng and Chen 2015; Wang et al. 2022a, b). Beyond documenting new seep systems, recent advances relate to seep footprints. Representative manifestations of seepage at the seafloor in the

Fig. 1.1 Upper left: The Cover image of the first issue of the 2002 journal *Acta Sedimentologica Sinica*. Upper right: Front page of the manuscript on which the term "冷泉" ("cold seeps" in Chinese) was created by the lead editor and his colleagues of this book. Lower: Duofu Chen and Xianpei Chen (with hat; deceased). Photo: Qinxian Wang

SCS are mud volcanoes, pockmarks, and carbonate deposits. The spatial distribution and morphology of these phenomena may provide information on the nature of the fluids, the conditions of their formation and evolution, and postformation secondary processes (e.g., Feng et al. 2018a).

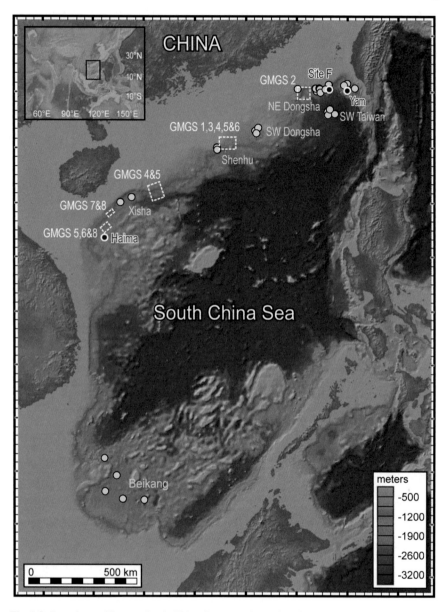

Fig. 1.2 Locations of known South China Sea seep sites (after Feng et al. (2018a); Wang et al. (2022a, b); and Tseng et al. (2023)). The active seep sites, Site F, Haima, and Yam are highlighted by red circles. The locations for gas hydrate drilling expeditions (GMGS 1–8) are marked by white boxes (after Yang et al. (2017) and Wei et al. (2020))

Fig. 1.3 The German research vehicle SONNE at the South China Sea in 2004. "Rocks on deck" on the RV SONNE during SO177. Photos curtesy of Jun Tao

The second major breakthrough on hydrocarbon seeps that involves detailed investigations using manned submersibles and remotely operated vehicles (ROV) started in the late 2000s (Lin et al. 2007; Machiyama et al. 2007). These underwater vehicles and robots have been routinely applied to examine seep systems since the early 2010s (Feng and Chen 2015; Feng et al. 2015, 2018a). At this stage, significant progress includes the detailed investigation of the known active seep Site F and the discovery of new active seeps: Haima and Yam (Table 1.1). The manned submersible *Jiaolong* was deployed in June–July 2013 to conduct a systematic survey and sampling of the active seep site, Site F, in the northern SCS. In 2015, another active cold seep, named the 'Haima seep', was discovered in the Qiongdongnan Basin of the SCS during the dives of the ROV *Haima* (Liang et al. 2017). These active seep sites were revisited annually from 2015 to 2022 by dives using remotely operated vehicles *Faxian* and *Haima* at Site F and the Haima seep, respectively (Liang et al. 2017; Zhang et al. 2017). The focus of these dives was to explore the interactions of fluids and chemosynthetic communities and to conduct in situ experiments on biogeochemical processes at hydrocarbon seeps (Fig. 1.4; Feng et al. 2018a, b; Wang et al. 2022a, b).

By linking subsurface and seafloor biospheres, seeps have become natural laboratories to enhance our knowledge of the processes of hydrocarbon migration and associated hydrocarbon-based ecosystems in the marine environment. Extensive efforts

Table 1.1 Information o the three active seeps in the South China Sea

Characteristic	Site F	Yam	Haima
Year of discovery	2007	2013	2015
Coordinates	22°07'N, 119°17'E	22°03'N, 119°48'E	16°43.8'N, 110°28.5'E
Water depth (m)	1120	1350	1350
Estimated area	32,000 m²	49,000 m²	350 km²
Timing of the seeps	since at least 57 ka BP	no data	since at least 21 ka BP
Origin of the methane	biogenic origin	biogenic origin	Mixture of biogenic and thermogenic origins
Seafloor manifestations	methane flare, mussel bed, carbonate pavement	methane flare, mussel bed, carbonate pavement	methane flare, mussel bed, clam bed, tubeworm bush, carbonate pavement
References	Machiyama et al. (2007); Feng and Chen (2015); Feng et al. (2015); Zhao et al. (2020); Wang et al. (2022a, b)	Klaucke et al. (2016); Tseng et al. (2023)	Liang et al. (2017); Yang et al. (2019); Wei et al. (2020); Wang et al. (2022a, b)

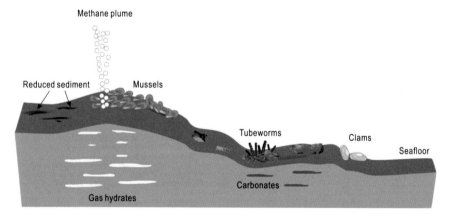

Fig. 1.4 Schematic illustration of hydrocarbon seeps in the South China Sea. Highly concentrated hydrate deposits provide a supply of methane that stimulates the entire ecosystem at cold seeps

on pore water studies from more than 250 sites have been made across the continental slope to detect hydrocarbon seeps in the SCS in the past two decades. The regional sulfate fluxes in the SCS highlight the importance of sulfate consumption fueled by deep-sourced methane, which is a factor that needs to be considered in any attempt to better quantify global fluxes of seawater sulfate in marine sediments (Hu et al. 2022, 2023; Chap. 8). In addition, drilling campaigns and coring programs have contributed greatly to our understanding of seep–hydrate dynamics on much larger timescales in the SCS (Fig. 1.2). These drilled cores merit further research to advance our understanding of the temporal evolution of seep systems (e.g., Chen et al. 2019; Wei et al. 2020). The macroecology of chemosynthesis-based ecosystems was first reviewed by Niu et al. (2017) and followed by Feng et al. (2018a) and Wang et al. (2022a, b). Documentation of which animals are present at the SCS seeps, how they interact with each other, and how the SCS seep animals are related to those from other seeps in a broader context have been important research areas during the last decade. Details about the macroecology of chemosynthesis-based ecosystems are presented in Chaps. 5 and 6.

Hydrocarbon seeps at the active margin of the SCS have also been extensively investigated in recent years (Fig. 1.5). For instance, elevated methane concentrations have been detected within seafloor sediments and bottom water (Chuang et al. 2010, 2013, 2019; Chen et al. 2017). Great progress was made in 2016, when Klaucke et al. (2016) investigated the Four-Way Closure Ridge and reported an active seep area, named the "Yam Seep", on top of the northern ridge (Tseng et al. 2023). Seafloor observations showed active gas emissions and the presence of extensive authigenic carbonate slabs and chemosynthetic bivalves (Klaucke et al. 2016). Different from the seep activities on the passive margin of the SCS that are closely related to gas hydrate dissociation during either changing bottom water temperature or seabed pressure (Tong et al. 2013; Chen et al. 2019; Deng et al. 2021; Wang et al. 2022a, b), hydrocarbon seeps on the active margins of the SCS are most likely caused by tectonic activity (Tseng et al. 2023).

In contrast to the northern SCS, seeps in the southern SCS are less well investigated. Feng et al. (2018c) and Li et al. (2018) provided the first report of cold seep activity on the southern continental slope of the SCS. The temporal variations in methane seepage were reconstructed using geochemical analyses of pore waters and sediments. Recently, Huang et al. (2022) provided a detailed description of deep-routed fluid seepage and inferred its indication for hydrates in the Beikang Basin. Zhang et al (2023) analyzed either seep carbonate samples collected from the Beikang Basin to reveal the fluid sources and sedimentary environments as well. More future work is needed to better characterize the seeps in the southern SCS.

The past two decades have witnessed significant progress in understanding a variety of aspects of the SCS cold seep systems. Cold seeps offer a great opportunity to examine complex geological, chemical, and biological processes from an interdisciplinary angle, thus requiring a philosophy that guides cold seep research from the perspective of Earth system science. Due to the scientific background of the editors and space limit, many interesting aspects, e.g., microbial ecology of the SCS seeps, are not discussed. The distinct microbial communities of the SCS seeps that are unique from other seeps and from other seafloor ecosystems are presented elsewhere (Niu et al. 2022). In addition, as seep studies rely heavily on innovative sampling and observation apparatuses, advancements in deep-sea technology will play a crucial role in future seep studies. In particular, *in-situ* (or on-site) observations on the seafloor have become an indispensable part of seep studies, e.g., they can further enhance the validity of the simulation model. We look forward to the next

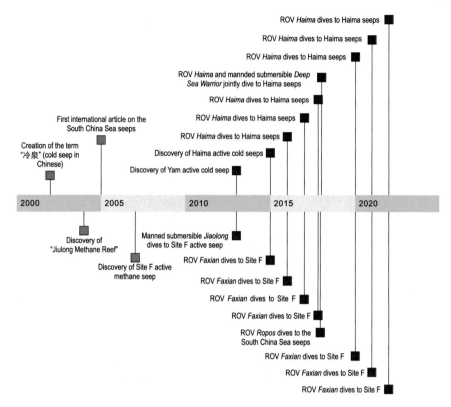

Fig. 1.5 Timeline of major milestones in South China Sea hydrocarbon seep research

decade of research on seep sites in the SCS and the advancement of knowledge that will result from it.

Acknowledgements Funding was provided by the NSF of China (Grants: 42225603 and 42176056). Zice Jia and Meng Jin are acknowledged for preparation of Fig. 1.4. Yu Hu and Min Luo are thanked for constructive comments, which have greatly improved the quality of the chapter.

References

Birgel D, Elvert M, Han XQ et al (2008) ^{13}C-depleted biphytanic diacids as tracers of past anaerobic oxidation of methane. Org Geochem 39(1):152–156

Boetius A, Wenzhöfer F (2013) Seafloor oxygen consumption fuelled by methane from cold seeps. Nat Geosci 6:725–734

Ceramicola S, Dupré S, Somoza L et al (2018) Cold seep systems. In: Micallef A, Krastel S, Savini A (Eds) Submarine Geomorphology. Springer, pp 367–387

Chen D, Huang Y, Yuan X et al (2005) Seep carbonates and preserved methane oxidizing archaea and sulfate reducing bacteria fossils suggest recent gas venting on the seafloor in the Northeastern South China Sea. Mar Pet Geol 22(5):613–621

Chen F, Hu Y, Feng D et al (2016) Evidence of intense methane seepages from molybdenum enrichments in gas hydrate-bearing sediments of the northern South China Sea. Chem Geol 443:173–181

Chen F, Wang X, Li N et al (2019) Gas hydrate dissociation during sea-level highstand inferred from U/Th dating of seep carbonate from the South China Sea. Geophys Res Lett 46(23):13928–13938

Chen J, Song H, Guan Y et al (2018) Geological and oceanographic controls on seabed fluid escape structures in the northern Zhongjiannan Basin, South China Sea. J Asian Earth Sci 168:38–47

Chen J, Song H, Guan Y et al (2015) Morphologies, classification and genesis of pockmarks, mud volcanoes and associated fluid escape features in the northern Zhongjiannan Basin, South China Sea. Deep-Sea Res Part II-Top Stud Oceanogr 122:106–117

Chen NC, Yang TF, Hong WL et al (2017) Production, consumption, and migration of methane in accretionary prism of southwestern Taiwan. Geochem Geophys Geosyst 18(8):2970–2989

Chen Z, Yan W, Chen M et al (2006) Discovery of seep carbonate nodules as new evidence for gas venting on the northern continental slope of South China Sea. Chin Sci Bull 51(10):1228–1237

Chuang PC, Yang T, Hong WL et al (2010) Estimation of methane flux offshore SW Taiwan and the influence of tectonics on gas hydrate accumulation. Geofluids 10(4):497–510

Chuang PC, Dale AW, Wallmann K et al (2013) Relating sulfate and methane dynamics to geology: Accretionary prism offshore SW Taiwan. Geochem Geophys Geosyst 14(7):2523–2545

Chuang PC, Yang TF, Wallmann K et al (2019) Carbon isotope exchange during anaerobic oxidation of methane (AOM) in sediments of the northeastern South China Sea. Geochim Cosmochim Acta 246:138–155

Deng Y, Chen F, Guo Q et al (2021) Possible links between methane seepages and glacialinterglacial transitions in the South China Sea. Geophys Res Lett 48(8): e2020GL091429

Feng D, Chen D (2015) Authigenic carbonates from an active cold seep of the northern South China Sea: New insights into fluid sources and past seepage activity. Deep-Sea Res Part II-Top Stud Oceanogr 122:74–83

Feng D, Cheng M, Kiel S et al (2015) Using Bathymodiolus tissue stable carbon, nitrogen and sulfur isotopes to infer biogeochemical process at a cold seep in the South China Sea. Deep-Sea Res Part I-Oceanogr Res Pap 104:52–59

Feng D, Qiu JW, Hu Y et al (2018a) Cold seep systems in the South China Sea: An overview. J Asian Earth Sci 168:3–16

Feng D, Peckmann J, Li N et al (2018b) The stable isotope fingerprint of chemosynthesis in the shell organic matrix of seep-dwelling bivalves. Chem Geol 479:241–250

Feng J, Yang S, Liang J et al (2018c) Methane seepage inferred from the porewater geochemistry of shallow sediments in the Beikang Basin of the southern South China Sea. J Asian Earth Sci 168:77–86

Ge L, Jiang SY, Swennen R et al (2010) Chemical environment of cold seep carbonate formation on the northern continental slope of South China Sea: Evidence from trace and rare earth element geochemistry. Mar Geol 277(1–4):21–30

Ge L, Jiang SY, Yang T et al (2011) Glycerol ether biomarkers and their carbon isotopic compositions in a cold seep carbonate chimney from the Shenhu area, northern South China Sea. Chin Sci Bull 56:1700–1707

Ge L, Jiang SY, Blumenberg M et al (2015) Lipid biomarkers and their specific carbon isotopic compositions of cold seep carbonates from the South China Sea. Mar Pet Geol 66:501–510

Gong S, Hu Y, Li N et al (2018) Environmental controls on sulfur isotopic compositions of sulfide minerals in seep carbonates from the South China Sea. J Asian Earth Sci 168:96–105

Gong S, Izon G, Peng Y et al (2022) Multiple sulfur isotope systematics of pyrite for tracing sulfate-driven anaerobic oxidation of methane. Earth Planet Sci Lett 597:117827

Guan H, Brigel D, Peckmann J et al (2018) Lipid biomarker patterns of authigenic carbonates reveal fluid composition and seepage intensity at Haima cold seeps, South China Sea. J Asian Earth Sci 168:163–172

Guan H, Feng D, Wu N et al (2016) Methane seepage intensities traced by biomarker patterns in authigenic carbonates from the South China Sea. Org Geochem 91:109–119

Guan H, Sun Y, Zhu X et al (2013) Factors controlling the types of microbial consortia in cold-seep environments: A molecular and isotopic investigation of authigenic carbonates from the South China Sea. Chem Geol 354:55–64

Han X, Suess E, Huang Y et al (2008) Jiulong methane reef: Microbial mediation of seep carbonates in the South China Sea. Mar Geol 249(3–4):243–256

Han X, Suess E, Liebetrau V et al (2014) Past methane release events and environmental conditions at the upper continental slope of the South China Sea: constraints from seep carbonates. Int J Earth Sci 103:1873–1887

Han X, Yang K, Huang Y (2013) Origin and nature of cold seep in northeastern Dongsha area, South China Sea: Evidence from chimney-like seep carbonates. Chin Sci Bull 58:3689–3697

Hu Y, Chen L, Feng D et al (2017) Geochemical record of methane seepage in authigenic carbonates and surrounding host sediments: A case study from the South China Sea. J Asian Earth Sci 138:51–61

Hu Y, Luo M, Chen L et al (2018) Methane source linked to gas hydrate system at hydrate drilling areas of the South China Sea: Porewater geochemistry and numerical model constraints. J Asian Earth Sci 168:87–95

Hu Y, Feng D, Peckmann J et al (2023) The crucial role of deep-sourced methane in maintaining the subseafloor sulfate budget. Geosci Front 14:101530

Hu Y, Zhang X, Feng D et al (2022) Enhanced sulfate consumption fueled by deep-sourced methane in a hydrate-bearing area. Sci Bull 67(2):122–124

Huang W, Meng M, Zhang W et al (2022) Geological, geophysical, and geochemical characteristics of deep-routed fluid seepage and its indication of gas hydrate occurrence in the Beikang Basin, Southern South China Sea. Mar Pet Geol 139:105610

Klaucke I, Berndt C, Crutchley G et al (2016) Fluid venting and seepage at accretionary ridges: the Four Way Closure Ridge offshore SW Taiwan. Geo-Mar Lett 36:165–174

Kunath P, Crutchley G, Chi WC et al (2022) Episodic Venting of a Submarine Gas Seep on Geological Time Scales: Formosa Ridge, Northern South China Sea. J Geophys Res-Solid Earth 127(9):e2022JB024668

Li N, Feng D, Chen LY et al (2016) Using sediment geochemistry to infer temporal variation of methane flux at a cold seep in the South China Sea. Mar Pet Geol 77:835–845

Li N, Yang X, Peng J et al (2018) Paleo-cold seep activity in the southern South China Sea: evidence from the geochemical and geophysical records of sediments. J Asian Earth Sci 168:106–111

Liang Q, Hu Y, Feng D et al (2017) Authigenic carbonates from newly discovered active cold seeps on the northwestern slope of the South China Sea: Constraints on fluid sources, formation environments, and seepage dynamics. Deep-Sea Res Part I-Oceanogr Res Pap 124:31–41

Lin S, Lim Y, Liu CS et al (2007) Formosa Ridge, A cold seep with densely populated chemosynthetic community in the passive margin, southwest of Taiwan. Geochim Cosmochim Acta 71:A582–A582 Suppl

Lin Q, Wang JS, Taladay K et al (2016a) Coupled pyrite concentration and sulfur isotopic insight into the paleo sulfate-methane transition zone (SMTZ) in the northern South China Sea. J Asian Earth Sci 115:547–556

Lin Q, Wang JS, Algeo TJ et al (2016b) Enhanced framboidal pyrite formation related to anaerobic oxidation of methane in the sulfate-methane transition zone of the northern South China Sea. Mar Geol 379:100–108

Lin Z, Sun X, Peckmann J et al (2016c) How sulfate-driven anaerobic oxidation of methane affects the sulfur isotopic composition of pyrite: A SIMS study from the South China Sea. Chem Geol 440:26–41

Lin Z, Sun X, Lu Y et al (2016d) Stable isotope patterns of coexisting pyrite and gypsum indicating variable methane flow at a seep site of the Shenhu area, South China Sea. J Asian Earth Sci 123:213–223

Lin Z, Sun X, Strauss H et al (2017) Multiple sulfur isotope constraints on sulfate-driven anaerobic oxidation of methane: evidence from authigenic pyrite in seepage areas of the South China Sea. Geochim Cosmochim Acta 211:153–173

Lu H, Liu J, Chen F et al (2005) Mineralogy and stable isotope composition of authigenic carbonates in bottom sediments on the offshore area of southwest Taiwan, South China Sea: evidence for gas hydrates occurrence. Earth Sci Frontiers 12(3):268–276 (in Chinese with English abstract)

Lu Y, Liu Y, Sun X et al (2017) Intensity of methane seepage reflected by relative enrichment of heavy magnesium isotopes in authigenic carbonates: a case study from the South China Sea. Deep-Sea Res Part I-Oceanogr Res Pap 129:10–21

Lu Y, Sun X, Xu H et al (2018) Formation of dolomite catalyzed by sulfate-driven anaerobic oxidation of methane: Mineralogical and geochemical evidence from the northern South China Sea. Am Mine 103(5):720–734

Luo M, Chen L, Tong H et al (2014) Gas hydrate occurrence inferred from dissolved Cl⁻ concentrations and $\delta^{18}O$ values of pore water and dissolved sulfate in the shallow sediments of the pockmark field in southwestern Xisha Uplift Northern South China Sea. Energies 7(6):3886–3899

Luo M, Dale AW, Wallmann K et al (2015) Estimating the time of pockmark formation in the SW Xisha Uplift (South China Sea) using reaction-transport modeling. Mar Geol 364:21–31

Machiyama H, Lin S, Fujikura K, Huang CY, Ku CY, Lin LH, Liu CS, Morita S, Nunoura T, Soh W, Toki T, Yang TF (2007) Discovery of "hydrothermal" chemosynthetic community in a cold seep environment, Formosa Ridge: seafloor observation results from first ROV cruise, off southwestern Taiwan. EOS Trans AGU 88 (52):OS23A-1041

Niu M, Liang Q, Feng D et al (2017) Ecosystems of cold seeps in the South China Sea. In: Kallmeyer J (Eds) Life at vents and seeps. Berlin, Boston, pp 139–160

Niu M, Deng L, Su L et al (2023) Methane supply drives prokaryotic community assembly and network at cold seeps of the South China Sea. Mol Ecol 32:660–679

Paull CK, Hecker B, Commeau R et al (1984) Biological communities at the Florida Escarpment resemble hydrothermal vent taxa. Science 226(4677):965–967

Suess E (2020) Marine cold seeps: background and recent advances. In: Wilkes (Ed) Hydrocarbons, oils and lipids: diversity, origin, chemistry and fate. Springer International Publishing, Cham, pp 747–767

Suess E, Huang Y, Wu N et al (2005) South China Sea: distribution, formation and effect of methane & gas hydrate on the environment. RV SONNE cruise report SO 177, Sino-German Cooperative Project, vol IFM-GEOMAR report no. 4. IFM-GEO-MAR, Kiel. https://oceanrep.geomar.de/id/eprint/1989

Sun Q, Wu S, Hovland M et al (2011) The morphologies and genesis of mega-pockmarks near the Xisha Uplift South China Sea. Mar Pet Geol 28(6):1146–1156

Sun Q, Wu S, Cartwright J et al (2012) Shallow gas and focused fluid flow systems in the Pearl River Mouth Basin, northern South China Sea. Mar Geol 315–318:1–14

Sun Q, Wu S, Cartwright J et al (2013) Focused fluid flow systems of the Zhongjiannan Basin and Guangle Uplift South China Sea. Basin Res 23(1):97–111

Sun Q, Cartwright J, Lüdmann T et al (2017) Three-dimensional seismic characterization of a complex sediment drift in the South China Sea: Evidence for unsteady flow regime. Sedimentology 64(3):832–853

Tong H, Feng D, Cheng H et al (2013) Authigenic carbonates from seeps on the northern continental slope of the South China Sea: new insights into fluid sources and geochronology. Mar Pet Geol 43:260–271

Tseng Y, Römer M, Lin S et al (2023) Yam Seep at four-way closure ridge–a prominent active gas seep system at the accretionary wedge SW offshore Taiwan. Int J Earth Sci 112:1043–1061

Wang S, Yan W, Chen Z et al (2014) Rare earth elements in cold seep carbonates from the southwestern Dongsha area, northern South China Sea. Mar Pet Geol 57:482–493

Wang M, Chen T, Feng D et al (2022a) Uranium-thorium isotope systematics of cold-seep carbonate and their constraints on geological methane leakage activities. Geochim Cosmochim Acta 320:105–121

Wang X, Guan H, Qiu J-W et al (2022b) Macro-ecology of cold seeps in the South China Sea. Geosyst Geoenviron 1(3):100081

Wang X, Li N, Feng D et al (2018) Using chemical compositions of sediments to constrain methane seepage dynamics: a case study from Haima cold seeps of the South China Sea. J Asian Earth Sci 168:137–144

Wang X, Barrat JA, Bayon G et al (2020) Lanthanum anomalies as fingerprints of methanotrophy. Geochem Perspect Lett 14:26–30

Wei J, Wu T, Zhang W et al (2020) Deeply Buried authigenic carbonates in the Qiongdongnan Basin, South China Sea: implications for ancient cold seep activities. Minerals 10(12):1135

Yang K, Chu F, Zhu Z et al (2018) Formation of methane-derived carbonates during the last glacial period on the northern slope of the South China Sea. J Asian Earth Sci 168:173–185

Yang M, Gong L, Sui J et al (2019) The complete mitochondrial genome of *Calyptogena marissinica* (Heterodonta: Veneroida: Vesicomyidae): insight into the deep-sea adaptive evolution of vesicomyids. PLoS ONE 14(9):e0217952

Yang S, Liang J, Lei Y et al (2017) GMGS4 gas hydrate drilling expedition in the South China Sea. Fire in the Ice 17:7–11

Yu X, Han X, Li H et al (2008) Biomarkers and carbon isotope composition of anaerobic oxidation of methane in sediments and carbonates of northeastern part of Dongsha, South China Sea. Acta Oceanol Sin 30:77–84 (in Chinese with English abstract)

Zhang X, Du Z, Luan Z et al (2017) In situ Raman detection of gas hydrates exposed on the seafloor of the South China Sea. Geochem Geophys Geosys 18(10):3700–3713

Zhang W, Chen C, Su P et al (2023) Formation and implication of cold-seep carbonates in the southern South China Sea. J Asian Earth Sci 241:105485

Zhao Y, Xu T, Law YS et al (2020) Ecological characterization of cold-seep epifauna in the South China Sea. Deep-Sea Res Part I-Oceanogr Res Pap 163:103361

Chapter 2
Distribution, Variability of Seeps

Wei Zhang

Abstract Various cold seep systems and related gas hydrate accumulations have been discovered in the South China Sea over the past two decades. Based on high-resolution seismic data, subbottom profiles, *in-situ* observations, deep drilling and coring, and hydrate gas geochemical analysis, the geological and geophysical characteristics of these cold seep systems and their associated gas hydrate accumulations in the Qiongdongnan Basin, Shenhu area and Dongsha area in the Pearl River Mouth Basin, Taixinan Basin, and Beikang Basin were investigated. Cold seep systems are present at different stages of evolution and exhibit various seabed microgeomorphic, geological, and geochemical features. Active cold seep systems with notable gas leakage, gas plumes, and microbial communities and inactive cold seep systems with authigenic carbonate pavements are related to the variable intensity of the gas-bearing fluid, which is usually derived from deep strata through mud volcanoes, mud diapirs, gas chimneys, and faults. This indicates a paragenetic relationship between the gas-bearing fluid and the seafloor morphology of cold seeps and deep-shallow coupling of gas hydrates, cold seeps, and deep petroleum reservoirs in the South China Sea.

2.1 Introduction

The South China Sea (SCS) is one of the largest marginal seas in the western Pacific Ocean and is located between the Eurasian plate, the Pacific Plate and the Indian Plate. The SCS is bounded by the South China mainland and Taiwan Island to the north, Indochina Peninsula and Malay Peninsula to the west, Sumatra and Kalimantan Islands of the Greater Sunda Islands to the south, and the Philippine Islands to the east (Fig. 2.1). The contour of the SCS is characterized by an irregular diamond shape, the longitudinal axis is approximately 3140 km long, the transverse axis is approximately 1250 km long, and the area reaches approximately 350×10^4 km^2. The average water depth is 1140 m, and the depth of the central basin is ~ 4200 m. The

W. Zhang (✉)
Guangzhou Marine Geological Survey, Ministry of Natural Resources, Guangzhou 510075, China
e-mail: zwgmgs@foxmail.com

© The Author(s) 2023
D. Chen and D. Feng (eds.), *South China Sea Seeps*,
https://doi.org/10.1007/978-981-99-1494-4_2

seabed topography of the SCS can be divided into the continental shelf, continental slope and deep-sea basin. There are many continental islands on the continental shelf, which are mainly composed of pre-Quaternary magmatic, metamorphic and sedimentary rocks adjacent to the mainland. Terrigenous materials originating from the Pearl River, the Red River, the Mekong River and other major rivers have all been imported into the SCS, which has also transported abundant organic materials into the SCS.

The northern SCS is a passive continental margin in which the Neogene tectonic activity intensity is high, and there are two main structural layers composed of Paleogene rift sequences and Neogene depression sequences. The abundant terrestrial clastic rocks derived from the Southern China and the Indochina Peninsula have resulted in Neogene sediment deposits up to ~ 10 km thick at the center of several petroliferous basins. The thick stratigraphic deposits and intense hydrocarbon generation and expulsion processes of the deep source kitchens have led to high fluid activity, forming a series of NE-trending basins that are rich in oil and gas, with widely distributed mud volcanoes, mud diapirs, gas chimneys, and submarine cold seeps (Chen et al. 2010; He et al. 2016; Su et al. 2016; Liang et al. 2019; Wan et al. 2019; Zhang et al. 2019a, b; 2020a). Numerous acoustic surveys, including 2D and 3D seismic surveys, subbottom profiles, and multibeam investigations, as well as extensive deep-sea diving observations, sampling, and gas hydrate drilling and coring (Matsumoto et al. 2011; Wang et al. 2014; Feng and Chen 2015; Zhang et al. 2017a, b; Liang et al. 2019; Wei et al. 2019; Zhang et al. 2020a, b), have been conducted in the SCS.

The number of cold seeps and associated gas hydrate accumulations discovered is continuously increasing. Since the discovery of the first cold seep system in the northern SCS in 2004, more than 40 seeps have been discovered in the Qiongdongnan Basin (QDNB), Xisha area, Shenhu area, Dongsha area, Taixinan Basin (TXNB), and Beikang Basin (BKB) (Fig. 2.1; Han et al. 2013; Berndt et al. 2014; Feng et al. 2018). Since 2007, the Guangzhou Marine Geological Survey (GMGS) has performed eight gas hydrate scientific drilling expeditions in the Shenhu area and the Xisha area (GMGS1, GMGS3 and GMGS4), the Dongsha area (GMGS2), and the Qiongdongnan Basin (GMGS5, GMGS6, GMGS7 and GMGS8) in the northern SCS (Yang et al. 2008; Wu et al. 2010, 2011; Wang et al. 2014; Sha et al. 2015a, b; Yang et al. 2015, 2017a, b, c; Zhang et al. 2015; Su et al. 2016; Liang et al. 2019; Wei et al. 2019; Zhang et al. 2020a, b). Except for the Xisha area, gas hydrate samples have been recovered in several regions where cold seep systems have been discovered. In addition, the Chinese Academy of Sciences has found exposed seafloor gas hydrates related to a submarine active cold seep in the Taixinan Basin (Feng and Chen 2015; Zhang et al. 2017b). These gas hydrate accumulations, which are related to the development and evolution of cold seep systems, are mainly distributed in the vicinity of the low uplift, submarine ridge, and channel–levee system and on the seabed above deep faults, large mud diapirs, and gas chimneys. Therefore, the development and evolution of cold seeps and the precipitation and accumulation of gas hydrates may be closely related to deep strata and petroleum systems. The objective of this chapter was to document the distribution and variability of hydrocarbon seeps in the SCS.

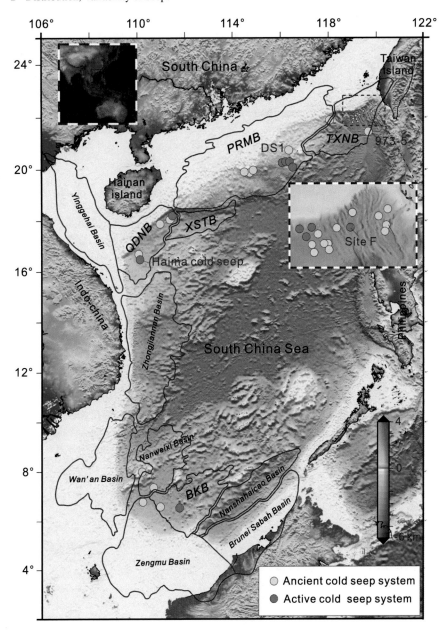

Fig. 2.1 Distribution of cold seeps in the marginal basins of the South China Sea. QDNB: Qiong-dongnan Basin; PRMB: Pearl River Mouth Basin; TXNB: Taixinan Basin; XSTB: Xisha Trough Basin; BKB: Beikang Basin

2.2 Distribution of Seeps

Based on comprehensive interpretation and analysis of 2D/3D seismic data, subbottom and multibeam data, and manned deep submersible and/or unmanned deep submersible (ROV) observations and sampling data, several regions of cold seep systems and associated gas hydrate accumulations have been discovered, and our understanding of cold seep systems in the SCS has been greatly improved (Schnürle et al. 2004; Suess 2005; Huang et al. 2006; Jiang et al. 2006; Liu et al. 2006; Han et al. 2008; Chen et al. 2014a, b; Feng and Chen 2015; Zhong et al. 2017; Feng et al. 2019; Wan et al. 2020; Cao et al. 2021).

In 2004, China and Germany jointly conducted the SO-177 Expedition in the deep water area near Dongsha in the eastern Pearl River Mouth Basin (PRMB), and for the first time, giant distributions of carbonate rocks, namely, the Jiulong Methane Reef, formed via seepage from cold seeps were discovered (Suess 2005; Han et al. 2008). In addition, the Jiulong Methane Reef was later confirmed as a gas hydrate accumulation area through deep gas hydrate drilling by the GMGS in 2013 (Sha et al. 2015a, b; Zhang et al. 2015; Zhong et al. 2017). In the same year, the *Jiaolong* manned deep submersible conducted its first scientific research voyage in the active cold seep area of the Site F in the TXNB, and intermittent activity of this cold seep system was initially identified (Feng and Chen 2015). In the deep waters of the Taixinan Basin, scholars have not only collected authigenic carbonates recording cold seep activity at five sites on the Gaoping Slope (Huang et al. 2006) but have also discovered very active cold seeps on the seafloor of the Formosa Ridge. In 2015, a giant active cold seep, namely, the Haima Cold Seep, was discovered by the GMGS in the southwestern QDNB. In 2018, a new submarine cold seep was discovered in the eastern QDNB. In addition, geophysical and geochemical evidence of cold seep activity was found in the BKB in the southern SCS through a geological survey (Huang et al. 2022).

Although dozens of cold seeps and associated gas seepages have been found in the SCS, active cold seeps have only been discovered in certain regions. Detailed results of the investigation of cold seeps in different regions are presented in Table 2.1 and addressed below.

2.3 Variability of Seeps

There are usually a series of microtopographic seabed features in cold seep and gas hydrate accumulation areas, such as slumps, pockmarks, mounds, mud volcanoes, depressions, and platforms (Kvenvolden 1993; Demirbas 2010; Matsumoto et al. 2011; Minshull et al. 2020). Through a comprehensive study of the seismic, subbottom profile, and ROV image data for the target area for cold seep and gas hydrate exploration in the SCS, various seismic and acoustic reflection features related to fluid seepage/leakage were recognized, including gas plumes, seabed

Table 2.1 Variability of the seeps discovered in the South China Sea

Location	Structure	Subsurface features	Seafloor features	Hydrate occurrence
Qiongdongnan Basin	Deep-sea plain Low uplift	Mud diapirs, gas chimneys, pipes faults, and bottom-simulating reflectors	Pockmarks, mounds, gas plumes, carbonate crusts, microbial communities, and bacterial mats	Massive, layers, nodules, fracture filling and disseminated
Shenhu area in the Pearl River Mouth Basin	Canyon Ridge Channel–levee	Mud diapirs, gas chimneys, pipes faults, and bottom-simulating reflectors	Pockmarks, gas plumes, and submarine slumps	Disseminated
Dongsha area in the Pearl River Mouth Basin	Canyon Ridge	Mud diapirs, gas chimneys, pipes faults, and bottom-simulating reflectors	Pockmarks, mounds, gas plumes, submarine slumps, and carbonate crusts	Massive, layers, nodules, fracture filling and disseminated
Taixinan Basin	Canyon Ridge	Mud diapirs, gas chimneys, pipes faults, and bottom-simulating reflectors	Pockmarks, mounds, gas plumes, submarine slumps, microbial communities, and carbonate crusts	Massive, layers, nodules, and fracture filling
Beikang Basin	Deep-sea plain	Mud volcanoes, mud diapirs, gas chimneys, pipes, and bottom-simulating reflectors	Pockmarks, mounds, possible gas plumes, and carbonate rocks	Nodules and disseminated

mounds, pockmarks, bright spots, acoustic blanking, and acoustic turbidity (Chen et al. 2010; Shang et al. 2013, 2014; Wang et al. 2014; Zhang et al. 2019b, 2020a, b; Cheng et al. 2020; Wu et al. 2020). In addition, microbial communities, bacterial mats, methane biochemical reefs, and carbonate crusts associated with cold seeps have been found in the QDNB, PRMB, and TXNB in the northern SCS and the BKB in the southern SCS (Fig. 2.1) (Chen et al. 2007; Feng and Chen 2015; Zhang et al. 2017b; Wei et al. 2020a, b; Zhang et al. 2023). These seafloor seepage phenomena, especially the discovery of gas plumes, suggest that a sufficient gas source exists in the study area. The appearance of many bivalves and bacterial mats is often a sign of a cold seep with active methane seepage, while carbonate crusts may indicate the cessation of methane seepage. These signs further indicate that gas hydrates are likely to accumulate in cold seep areas.

2.3.1 Qiongdongnan Basin

Based on fine interpretation of a 3D seismic profile in the QDNB, many geophysical anomalies closely related to gas seepage were identified in the deep-water area (Fig. 2.2). In addition, a large number of seabed pockmarks, mounds, and acoustic blanking reflections, which indicate the presence of free gas and gas hydrates, were identified in subbottom and seismic profile (Fig. 2.2a, b). The reflection of acoustic blanking suggests gas-bearing strata, which is consistent with the local structural high, indicating the accumulation characteristics of gas at high points. Small mounds, i.e., tens to hundreds of meters in diameter, were interpreted atop the observed acoustic blanking, indicating the accumulation and seepage of gas from depth. The presence of large flourishing biological communities often indicates the occurrence of current methane seepage, which is likely related to the dissociation of gas hydrates precipitated in shallow strata.

The bottom-simulating reflector (BSR) was clearly identified by the seismic anomalies resulting from gas migration along the low uplift, with a fuzzy seismic reflection zone below the BSR (Fig. 2.2c, f). Gas seepage pathways extending from the BSR to the seafloor were identified in the high-precision 3D seismic profile, which shows the pulled-up features of the events on both sides of the vertical pathways (Fig. 2.2c). Pulled-up reflectors are usually caused by the presence of a material that is harder than the surrounding strata since this hard material could cause velocity anomalies (Yoo et al. 2013). The gas hydrate drilling and sampling campaigns conducted during GMGS expeditions 5 and 6 demonstrated that hard materials, including gas hydrates and authigenic carbonate rocks precipitated within seepage/migration pathways, corresponding to pulled-up features (Liang et al. 2019). Piston and push coring processes also recovered gas hydrates and authigenic carbonate rocks from subsurface sediments at seepage sites in the QDNB (Liang et al. 2017). In addition, seepage pathways usually correspond to cold seep vents on the seafloor, and this phenomenon has been commonly observed in the Haima Cold Seep (Fig. 2.2c–e). Anomalies with various amplitudes associated with seafloor seepage were observed in the seismic profiles, mainly including high-amplitude bright spots associated with BSRs and large areas of blanking or chaotic reflection zones due to mud diapirs and/or gas chimneys, which promote gas migration and accumulation in deep strata (Zhang et al. 2019b).

The cold seeps in the QDNB developed atop strata overlying the low uplift of the pre-Paleogene basement, which resulted from intense magmatic intrusion (Wang et al. 2018a; Wei et al. 2020a). In addition, the faults and fractures formed by the uplift of structures generated active plumping systems (Wang et al. 2018a; Wei et al. 2020a) connected to the deep Paleogene source rocks, transporting deep hydrocarbons upward and resulting in the formation of gas chimneys and associated fluid leakage under overpressure conditions. Many thermogenic and biogenic gases migrated along these gas chimneys into the GHSZ to form fracture-filling gas hydrates in mass transport deposits (Liang et al. 2019). Analysis of the gas hydrate petroleum system revealed the close relationship between these shallow gas

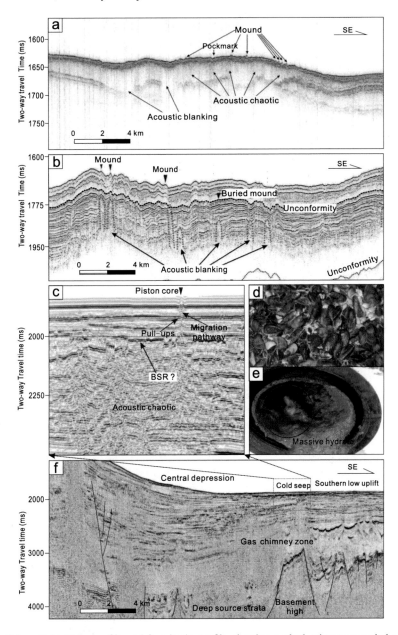

Fig. 2.2 a Subbottom profile and **b** seismic profile showing seabed microgeomorphological features related to gas seepage in the deep-water Qiongdongnan Basin. **c** Seismic profile showing hydrocarbon migration and seepage features of the Haima Cold Seep. **d** Microbial communities and **e** massive gas hydrates recovered from the Haima Cold Seep (modified from Liang et al. 2017). **f** Seismic profile showing deep structural features below the Haima Cold Seep in the southern low uplift area of the Qiongdongnan Basin. A large range of acoustic fuzzy zones resulting from gas chimneys can be observed in the seismic profile, indicating upward migration of deep hydrocarbons, supplying materials for gas seepage and gas hydrate accumulation

hydrates and deep conventional petroleum reservoirs. The deep central channel sand gas reservoir and deep source strata exhibit a direct vertical coupling relationship with the distribution of gas hydrates. The geochemical characteristics of the gas hydrates drilled during the GMGS5 expedition indicate that the thermogenic gas in the shallow gas hydrate accumulation area is consistent with the origin of natural gas in deep reservoirs such as LS17-2, LS22-1, and Y8-1 (Zhang et al. 2017c, 2019c; Lai et al. 2021; Zhang et al. 2020a, c). Therefore, it is believed that shallow gas hydrates are homologous with deep gas reservoirs. The gas-bearing fluid provided by the development and evolution of source rocks in the deep depression, the migration system comprising the low uplift due to tectonic activity, the associated faults, and the gas chimneys jointly controlled submarine cold seep system formation and gas hydrate accumulation in the QDNB.

2.3.2   Shenhu Area in the Middle PRMB

Similar to the QDNB, the seismic profiles of the Shenhu area indicated that there exists a large range of fuzzy reflection zones formed by diapirs and/or gas chimneys, which have a columnar or mushroom-shaped appearance along the vertical direction (Wang et al. 2014; Su et al. 2017; Liang et al. 2019; Zhang et al. 2020b). Bright spots were observed at the top and edges of the gas chimneys, indicating the presence of active gas-bearing fluids in the Shenhu area. BSRs were widely distributed in the study area, and very high-amplitude BSRs occurred in the upper parts of the gas chimneys (Wang et al. 2014; Chen et al. 2016; Cheng et al. 2020; Zhang et al. 2020b). A set of amorphous enhanced reflections oblique to the BSR usually occurs immediately above the BSR. Drilling confirmed that these enhanced reflections indicate the presence of gas hydrates (Wang et al. 2014; Zhang et al. 2020b) (Fig. 2.3). With the acquisition and interpretation of high-resolution 3D seismic data, large-scale listric faults connecting deep source kitchens, medium-deep petroleum reservoirs, and shallow GHSZ were observed in the gas hydrate accumulation zone (Chen et al. 2016; Su et al. 2017; Cheng et al. 2020). In addition, it was found that the gliding faults in the GHSZ are partially connected with the BSRs and extended upward to the seabed, constituting possible pathways along which gas could enter the GHSZ or escape and leak into the water column after gas hydrate dissociation due to slope failure. Although no active cold seeps have been confirmed by ROV observations at present, gas seepage and paleo-cold seep activity have been confirmed through geophysical detection and geochemical analysis of sediments recovered from the Shenhu area (Xu et al. 2012; Chen et al. 2014a, b; Lin et al. 2016; Hu et al. 2020).

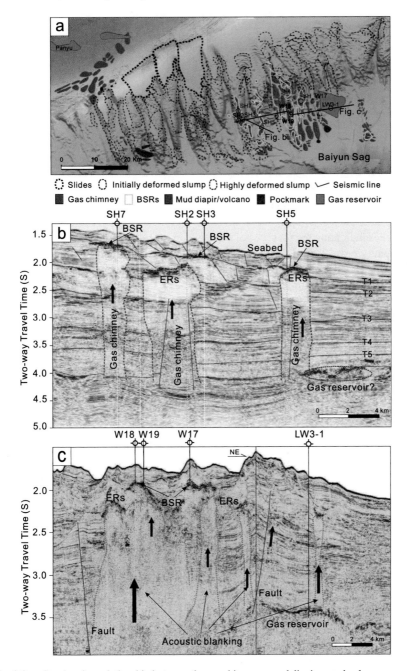

Fig. 2.3 **a** Overlapping relationship between the gas chimneys, mud diapirs, mud volcanoes, pockmarks, BSRs, and gas reservoirs in the Shenhu area (modified from Chen et al. 2016). **b** and **c** Seismic profiles showing geophysical reflection characteristics of the gas hydrate drilling sites in the Shenhu area. ERs: Enhanced reflections

2.3.3 Dongsha Area in the Eastern PRMB

Gas chimneys with chaotic seismic reflections and pull-down features indicating hydrocarbon migration and accumulation, as well as BSRs and associated seismic anomalies indicating seeps, have been identified by interpreting quasi-3D seismic data for the Dongsha area, northern PRMB (Sha et al. 2015a, b; Zhang et al. 2015). In addition, a large number of slumps and listric faults have been identified in shallow strata. The above characteristics indicate the occurrence of gas seepage in the area and the possibility of the development of various submarine microgeomorphologies associated with gas seepage. A large range of blanking zones was recognized in subbottom profiles (Fig. 2.4a; Shang et al. 2013, 2014), indicating that there may exist mud diapirs with a high strata pressure or faults conducive to fluid migration in this area. Based on analysis of subbottom profiles, acoustic blanking reflection anomalies indicating the presence of mud volcanoes and gas accumulation in shallow sediments were observed (Fig. 2.4b), and it was found that many pockmarks that may be caused by submarine gas leakage (Fig. 2.4c–h; Shang et al. 2013, 2014). When the deep gas-bearing fluid is blocked by shallow strata and cannot reach the seafloor, the upper strata are deformed due to overpressure and can form dome features in the seafloor, which is also one of the macroscopic manifestations of gas seepage below the sea floor (Fig. 2.4d).

Gas chimneys are commonly observed in the seismic profile of the Dongsha area (Kuang et al. 2018; Wang et al. 2018b). In general, the internal reflections of these gas chimneys are disordered and chaotic or are characterized by acoustic blanking. In addition, continuous reflection events on both flanks are suddenly interrupted and terminate at the edge of the chimney zone. The pulled-down phenomenon of events can commonly be observed at the top of the gas chimney, and the bright spots on both sides and the top of the upper part are distinct (Fig. 2.4). Low-amplitude, pulled-down features are mostly caused by low-velocity anomalies created by gas charging. Bright spots are the enhanced reflections of free gas accumulations. Most of the gas chimneys extend from Miocene to Quaternary strata. In plan view, the gas chimneys are mainly located along the eastern and western submarine ridges, indicating that the local structural high controls hydrocarbon migration and accumulation (Fig. 2.5). Vertically, the large gas chimneys in the drilling area nearly terminate below the BSR, indicating that the gas-bearing fluid is enriched at the base of the gas hydrate stability zone (BGHSZ) during upward migration and that gas hydrates are precipitated under the appropriate temperature and pressure conditions (Liu et al. 2006; Chen et al. 2010; Wang et al. 2018b). Notably, gas chimneys control hydrocarbon migration and gas hydrate formation and accumulation in the Dongsha area.

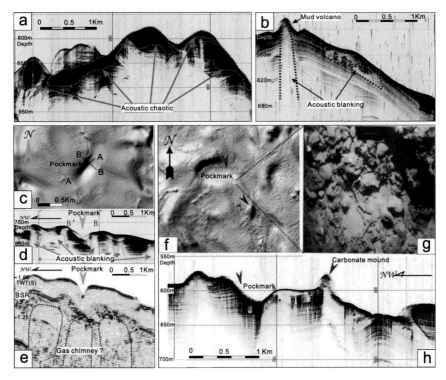

Fig. 2.4 Subbottom profile and seismic profile showing the microgeomorphological features associated with the cold seep and gas hydrate accumulation in the Dongsha area (modified from Shang et al. 2013, 2014). The pockmarks are closely related to gas-bearing fluid migration from the deep strata, which are indicated by acoustic blanking and chaotic features in the subbottom and seismic profiles. The submarine mound with authigenic carbonate pavement resulted from gas hydrate dissociation and hydrocarbon seepage

2.3.4 Taixinan Basin

Mud diapirs and mud volcanoes also developed in the deep-water TXNB offshore southwest of Taiwan (Schnürle et al. 2011). They exhibit a large range of vertical acoustic blanking and chaotic reflections in seismic profiles, and domes and cone-shaped structures are commonly observed in the seafloor. Continuous BSRs have been interpreted in the vicinity of mud volcanoes, mud diapirs, and gas chimneys, indicating the presence of gas hydrates (Fig. 2.6). Moreover, gas seepages and associated cold seeps have been confirmed via multibeam profiles and in situ ROV observations.

There are cold seeps with variable activity intensities and associated gas hydrate accumulations in the Dongsha area. The direct evidence obtained from geochemical analysis of available recovered hydrate gas samples indicates that the hydrate gas is microbial in origin and is unrelated to deep source rocks. No commercial conventional

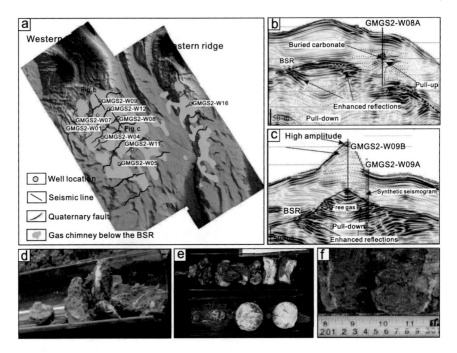

Fig. 2.5 a Distribution of gas chimneys, faults, and gas hydrate drilling sites in the GMGS2 expedition drilling zone in the Dongsha area. The locations of the drilling sites are retrieved from Zhang et al. 2015. **b** and **c** Seismic profiles showing the gas hydrate accumulation features at the drilling and coring sites, where gas hydrates and buried carbonates were confirmed via logging and pressure coring (modified from Wang et al. 2018b). **d** and **e** Massive gas hydrate samples collected at site GMGS2-W08 (Sha et al. 2015b). **f** Carbonate rock recovered at site GMGS2-W08 (modified from Zhong et al. 2017)

oil and gas reservoirs have been discovered in the Dongsha area, making it more difficult to prove the relationship between shallow gas hydrates and deep petroleum reservoirs. However, recent basin modeling and geochemical studies have shown that deep source rocks and possible oil and gas reservoirs may contribute to the formation of shallow gas hydrates (Gong et al. 2017; Li et al. 2021). In addition, thermogenic gas associated with mud volcanoes has been detected in the nearby central-eastern TXNB, suggesting a potential coupling relationship between shallow gas hydrates and potential deep petroleum reservoirs (Chen et al. 2017, 2020).

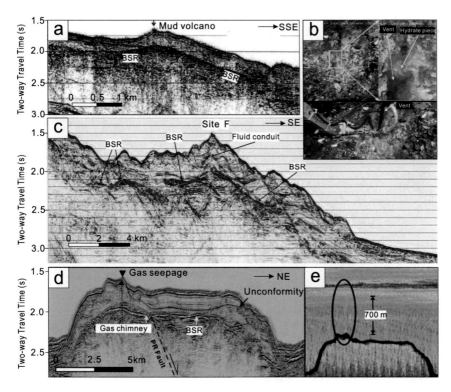

Fig. 2.6 a Seismic profile showing the BSRs observed in the vicinity of a mud volcano developed offshore of southwestern Taiwan (modified from Liu et al. 2006). **b** Cold seep vents and gas hydrates exposed on the seafloor at site F on the Formosa Ridge (Zhang et al. 2017b). **c** Seismic profile showing gas seepage and BSRs crossing Site F on the Formosa Ridge (Hsu et al. 2018). **d** Seismic profile showing gas seepage and BSRs on Pointer Ridge (Han et al. 2019). **e** Gas plume detected on Pointer Ridge (Han et al. 2019). Fault PR is the Pointer Ridge fault

2.3.5 Beikang Basin in the Southern SCS

Based on high-resolution 2D seismic profiles and multibeam data, multiple types of deep-routed conduits—typically manifested as mud volcanoes, diapirs, fluid-escape pipes, and paleo-uplift-associated faults—were identified. These manifestations were generally accompanied by BSRs, blanking zones, locally enhanced reflections, pulled-up reflections, pockmarks, mounds, and possible gas plumes, which indicated fluid seepage and potential gas hydrate accumulation (Fig. 2.7). Arching deformation of deep mud source rock strata and accompanying gas-bearing fluid invasion yielded blanking zones, fuzzy zones, and enhanced reflections inside the mud volcanoes and diapirs. The sulfate methane transition zone (SMTZ) at the fluid seepage sites occurred at relatively shallow depths (325–660 cm below the seafloor (cmbsf)), suggesting a high methane flux and sufficient gas supply. The carbon isotope composition of methane ($-92.6‰ < \delta^{13}C < -50.2‰$) and the

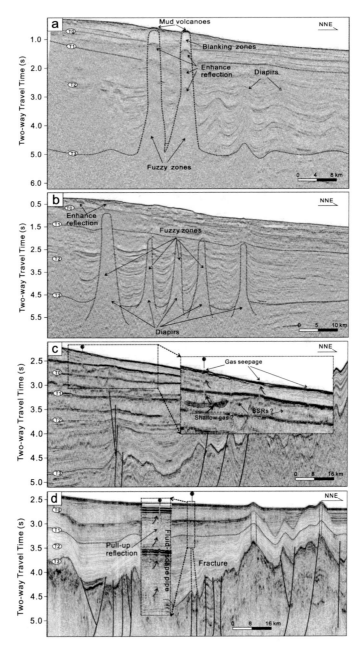

Fig. 2.7 Seismic profile showing geophysical characteristics of the **a** mud volcanoes and **b** mud diapirs. Seismic profiles showing geological features of the fluid escape pipes. **c** Fluid escape pipes originating at the top of the shallow gas zone. **d** Fluid escape pipes originating in the depressions with their roots connected to the fault

detection of ethane, propane, and n-butane, as well as multiple geological configuration biomarkers, all indicated that deep-sourced hydrocarbons and thermogenic gas migrated into the shallow layer through deep-routed conduits. Deep-routed fluid seepage and the accompanying hydrocarbon gas supply may have significantly influenced cold seeps and potential thermogenic gas hydrates (Huang et al. 2022).

2.4   Gas and Fluid Migration Paths and Seafloor Seepage

Although the majority of the gas seeps discovered in the SCS are related to preexisting pathways composed of faults, gas chimneys, and mud diapirs, there are gas seeps that formed due to the behavior of the gases themselves. Many high-resolution seismic profiles exhibit indicators of the presence of shallow gas seeps in Quaternary sediments in the SCS, such as pipe structures, bright spots, seabed mounds, and pockmarks. These subsurface gas seepage/leakage areas are probably caused by gas hydrate dissociation and exhibit a close relationship with fluid pathways and deep petroleum systems.

The drilling sites of the GMGS1, GMGS3, and GMGS4 expeditions in the Shenhu area were located atop gas chimneys or mud diapirs, which provided pathways by which deep gas-bearing fluids could migrate into shallow strata and transport thermogenic gas generated by Paleogene source rocks and shallow biogenic gas in the Baiyun Sag to the GHSZ to form gas hydrates (Wang et al. 2014; Su et al. 2017). At the drilling sites where high-saturation gas hydrates (~60%) were recovered through coring, such as sites W11, W17, W18 and W19 (Fig. 2.8), gas-bearing fluids migrated upward from the deep Paleogene strata along large-scale mud diapirs and gas chimneys, resulting in large fuzzy seismic zones in the seismic profiles (Zhang et al. 2019b, 2020b). In plan view, the development range of the diapirs and gas chimney attained a very obvious relationship with the distribution of the BSR and gas hydrate accumulations, which are mainly distributed along the submarine ridge (Fig. 2.3). This demonstrates that the migration pathways of these gas-bearing fluids controlled gas hydrate accumulation in the Shenhu area (Wang et al. 2014; Su et al. 2017; Cheng et al. 2020). Based on high-resolution 3D seismic data (Fig. 2.8), it was found that there are also large, deep faults in the Shenhu area that connect the Paleogene source rocks and the shallow GHSZ, which also served as vertical pathways for fluid migration, especially thermogenic gas derived from deep source rocks (Wang et al. 2014; Cheng et al. 2020; Jin et al. 2020). For example, the deep thermogenic gas in gas field LW3-1 was transported to the GHSZ via large faults and mud diapirs to form high-saturation gas hydrates (Fig. 2.3), which was confirmed based on the isotopic compositions of the hydrate gas and conventional gas in the Baiyun Sag (Li et al. 2019; Zhang et al. 2019a; Jin et al. 2020). Within the GHSZ, due to sediment gliding and/or slope failure, some of the gas hydrate-bearing strata formed slump faults, which connect the base of the GHSZ with the seafloor, constituting pathways along which gas entering the GHSZ could move further upward, and some of the gas may leak from the sea floor after gas hydrate dissociation, producing upward fluid migration (Fig. 2.8c). Additionally, gas-bearing fluids commonly migrate along the preexisting slump faults developed above the BGHSZ, and this fluid escape may in turn cause seafloor collapse and gas hydrate dissociation.

Medium-strong seafloor gas seepage occurs in the deep-water QDNB, and many seepage pathways related to gas hydrate formation and accumulation have developed, including mud diapirs, gas chimneys, and faults of variable scales

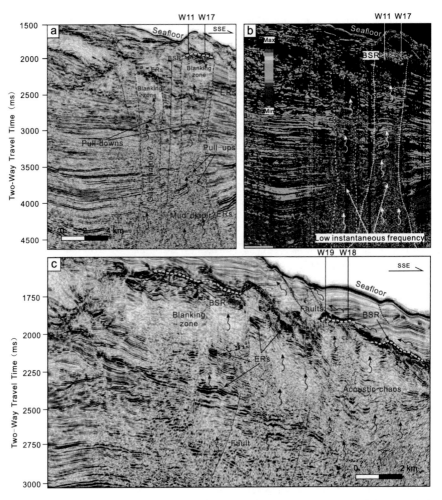

Fig. 2.8 Seismic characteristics of the various types of hydrocarbon migration pathways and their coupling relationships with gas hydrate stability zones in the Shenhu area. Acoustic blanking (BZ) and acoustic chaos reflections below the BSR. Seismic events in the upper part of the abnormal reflection zone exhibit pull-down characteristics. Pull-downs and enhanced reflections (ERs) suggest the presence of gas. Distinct low-frequency zones in the instantaneous frequency image indicate free gas charging and trapping in sediments. The dark blue and white curved arrows indicate possible gas-bearing fluid migration

(Zhang et al. 2019b, 2020a, c). Most significantly, the seismic reflections below the Haima Cold Seep discovered in 2015 are chaotic, creating a large fuzzy zone due to the migration of deep gases into shallow strata (Wang et al. 2018a). In addition, several faults and fractures have been identified within the cold seep area, constituting vertical pathways for deep fluid migration and seepage (Wei et al. 2020a). In 2018 and 2019, massive gas hydrate samples were successfully obtained within a large gas chimney development area in the eastern QDNB (Liang et al. 2019; Wei et al. 2019). Seismic interpretation and gas hydrate drilling demonstrated that this gas hydrate accumulation is closely related to the underlying gas chimneys and deeply buried low uplift. The large faults developed on both sides of the uplift directly connect the deep source kitchens with the vertical gas chimneys and GHSZ, functioning as thermogenic hydrocarbon migration pathways (Liang et al. 2019; Lai et al. 2021).

Since the late Miocene, neotectonic activity has occurred in the cold seep and gas hydrate accumulation area of the Dongsha area. Many gas chimneys, mud diapirs, and mud volcanoes, which function as gas-bearing fluid migration pathways, were developed in this area. Clusters of gas chimneys were identified in seismic profiles of the GMGS2 gas hydrate drilling zone (Kuang et al. 2018; Wang et al. 2018b; Wu et al. 2020). Another type of hydrocarbon migration pathway in the Dongsha drilling area entails faults, which are mainly inherited faults and small active faults at the top and/or flanks of the large gas chimneys, and some of the inherited faults extend downward into the deeply buried basement (Kuang et al. 2018). Fault development is usually accompanied by gas chimneys or directly occurs within the gas chimneys. These faults have remained active since the late Miocene and are mainly distributed along the western ridge in the drilling area. However, the faults developed along the eastern ridge are less abundant and are smaller in size (Fig. 2.5). Most of the identified faults are subvertical and steeply dipping faults and were developed from the late Miocene to the Quaternary. Some of these faults even reach the seafloor. Many of the faults developed at the top and/or flanks of the gas chimney are steep in occurrence, small in size, and variable in strike. They often cut the BSR and are associated with enhanced reflections, which are the result of upward hydrocarbon migration along gas chimneys and deep large-scale faults. When the gas flux is sufficient, gas hydrates can form and accumulate in the GHSZ. In addition, the gas hydrate accumulation area corresponds well with the strike of the faults and the extension direction of the gas chimneys (Fig. 2.5; Zhang et al. 2015). The gas hydrate drilling and coring sites in the Dongsha area are basically located near the fault at the top of a gas chimney.

In the eastern Dongsha area, the mud volcanoes and mud diapirs near the BSRs are well developed in the TXNB, indicating that they constitute the main pathways for the migration of deep gas-bearing fluids needed for gas hydrate formation (Liu et al. 2006). Gas chimneys, pipe structures, and unconformities, which constitute the pathways by which gas-bearing fluids can migrate into the GHSZ, were also observed in this area (Fig. 2.6). Based on the seismic profile crossing the Site F (Formosa ridge) cold seep in the Taixinan Basin, gas chimneys constituting the migration pathway by which deep gas leaks into the seabed and forms a cold seep were identified (Hsu et al. 2018). In addition, mud diapirs and gas chimneys that directly connect the BSR with the seabed and constitute gas migration and leakage pathways were identified along

Cheng C, Jiang T, Kuang Z et al (2020) Characteristics of gas chimneys and their implications on gas hydrate accumulation in the Shenhu area, northern South China Sea. J Nat Gas Sci Eng 84:103629

Demirbas A (2010) Methane gas hydrate: as a natural gas source. Green Energy and Technology. Springer, London, pp 113–160

Feng D, Chen D (2015) Authigenic carbonates from an active cold seep of the northern South China Sea: new insights into fluid sources and past seepage activity. Deep-Sea Res Part II-Top Stud Oceanogr 122:74–83

Feng D, Qiu J, Hu Y et al (2018) Cold seep systems in the South China Sea: an overview. J Asian Earth Sci 168:3–16

Feng J, Yang S, Wang H et al (2019) Methane source and turnover in the shallow sediments to the west of Haima cold seeps on the northwestern slope of the South China Sea. Geofluids 2019:1–18

Gong J, Sun X, Xu L et al (2017) Contribution of thermogenic organic matter to the formation of biogenic gas hydrate: evidence from geochemical and microbial characteristics of hydrate-containing sediments in the Taixinan Basin, South China Sea. Mar Pet Geol 80:432–449

Han X, Suess E, Huang Y et al (2008) Jiulong methane reef: microbial mediation of seep carbonates in the South China Sea. Mar Geol 249(3–4):243–256

Han X, Yang K, Huang Y (2013) Origin and nature of cold seep in northeastern Dongsha area, South China Sea: evidence from chimney-like seep carbonates. Chin Sci Bull 58(30):3689–3697

Han W, Chen L, Liu C et al (2019) Seismic analysis of the gas hydrate system at Pointer Ridge offshore SW Taiwan. Mar Pet Geol 105:158–167

He J, Wang S, Zhang W et al (2016) Characteristics of mud diapirs and mud volcanoes and their relationship to oil and gas migration and accumulation in a marginal basin of the northern South China Sea. Environ Earth Sci 75(15):1122

Hsu H, Liu C, Morita S et al (2018) Seismic imaging of the Formosa ridge cold seep site offshore of southwestern Taiwan. Mar Geophys Res 39(4):523–535

Hu Y, Feng D, Peckmann J et al (2020) The impact of diffusive transport of methane on pore-water and sediment geochemistry constrained by authigenic enrichments of carbon, sulfur, and trace elements: a case study from the Shenhu area of the South China Sea. Chem Geol 553:119805

Huang C, Chien C, Zhao M et al (2006) Geological study of active cold seeps in the syn-collision accretionary prism Kaoping slope off sw Taiwan. Terr Atmos Ocean Sci 17(4):679–702

Huang W, Meng M, Zhang W et al (2022) Geological, geophysical, and geochemical characteristics of deep-routed fluid seepage and its indication of gas hydrate occurrence in the Beikang Basin, Southern South China Sea. Mar Pet Geol 139:105610

Jiang W, Chen J, Huang B et al (2006) Mineralogy and physical properties of cored sediments from the gas hydrate potential area of offshore southwestern Taiwan. Terr Atmos Ocean Sci 17(4):981–1007

Jin J, Wang X, Guo Y et al (2020) Geological controls on the occurrence of recently formed highly concentrated gas hydrate accumulations in the Shenhu area. South China Sea. Mar Pet Geol 116:104294

Kuang Z, Fang Y, Liang J et al (2018) Geomorphological-geological-geophysical signatures of high-flux fluid flows in the Easter Pearl River Mouth Basin and effects on gas hydrate accumulation. Sci China-Earth Sci 61:914–924

Kvenvolden K (1993) Gas hydrates as a potential energy resource-a review of their methane content. the future of energy gases. USGS Prof Paper 1570

Lai H, Fang Y, Kuang Z et al (2021) Geochemistry, origin and accumulation of natural gas hydrates in the qiongdongnan basin, South China Sea: implications from site GMGS5-W08. Mar Pet Geol 123:104774

Li J, Lu J, Kang D et al (2019) Lithological characteristics and hydrocarbon gas sources of gas hydrate bearing sediments in the Shenhu area, South China Sea: Implications from the W01B and W02B sites. Mar Geol 408:36–47

Li Y, Fang Y, Zhou Q et al (2021) Geochemical insights into contribution of petroleum hydrocarbons to the formation of hydrates in the Taixinan Basin, the South China Sea. Geosci Front 12(6):394–403

Liang Q, Hu Y, Feng D et al (2017) Authigenic carbonates from newly discovered active cold seeps on the northwestern slope of the South China Sea: Constraints on fluid sources, formation environments, and seepage dynamics. Deep Sea Res Part i: Ocea Res Papers 124:31–41

Liang J, Zhang W, Lu J et al (2019) Geological occurrence and accumulation mechanism of natural gas hydrates in the eastern Qiongdongnan Basin of the South China Sea: insights from site GMGS5-W9-2018. Mar Geol 418:1–19

Lin Z, Sun X, Lu Y et al (2016) Stable isotope patterns of coexisting pyrite and gypsum indicating variable methane flow at a seep site of the Shenhu area, South China Sea. J Asian Earth Sci 123:213–223

Liu C, Schnurle P, Wang Y et al (2006) Distribution and Characters of Gas Hydrate Offshore of Southwestern Taiwan. Terr Atmos Ocean Sci 17(4):615–644

Matsumoto R, Ryu B, Lee S (2011) Occurrence and exploration of gas hydrate in the marginal seas and continental margin of the Asia and oceania region. Mar Pet Geol 28(10):1751–1767

Minshull T, Marín-Moreno H, Betlem P et al (2020) Hydrate occurrence in Europe: A review of available evidence. Mar Pet Geol 111:735–764

Schnürle P, Liu C, Hsiuan T et al (2004) Characteristics of gas hydrate and free gas offshore southwestern Taiwan from a combined MCS/OBS data analysis. Mar Geophys Res 25(1–2):157–180

Schnürle P, Liu C, Lin A (2011) Structural controls on the formation of BSR over a diapiric anticline from a dense MCS survey offshore southwestern Taiwan. Mar Pet Geol 28(10):1932–1942

Sha Z, Liang J, Su P et al (2015a) Natural gas hydrate accumulation elements and drilling results analysis in the eastern part of the Pearl River Mouth Basin. Earth Sci Frontiers 22(6):125–135

Sha Z, Liang J, Zhang G et al (2015b) A seepage gas hydrate system in northern South China Sea: seismic and well log interpretations. Mar Geol 366:69–78

Shang J, Sha Z, Liang J et al (2013) Acoustic reflections of shallow gas on the northern slope of South China Sea and implications for gas hydrate exploration. Marine Geological Frontiers 29(10):23–30

Shang J, Wu L, Liang J et al (2014) The microtopographic features and gas seep model on the slope in the northeastern South China Sea. Mar Geol Q Geol 34(1):129–136

Su M, Yang R, Wang H et al (2016) Gas hydrates distribution in the Shenhu area, northern South China Sea: comparisons between the eight drilling sites with gas-hydrate petroleum system. Geol Acta 14(2):79–100

Su M, Sha Z, Zhang C et al (2017) Types, characteristics and significances of migration pathways of gas-bearing fluids in the Shenhu area, northern continental slope of the South China Sea. Acta Geol Sin-Engl Ed 91(1):219–231

Suess E (2005) RV Sonne cruise report SO 177, Sino-German Cooperative Project, South China Sea Continental Margin: Geological Methane Budget and Environmental Effects of Methane Emissions and Gas Hydrates IFM-GEOMAR Report. DOI https://doi.org/10.3289/ifm-geomar_rep_4_2005

Wan Z, Yao Y, Chen K et al (2019) Characterization of mud volcanoes in the northern Zhongjiannan basin, western South China Sea. Geol J 54(1):177–189

Wan Z, Chen C, Liang J et al (2020) Hydrochemical Characteristics and Evolution Mode of Cold Seeps in the Qiongdongnan Basin, South China Sea. Geofluids 2020:1–16

Wang X, Collett T, Lee M et al (2014) Geological controls on the occurrence of gas hydrate from core, downhole log, and seismic data in the Shenhu area, South China Sea. Mar Geol 357:272–292

Wang J, Wu S, Kong X et al (2018a) Subsurface fluid flow at an active cold seep area in the Qiongdongnan Basin, northern South China Sea. J Asian Earth Sci 168:17–26

Wang X, Liu B, Qian J, et al (2018b) Geophysical evidence for gas hydrate accumulation related to methane seepage in the Taixinan Basin, South China Sea. J Asian Earth Sci 168:27–37

Wei J, Liang J, Lu J et al (2019) Characteristics and dynamics of gas hydrate systems in the northwestern South China Sea-Results of the fifth gas hydrate drilling expedition. Mar Pet Geol 110:287–298

Wei J, Li J, Wu T et al (2020a) Geologically controlled intermittent gas eruption and its impact on bottom water temperature and chemosynthetic communities—A case study in the "Haima" cold seeps. South China Sea. Geol J 55(9):6066–6078

Wei J, Wu T, Zhang W et al (2020b) Deeply buried authigenic carbonates in the Qiongdongnan Basin, South China Sea: Implications for ancient cold seep activities. Minerals 10(12):1–19

Wu N, Yang S, Zhang H (2010) Gas hydrate system of Shenhu area, northern South China Sea: wire-line logging, geochemical results and preliminary resources estimates. Proceedings of the Annual Offshore Technology Conference 1(1):654–666

Wu T, Wei J, Liu S et al (2020) Characteristics and formation mechanism of seafloor domes on the north-eastern continental slope of the South China Sea. Geol J 55(1):1–10

Wu N, Zhang H, Yang S (2011) Gas hydrate system of Shenhu area, northern South China Sea: Geochemical results. Journal of Geological Research 1–10

Xu H, Xing T, Wang J et al (2012) Detecting seepage hydrate reservoir using multi-channel seismic reflecting data in Shenhu Area. Earth Sci-J China Univ Geosci. 37(z1):195–202

Yang S, Zhang H, Wu N et al (2008) High concentration hydrate in disseminated forms obtained in Shenhu area, North Slope of South China Sea. Proceedings of the 6th International Conference on Gas Hydrate (ICGH2008), Vancouver, British Columbia, Canada. DOI.org/https://doi.org/10.14288/1.0041052

Yang S, Zhang M, Liang J et al (2015) Preliminary results of China's third gas hydrate drilling expedition: A critical step from discovery to development in the South China Sea. Fire in the Ice 15(2):1–5

Yang S, Liang J, Lei Y et al (2017a) GMGS4 gas hydrate drilling expedition in the South China Sea. Fire in the Ice: Methane Hydrate Newsletter 17(1):7–11

Yang S, Liang J, Lu J et al (2017b) New understandings on the characteristics and controlling factors of gas hydrate reservoirs in the Shenhu area on the northern slope of the South China Sea. Earth Sci Frontiers 2(4):1–14 (in Chinese with English abstract)

Yang S, Lei Y, Liang J et al (2017c) Concentrated gas hydrate in the Shenhu area, South China Sea: results from drilling expeditions GMGS3 & GMGS4. In Proceedings of 9th International Conference on Gas Hydrates, Denver, 105:25–30

Yoo D, Kang N, Yi B et al (2013) Occurrence and seismic characteristics of gas hydrate in the Ulleung Basin, East Sea. Mar Pet Geol 47:236–247

Zhang G, Liang J, Lu J et al (2015) Geological features, controlling factors and potential prospects of the gas hydrate occurrence in the east part of the Pearl River Mouth Basin, South China Sea. Mar Pet Geol 67:356–367

Zhang W, Liang J, Lu J et al (2017a) Accumulation features and mechanisms of high saturation natural gas hydrate in Shenhu Area, northern South China Sea. Petroleum Explor Dev 44(5):708–719

Zhang X, Du Z, Luan Z et al (2017b) *In situ* Raman detection of gas hydrates exposed on the seafloor of the South China Sea. Geochem Geophys Geosyst 18(10):3700–3713

Zhang Y, Xu X, Gan J et al (2017c) Study on the geological characteristics, accumulation model and exploration direction of the giant deepwater gas field in the Qiongdongnan Basin. Acta Geological Sinica 91(7):1620–1633

Zhang W, Liang J, Wei J et al (2019a) Origin of natural gases and associated gas hydrates in the Shenhu area, northern South China Sea: Results from the China gas hydrate drilling expeditions. J Asian Earth Sci 183:103953

Zhang W, Liang J, Su P et al (2019b) Distribution and characteristics of mud diapirs, gas chimneys, and bottom simulating reflectors associated with hydrocarbon migration and gas hydrate accumulation in the Qiongdongnan Basin, northern slope of the South China Sea. Geol J 54(6):3556–3573

Zhang Y, Gan J, Xu X et al (2019c) The source and natural gas lateral migration accumulation model of Y8–1 gas bearing structure, east deep water in the Qiongdongnan Basin. Earth Sci 44(8):2609–2618

Zhang W, Liang J, Yang X et al (2020a) The formation mechanism of mud diapirs and gas chimneys and their relationship with natural gas hydrates: insights from the deep-water area of Qiongdongnan Basin, northern South China Sea. Int Geol Rev 62(7–8):789–810

Zhang W, Liang J, Wei J et al (2020b) Geological and geophysical features of and controls on occurrence and accumulation of gas hydrates in the first offshore gas-hydrate production test region in the Shenhu area, Northern South China Sea. Mar Pet Geol 114:104191

Zhang W, Liang J, Lu J et al (2020c) Characteristics and controlling mechanism of typical leakage gas hydrate reservoir forming system in the Qiongdongnan Basin, northern South China Sea. Nat Gas Ind 40(8):90–99

Zhang W, Chen C, Su P et al (2023) Formation and implication of cold-seep carbonates in the southern South China Sea. J Asia Earth Sci 241:105485

Zhong G, Liang J, Guo Y et al (2017) Integrated core-log facies analysis and depositional model of the gas hydrate-bearing sediments in the northeastern continental slope, South China Sea. Mar Pet Geol 86:1159–1172

Chapter 3
Gas Seepage Detection and Gas Migration Mechanisms

Kun Zhang, Haibin Song, Jiangxin Chen, Minghui Geng, and Boran Liu

Abstract Gas seepages are often observed at the seafloor and can form cold seep systems, which are important for climate change, geohazards, and biogeochemical cycles. Many kinds of methods have been used to detect gas seepages, e.g., video imaging, active acoustic, passive acoustic, and direct gas sampling. In this chapter, we introduce the characteristics of these methods and show their applications in the South China Sea (SCS). Particle image velocimetry (PIV) technology is used here to quantitatively detect gas seepage in the northwestern SCS and visualize the seepage flow field. The gas migration mechanisms are also discussed. With the development of technologies, long-term, three-dimensional, and comprehensive observations is permitted to quantitatively characterize gas seepages, which can help us understand the formation and mechanism of gas seepages further in the future. Physical and numerical simulations of gas migration and geohazard processes would also be helpful in the future for understanding the fate of gas seepages.

K. Zhang · H. Song (✉)
State Key Laboratory of Marine Geology, School of Ocean and Earth Science, Tongji University, Shanghai 20092, China
e-mail: hbsong@tongji.edu.cn

K. Zhang
e-mail: kunzhang@tongji.edu.cn

J. Chen
Key Laboratory of Gas Hydrate, Ministry of Natural Resources, Qingdao Institute of Marine Geology, Qingdao 266071, China
e-mail: jiangxin_chen@sina.com

M. Geng
Key Laboratory of Marine Mineral Resources, Ministry of Natural Resources, Guangzhou Marine Geological Survey, Guangzhou 510760, China
e-mail: gengminghui5591788@163.com

B. Liu
Laboratory of Coastal and Marine Geology, Ministry of Natural Resources, Third Institute of Oceanography, Xiamen 361005, China
e-mail: liuboran@tio.org.cn

© The Author(s) 2023
D. Chen and D. Feng (eds.), *South China Sea Seeps*,
https://doi.org/10.1007/978-981-99-1494-4_3

3.1 Introduction

Gas seepage is a widely distributed geological feature around the world's oceans and lakes. It can transport larger amounts of natural gas (mainly methane) into the water column and even the atmosphere, which has been invoked as an important factor in climate change (Judd and Hovland 2007), submarine slope failures (Kvenvolden 1993), and biogeochemical cycles (Feng et al. 2010).

Gas seepages can not only influence the physical properties of the sediments near the seafloor but also significantly change the geomorphology of the seafloor, e.g., the formation of pockmarks, mud volcanoes, authigenic carbonates and cold seep biochemistry communities (Fig. 3.1). These features show different characteristics in geological, biogeochemical, and geophysical data (Judd and Hovland 2007), and many gas seepages have been detected in the South China Sea (SCS) through these features. Most of the gas seepages are located in the water depth range of 200–3000 m in the northern SCS, e.g., Southwest Taiwan Island, Dongsha Uplift, and Qiongdongnan Basin. Haima and Site F are the most famous and well-studied active cold seep areas in the SCS (Fig. 3.2). The geomorphological properties, subsurface structures, and water column characteristics of gas seepages have been studied and summarized to show the features and mechanisms in the SCS (Feng et al. 2018), and some quantitative studies have also attempted to show the fate and effect of these seepages (Di et al. 2020).

In this chapter, we first introduce the progress of gas seepage detection in the SCS and then demonstrate the methods for the quantitative characterization of gas seepage. Subsequently, gas migration mechanisms are discussed. Finally, we provide some ideas for detecting gas seepages and studying gas migration mechanisms.

3.2 Detecting Gas Seepage in the South China Sea

3.2.1 Bubbles

The bubbles emitted from gas seepages are the most significant characteristic of gas seepages. Because gas bubbles induce a strong change in acoustic impedance, they can be detected in acoustic water column data by their typical "flare" shape or as rising lines when single bubbles/bubble clouds are emitted (Urban et al. 2017).

Gas plumes have been identified for their significant "flare" shape on multibeam echosounder system (MBES) water column images (Fig. 3.3) in offshore SW Taiwan Island (Hsu et al. 2013, 2018a), the Pearl River Mouth Basin (Zhu et al. 2018), and the Qiongdongnan Basin (Yang et al. 2018; Wei et al. 2020; Liu et al. 2021). The gas plumes in these regions have significant spatiotemporal variations. The heights of the plumes range from tens of meters to over 750 m, some of which are higher than the upper boundary of the Gas Hydrate Stability Zone (GHSZ) in the Qingdongnan Basin. The hydrated skin of bubbles may extend the lifetime of gas bubbles and cause

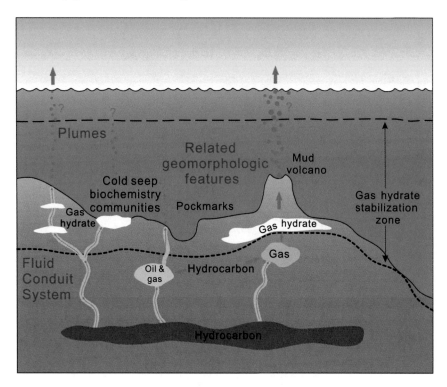

Fig. 3.1 Main elements of gas seepages (modified from Marine and Petroleum Geology, 22(4), Whelan et al. Surface and subsurface manifestations of gas movement through a N–S transect of the Gulf of Mexico Sea, 479–497.Copyright (2005), with permission from Elsevier)

the gas plumes to reach higher (Liu et al. 2021). The shapes of these plume usually vary with time due to the bottom current (Liu et al. 2021). Ocean tides may also influence the strength of gas seepages and cause emission tremors (Hsu et al. 2013). Due to periodic variations in subsurface pressure, the locations of gas emission points are not fixed, and gas flux is unstable in the Qiongdongnan Basin (Wei et al. 2020).

Gas bubbles also show unique features on water column images of the chirp subbottom system and multichannel seismic system (Xu et al. 2012; Liu et al. 2015; Chen et al. 2017, 2020). Liu et al. (2015) identified an acoustic plume on subbottom water column images in the northeast SCS, and they suspected that this plume was caused by gas seepage (Fig. 3.4). Acoustic turbidity, acoustic curtain, acoustic blanket, and enhanced reflection were identified on subbottom profiles and considered to be indicators of shallow gas below the seafloor. On seismic sections of the water column, bubble plumes are characteristic of staggered boundary events, strong amplitude, and high frequency (Chen et al. 2017, 2020). For example, Chen et al. (2017) processed the multichannel seismic data near the Dongsha uplift and the northern Zhongjiannan Basin using seismic oceanography methods. Their analysis

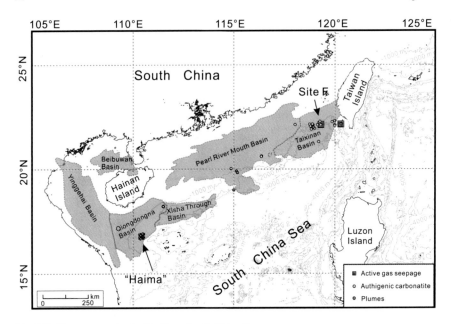

Fig. 3.2 Gas seepages in the Northern South China Sea

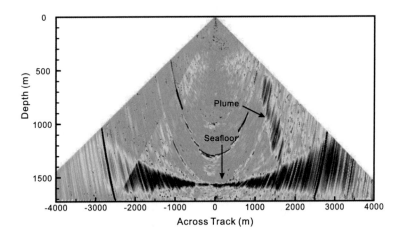

Fig. 3.3 Multibeam Echosounder Systems water column images. The flare shape reflection indicates the gas seepage

showed that gas seepages primarily present plume, broom, and/or irregular shapes that have weak and chaotic seismic reflections in the water column (Fig. 3.5).

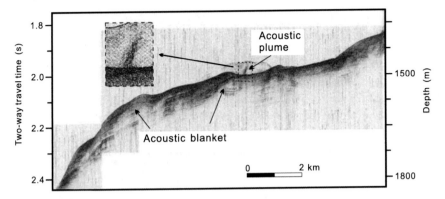

Fig. 3.4 Acoustic plume and blanket on chirp subbottom profiles (redrawn from Chinese Journal of Geophysics-Chinese Edition, 58(1), Liu et al. Characteristics and formation mechanism of cold seep system in the northeastern continental slope of South China Sea from sub-bottom profiler data, 247–256. Copyright (2015), with permission from Science Press)

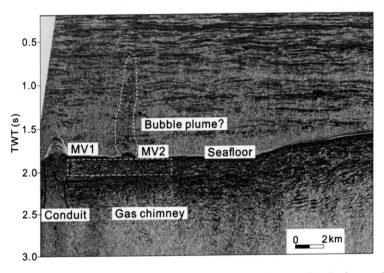

Fig. 3.5 Possible bubble plumes in seismic sections of the water column, and mud volcanoes (MV) and gas chimneys in seismic sections of the sedimentary strata near the Dongsha uplift, northern South China Sea. (Reprinted from Chinese Journal of Geophysics-Chinese Edition, 60(2), Chen et al. A preliminary study of submarine cold seeps applying seismic oceanography techniques, 604–616. Copyright (2017), with permission from Science Press)

3.2.2 Pockmarks

Pockmarks consist of depressions on the seafloor and are often considered to originate from fluid escape activities (Dandapath et al. 2010; Hovland et al. 2010). Different kinds of pockmarks have been discovered in the Taixinan Basin (Chen et al. 2010),

Pearl River mouth basin (Chen et al. 2015a; Zhu et al. 2020), Yinggehai Basin (Di et al. 2012), Qiongdongnan Basin (Bai et al. 2014; Wang et al. 2019), Xisha Massif (Sun et al. 2011; Luo et al. 2015), Zhongjiannan Basin (Chen et al. 2015a; Yu et al. 2021), Beikang Basin (Zhang et al. 2020a), and Reed Basin (Zhang et al. 2019; Zhu et al. 2020) of the northern, western, and southern margins of the SCS. According to the characteristics of pockmarks in the SCS, Chen et al. (2015b) proposed a general pockmark classification system based on three main pockmark characteristics: their shape, size, and composite pattern (Fig. 3.6). According to the geometrical shapes, pockmarks are classified as circular, elliptical, crescent, comet, elongated, and irregular pockmarks. By size, they can be classified as small pockmarks, normal pockmarks, giant pockmarks, and mega pockmarks, which have diameters on the order of several meters, tens of meters, hundreds of meters, and thousands of meters, respectively. A composite pattern describes how groups of pockmarks are organized. They can be classified as composites of pockmarks, pockmark strings, and pockmark groups.

Different geometrical parameters of pockmarks were also presented, e.g., diameter, depth, area, filled volume, and slopes (Fig. 3.7; Chen et al. 2015a; Zhang et al. 2020a). Many giant and mega pockmarks were found in the western and southern SCS margins, and the pockmarks reported in the northern SCS margin were relatively small. The diameter and depth of pockmarks usually have a positive correlation in the SCS (Chen et al. 2015a; Zhang et al. 2020a). However, the slopes and coefficients

Types	Circle	Elliptical	Crescent	Comet	Elongated	Irregular
3D View	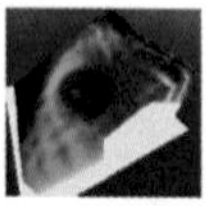		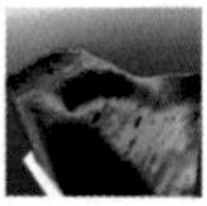	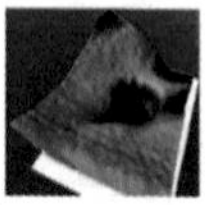		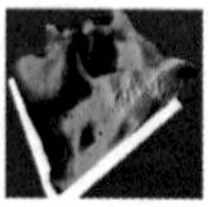
Types	Small		Normal		Gaint	Mega
Sizes	Several meters		Tens of meters		Hundreds of meters	Thousands of meters
Types	Composite		String		Group	
3D view						

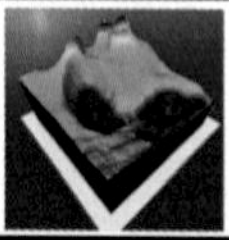

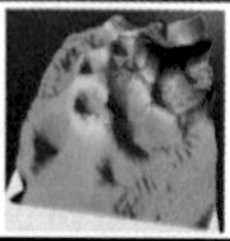

Fig. 3.6 Classification of pockmarks according to individual standard (modified from Deep Sea Research Part II: Topical Studies in Oceanography, 122, Chen et al. Morphologies, classification and genesis of pockmarks, mud volcanoes and associated fluid escape features in the northern Zhongjiannan Basin, South China Sea, 106–117.Copyright (2015), with permission from Elsevier, and Marine Geophysical Research, 41(2), Zhang et al. A preliminary study on morphology and genesis of giant and mega pockmarks near Andu Seamount, Nansha Region (South China Sea), 1–12. Copyright (2015), with permission from Springer Nature)

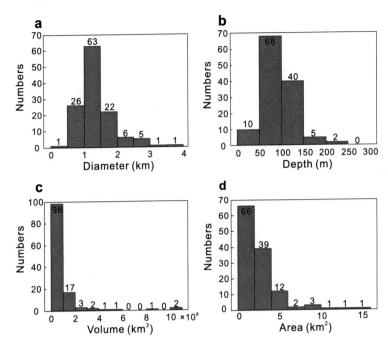

Fig. 3.7 Histograms and rose diagram of morphological parameters of pockmarks in the Beikang Basin, southern South China Sea. **a** Diameter, **b** depth, **c** volume, and **d** surface area (modified from Marine Geophysical Research, 41(2), Zhang et al. A preliminary study on morphology and genesis of giant and mega pockmarks near Andu Seamount, Nansha Region (South China Sea), 1–12. Copyright (2015), with permission from Springer Nature)

of determination of the pockmark fitting curves are different in each area, showing a relatively poor linear relationship (Chen et al. 2015a).

High-angle faults, high-amplitude reflections, filled-up structures, and other fluid escape features are usually found beneath pockmarks (Sun et al. 2011; Zhang et al. 2020a). These underlying geological structures controls the formation and evolution of pockmarks (Gay et al. 2006). The seabed gradient, bottom currents, sediment types, and gas hydrate evolution also contribute to the development of pockmarks (Pilcher and Argents 2007; Dandapath et al. 2010; Bai et al. 2014; Sultan et al. 2014). In the western SCS, the development of pockmarks was found to be related to the formation of submarine channels (Chen et al. 2015b; Yu et al. 2021).

3.2.3   Mud Volcanoes

In contrast with pockmarks, mud volcanoes are positive geomorphological features on the seafloor and have flat, conical, or concave-down tops (Dupré et al. 2008; Ceramicola et al. 2018). Due to the rough seafloor and breccia deposits around mud volcanoes, the acoustic backscatter strength near mud volcanoes is high (Chen and Song 2005). On chirp subbottom profiles, mud volcanoes show positive geomorphological features and acoustic blanket, acoustic turbidity, and enhanced reflections because of the clay, water, and gas inside the mud volcanoes (Schroot et al. 2005). On seismic profiles, weak reflections, chaotic reflections, and blank reflections are distributed inside mud volcanoes, and mud volcanoes also show positive relief (He et al. 2010; Wan et al. 2019).

Many mud volcanoes have been reported on the northern (Fig. 3.8) and western SCS margins (Chen et al. 2015a; Geng et al. 2019; Wan et al. 2019). The diameters of them are distributed between hundreds of meters and two kilometers, and their heights range from several meters to two hundred meters. Chen et al. (2015a) presented the statistical results on the relationship between the diameter and height of mud volcanoes in the northern and western margins of the SCS. They found that the mud volcanoes in individual areas showed good linear relationships, but all of them showed a relatively bad linear relationship. The classification of mud volcanoes is also diverse. Based on their morphology, structure, and kinetic characteristics, mud volcanoes are classified into deep source high-energy large-size mud volcanoes, shallow source-low energy-small mud volcanoes, budding mud volcanoes, canyon- and bottom current channel-related mud volcanoes, and pockmark- and gas chimney-related mud volcanoes in the western SCS margin (Wan et al. 2019). The formation mechanisms of mud volcanoes are distinct in different regions (Chen et al. 2015a), but the abundant sediments and overpressure usually provide the necessary conditions for the formation of mud volcanoes.

3.2.4 Authigenic Carbonates

The authigenic carbonates and chemosynthetic communities formed by gas seepages can change the acoustic impedance and roughness of the seafloor, which can be identified by enhanced backscatter on Sidescan Sonar Images (Chen and Song 2005; Dumke et al. 2014). For example, Site F is characterized by irregular and hummocky topography on a seabed relief map and by patches with high to medium backscatter intensities on side-scan sonar (Wang et al. 2021). On seismic and subbottom profiles, acoustic blanking usually appears beneath gas seepages because of acoustic shielding from carbonate rocks (Liu et al. 2015; Liu 2017).

3.2.5 Subsurface Features Beneath a Gas Seepage

Subsurface features connect a gas seepage at the seafloor with the source of gas in the subsurface layers and provide conduits for gas migration (Talukder 2012). High-angle faults, pipes, gas chimneys, mud diapirs, and bottom simulating reflections (BSRs) are often found beneath gas seepages (Sun et al. 2012, 2013; Wang et al. 2018a;

Fig. 3.8 Mud volcanos discovered in the northeastern South China Sea (modified from Marine Geology Frontiers, 35(10), Geng et al. The distribution and characteristics of mud volcanoes in the northeastern South China Sea.: 1–10, Copyright (2015), with permission from Science Press)

Zhang et al. 2020b). These features are characterized as chaotic, disrupted, blanking, pull-up, pull-down, and/or negative reflections on seismic and subbottom profiles (Fig. 3.9). These features are helpful for identifying the location and determining the formation mechanisms of gas seepages. High-angle faults, pipes, gas chimneys, and mud diapirs are usually good conduits for gas migration; they permits hydrocarbons to migrate from the deep strata to shallow strata and seep into the water column (Sun et al. 2012; Zhang et al. 2020b).

3.3 Quantitative Characterization of Gas Seepage

Quantitatively characterizations of gas seepage facilitate an understanding the fate and effect of gas seepage. The approaches for the estimation of gas seepage fluxes include acoustic methods (von Deimling et al. 2011; Li et al. 2020; Turco et al. 2022), chemical methods (Tryon and Brown 2004; Di et al. 2014; Leifer 2015), and optical methods (Römer et al. 2012; Wang et al. 2016; Higgs et al. 2019).

Optical methods usually rely on the manual and frame-by-frame analysis of the behavior and parameters of bubbles from video data acquired by a camera (Römer et al. 2012; Wang et al. 2016). Some semiautomatic and automatic analysis techniques

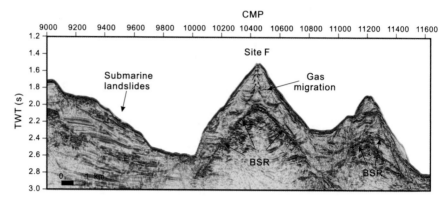

Fig. 3.9 Gas migration pathway beneath the gas seepage station and bottom simulating reflection (BSR) at Site F in the northeastern South China Sea

have also been applied for bubble behavior and parameter measurements (Johansen et al. 2017; Higgs et al. 2019; Di et al. 2020; Veloso-Alarcón et al. 2022). However, most studies only show the behavior of single bubble streams and do not provide flow field images of gas seepages. Particle image velocimetry (PIV) technology is commonly used for visualizing fluid motion (Westerweel 2000; Adrian 2005). For example, Zhang et al. (2018) used PIV to image flow field of the hydrothermal plumes at mid-ocean ridge. Zhang et al. (2020c) and Li et al. (2021) tried to use PIV to image the flow field of cold seep in the SCS and Cascadia margin.

During the Deep-sea Exploration Shared Cruise (2019.4.10–2019.5.16, Northwestern South China Sea) with R/V Haiyang-6, a new and active cold seep (called "Haima 2019" in this chapter) was found approximately 50 km northeast of the "Haima" cold seeps (Geng et al. 2021). In situ video data from Haima 2019 were acquired using high-speed cameras mounted on the ROV Haima. The video data were converted to full resolution (1920 × 1080 pixels) images frame by frame using the open source software FFmpeg (https://www.ffmpeg.org). The pixel-to-mm parameter can be acquired separately for each image sequence according to a known reference measurement. After the conversion of pixels to mm, these high-resolution image sequences allow us to make detailed and accurate observations of gas seepage behaviors and parameters in Haima 2019 in the SCS.

3.3.1 Manual Observations

According to the video image sequences, there were two types of gas seepages in Haima 2019 (Fig. 3.10). One consisted of gas bubbles, which were formed through gas hydrate decomposition and the slow escape of gas from the seafloor (seeping structure; Fig. 3.10a). The other was via plumes of gas-containing fluids, which erupted rapidly from the vent (erupting structure; Fig. 3.10b). As Fig. 3.10 shows, there were some bubble streams near the seafloor, with abundant gas hydrate and cold seep biology, e.g., mussels (Fig. 3.10a) developed on the seafloor. The dissolved methane concentrations at Haima 2019 reached a value of 91 nmol/L, which is much higher than that of the normal bottom water (0.5–2.0 nmol/L, Di et al. 2020). It is speculated that most of the bubbles were methane bubbles formed through gas hydrate decomposition (Fig. 3.10c); these bubbles then slowly escaped from the seafloor and became methane bubble streams (Fig. 3.10a). As the bubbles rose, they became yellow and sheet-like (Fig. 3.10d). From this observation, we inferred that hydrate skins formed around the bubbles as they rose (Fig. 3.10d), in agreement with observations by Rehder et al. (2002) and Römer et al. (2012).

A gas-containing fluid rapidly erupted from the seafloor and formed three plumes (Fig. 3.10b). The plumes on the left and middle of this site are in gray, and the plume on the right is in white. The white particles in the plumes are inferred to be gas hydrates that erupted from the seafloor because of the high reservoir pressure in the study area (Wang et al. 2018a).

From the video observation, the average equivalent spherical bubble radius (r) was estimated to be 2.924 mm. Thus, the average bubble volume, calculated as the mean

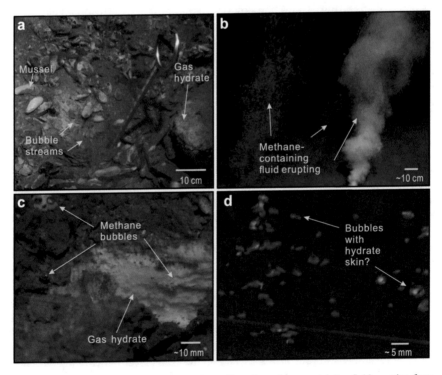

Fig. 3.10 **a** Bubble streams seeping from the seafloor; **b** methane-containing fluid erupting from the seafloor; **c** gas hydrate decomposing and methane bubbles forming; **d** bubbles with hydrate skin. The scale of (**a**) is calculated according to the ruler in the image, while the (**b**), (**c**), and (**d**) is estimated according to experience point

of all individually estimated bubble volumes, was 0.128 ml. During an observation period of 2 min, the average bubble emission frequency was 12.34 bubbles/s. Using the average bubble emission frequency and the average bubble volume, a total bubble flux of approximately 94.8 ml/min for an individual bubble stream was calculated. The average rising velocity was 0.263 m/s, estimated from 23 bubbles within the size range of 2.556–4.624 mm in the upper region of the bubble streams.

3.3.2 PIV Observations

A GUI-based open-source tool (PIVlab) in MATLAB (MathWorks, Natick, Massachusetts) was used for PIV analyses of these gas seepage image sequences. The initial and final steps were 64 × 64 pixels and 32 × 32 pixels, or 48 × 48 pixels and 24 × 24 pixels, respectively, with an overlap of 50%. After the velocity field was acquired, the vorticity (ω) was also calculated according to the curl of the velocity. More detailed information can be found in Thielicke and Stamhuis (2014).

As the vector field images show (Fig. 3.11a and b), both the outlines of the seeping structure (Fig. 3.11a) and erupting structure (Fig. 3.11b) flow field had a plume structure, which is smaller at the bottom and gradually enlarges upward. The plume boundaries were easier to identify in velocity field images (Fig. 3.11c). The streamlines of the cold seep flow were zigzags, which indicated that the bubbles and other fluids did not rise vertically (Fig. 3.11a and b). The velocity magnitude was higher at the center region of the plume and gradually decreased to almost zero at the boundary region of the plume (Fig. 3.11c and d). Many high-velocity magnitude points are present in the velocity field of the bubble streams (Fig. 3.11c). These points can be considered methane bubbles, and the velocity decreased with distance from the core of the bubbles (Gong et al. 2009). Thus, we obtained not only the single rise velocity of the bubbles but also the velocity of the bubble-induced flow. Turbulent motion is visible within the flow field images (Fig. 3.11), and they are more distinct in the velocity field (Fig. 3.11c and d) and vorticity field (Fig. 3.11e and f). At the edge of the plumes, seawater was carried into the plumes by the turbulent eddies, which may have been responsible for the upward enlargement of the plumes (Fig. 3.11d). Within the plumes, the turbulent eddies were more developed (Fig. 3.11e and f).

These results also provide quantitative information on the cold seep flow field. The seeping velocity flow field ranged from 0 to 0.337 m/s, while the erupting velocity flow field ranged from 0 to approximately 6.461 m/s. The maximum velocity of the erupting structure was approximately 19 times larger than the maximum velocity of the seeping structures. The vorticity of the seeping structure ranged from −27.073 to 20.394 1/s. The vorticity of the erupting structure ranged from −809.324 to 910.307 1/s. The quantitative results also showed that the vorticity of the erupting structure was much larger than that of the seeping structure.

Figure 3.12a and b show velocity field changes. The velocity field change in Fig. 3.12b was recorded 0.0334 s after that in Fig. 3.12a. It is clear that the velocity field changed, even within this short time interval. Both the direction and velocity of

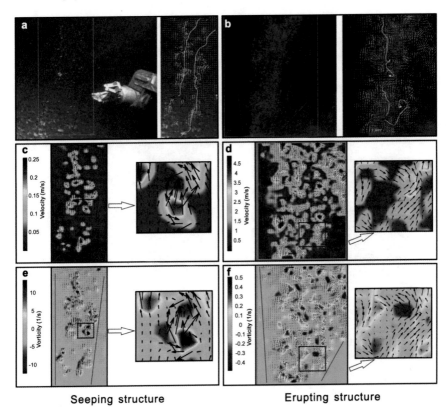

Fig. 3.11 Images of the cold seep flow field. **a** Original (left) and vector (right) images of the seeping structure; **b** original (left) and vector (right) images of the erupting structure; **c** velocity field of the seeping structure; **d** velocity field of the erupting structure; **e** vorticity image of the seeping structure; **f** vorticity image of the erupting structure. The red lines show the boundaries of the cold seep flow. The streams are indicated by yellow lines

the cold seep flow changed with time (Fig. 3.12a and b). For example, the velocity field direction changed from the upper righter to vertically upward in the red box region of Fig. 3.12a and b. In addition, the velocity values became slightly lower. The time-dependent variations in maximum velocity are mapped in Fig. 3.12c. A short-period cycle of 6.6 s was observed (Fig. 3.12c), which is far shorter than the general tidal cycle (~24 h) of cold seeps (Römer et al. 2016).

These results show that the PIV method is a feasible and powerful tool to visualize cold seep flow properties within the water column. It is a nonintrusive measurement with high sampling frequency and can show overall changes in field images, not simply individual static values. Regarding the possibilities for the use of PIV in cold seep systems, it can be stated that the images of cold seep flow fields provide a new approach for observing cold seep flows, and the scope of their future research applications is broad.

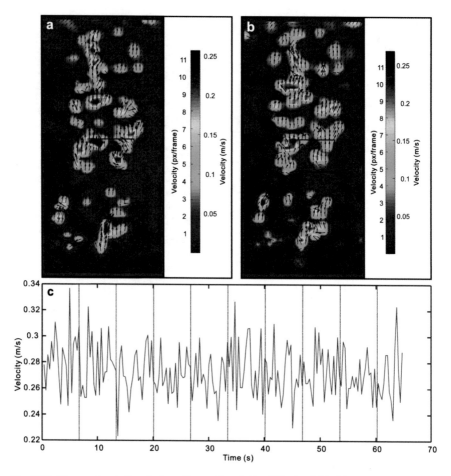

Fig. 3.12 Temporal variations in the cold seep flow field. **a** Velocity field of frame 1152; **b** velocity field of frame 1153; **c** time series of velocity

3.4 Gas Migration Mechanisms

Overpressure is a fundamental condition for gas migration and seepage. However, differences in gas sources, conduit systems and patterns induce different gas migration mechanisms in different regions of the SCS. At Site F, continuous and strong BSRs, sediment waves, mass transport deposits, and gas chimneys were discovered in seismic profiles (Hsu et al. 2018b; Wang et al. 2018b). Methane-rich fluids present inside sediment ridges migrated upward along gas chimneys and other conduits, and sulfate carried by cold seawater flowed into the fluid systems from both flanks of the sediment ridge, and this formed the gas seepages at Site F.

For the Haima cold seep area and Haima 2019 in the Qiongdongnan Basin, polarity reverses, pull-downs, deep faults, minor faults, gas chimneys, and BSRs were also discovered in the seismic profiles (Wei et al. 2020; Geng et al. 2021). The gas seepages showed spatiotemporal variations, which are controlled by the gas migration process from the deep strata to the seafloor. The gas migration originating from the deep reservoir migrates along the deep faults, the slope of the basal uplift, and the gas chimney to the shallow sediments. Gas accumulated at shallow depths forms gas hydrates and increases pore pressure. When the pore pressure overcomes the overburdened sediment, fractures are generated for gas migration to the water column. Gas emission and eruption decrease the pressure in the shallow sediments, which stops the fractures from reaching the seafloor. However, gas seepage along the fractures might cease with gas hydrate development and authigenic carbonate precipitation, which decreases the permeability of the original fractures. Consequently, pressure rebuilds in the shallow sediments, and new fractures are generated.

However, most gas migration mechanism studies in the SCS are limited to isolated stations. Comparison and holistic studies are needed to summarize the gas migration mechanisms in the SCS in the future. Physical and numerical simulations of the gas migration process are scarce in the SCS. The relationships between gas seepage and geohazards, e.g., submarine landslides and tsunamis, also need to be studied in the future.

3.5 Summary and Perspectives

Gas seepage is characterized as bubble plumes in water, which can be detected by several kinds of acoustic devices and optical investigations. The formation and evolution of gas seepages can alter the seafloor and produce various kinds of features on and below the seafloor, e.g., chemosynthetic communities, authigenic carbonate, pockmarks, and mud volcanoes on the seafloor, and BSRs, gas chimneys, high-angle faults, pipes, and polarity reversals in the subsurface layers. These features have unique manifestations in topographical data, seismic data, sidescan sonar data, and subbottom data.

The research on gas seepages has gradually shifted from qualitative descriptions to quantitative characterizations. Generally, the parameters for gas seepages are estimated according to manual observation or acoustic devices. A nonintrusive measurement, called the PIV method, conducted with a high sampling frequency optical device is introduced in this chapter and provides a new approach to study gas seepages quantitatively. Gas migration mechanisms are discussed in this chapter. Overpressure in marine sediments is a significant gas transport mechanism. The subsurface structures, e.g., faults, gas chimneys, pipes, and other conduits, also influence gas migration.

With the development of research equipment and underwater positioning and navigation systems, the comprehensive detection of gas seepages has become feasible, which can help us understand the formation and mechanism of gas seepages further.

Long-term, three-dimensional, and comprehensive observations are needed to quantitatively characterize gas seepages. Physical and numerical simulations of gas migration and hazard processes could also facilitate understanding of the fate of gas seepages in the future.

Acknowledgements We thank the Guangzhou Marine Geological Survey for releasing these data for scientific research. Mr. Yongxian Guan and Dr. Baojin Zhang are thanked for their kind help and support during our visit in Guangzhou.

References

Adrian RJ (2005) Twenty years of particle image velocimetry. Exp Fluids 39(2):159–169

Bai Y, Song H, Guan Y et al (2014) Structural characteristics and genesis of pockmarks in the Northwest of the South China Sea derived from reflective seismic and multibeam data. Chinese J Geophys-Chinese Ed 57(7):2208–2222 (in Chinese with English abstract)

Ceramicola S, Dupré S, Somoza L et al (2018) Cold seep systems. In: Micallef A, Krastel S, Savini A (eds) Submarine Geomorphology. Springer International Publishing, Cham, pp 367–387

Chen L, Song H (2005) Geophysical features and identification of natural gas seepage in marine environment. Prog Geophys 20(4):1067–1073 (in Chinese with English abstract)

Chen SC, Hsu SK, Tsai CH et al (2010) Gas seepage, pockmarks and mud volcanoes in the near shore of SW Taiwan. Mar Geophys Res 31(1–2):133–147

Chen J, Guan Y, Song H et al (2015a) Distribution characteristics and geological implications of pockmarks and mud volcanoes in the northern and western continental margins of the South China Sea. Chinese J Geophys-Chinese Ed 58(3):919–938 (in Chinese with English abstract)

Chen J, Song H, Guan Y et al (2015b) Morphologies, classification and genesis of pockmarks, mud volcanoes and associated fluid escape features in the northern Zhongjiannan Basin, South China Sea. Deep-Sea Res Part II-Top Stud Oceanogr 122:106–117

Chen J, Song H, Guan Y et al (2017) A preliminary study of submarine cold seeps by seismic oceanography techniques. Chinese J Geophys-Chinese Ed 60(1):117–129

Chen J, Tong S, Han T et al (2020) Modelling and detection of submarine bubble plumes using seismic oceanography. J Mar Syst 209(103375):1–11

Dandapath S, Chakraborty B, Karisiddaiah SM et al (2010) Morphology of pockmarks along the western continental margin of India: employing multibeam bathymetry and backscatter data. Mar Pet Geol 27(10):2107–2117

Di P, Huang H, Huang B et al (2012) Seabed pockmark formation associated with mud diapir development and fluid activities in the Yinggehai Basin of the South China Sea. J Trop Oceanogr 31(5):26–36 (in Chinese with English abstract)

Di P, Feng D, Chen D (2014) In-situ and on-line measurement of gas flux at a hydrocarbon seep from the northern South China Sea. Cont Shelf Res 81:80–87

Di P, Feng D, Tao J et al (2020) Using time-series videos to quantify methane bubbles flux from natural cold seeps in the South China Sea. Minerals 10(3):1–17

Dumke I, Klaucke I, Berndt C et al (2014) Sidescan backscatter variations of cold seeps on the Hikurangi margin (New Zealand): indications for different stages in seep development. Geo-Mar Lett 34(2–3):169–184

Dupré S, Buffet G, Mascle J et al (2008) High-resolution mapping of large gas emitting mud volcanoes on the Egyptian continental margin (Nile Deep Sea Fan) by AUV surveys. Mar Geophys Res 29(4):275–290

Feng D, Chen D, Peckmann J et al (2010) Authigenic carbonates from methane seeps of the northern Congo Fan: microbial formation mechanism. Mar Pet Geol 27(4):748–756

Feng D, Qiu JW, Hu Y et al (2018) Cold seep systems in the South China Sea: an overview. J Asian Earth Sci 168:3–16

Gay A, Lopez M, Cochonat P et al (2006) Isolated seafloor pockmarks linked to BSRs, fluid chimneys, polygonal faults and stacked Oligocene-Miocene turbiditic palaeochannels in the Lower Congo Basin. Mar Geol 226(1–2):25–40

Geng M, Song H, Guang Y et al (2019) The distribution and characteristics of mud volcanoes in the northeastern South China Sea. Mar Geol Front 35(10):1–10 (in Chinese with English abstract)

Geng M, Zhang R, Yang S et al (2021) Focused fluid flow, shallow gas hydrate, and cold seep in the Qiongdongnan Basin, northwestern South China Sea. Geofluids 2021:1–11

Gong X, Takagi S, Matsumoto Y (2009) The effect of bubble-induced liquid flow on mass transfer in bubble plumes. Int J Multiph Flow 35(2):155–162

He J, Zhu Y, Weng R et al (2010) Characters of north-west mud diapirs volcanoes in South China Sea and relationship between them and accumulation and migration of oil and gas. Earth Sci 35(1):75–86 (in Chinese with English abstract)

Higgs B, Mountjoy J, Crutchley GJ et al (2019) Seep-bubble characteristics and gas flow rates from a shallow-water, high-density seep field on the shelf-to-slope transition of the Hikurangi subduction margin. Mar Geol 417:105985

Hovland M, Heggland R, De Vries MH et al (2010) Unit-pockmarks and their potential significance for predicting fluid flow. Mar Pet Geol 27(6):1190–1199

Hsu SK, Wang SY, Liao YC et al (2013) Tide-modulated gas emissions and tremors off SW Taiwan. Earth Planet Sci Lett 369–370:98–107

Hsu HH, Liu CS, Morita S et al (2018a) Seismic imaging of the Formosa Ridge cold seep site offshore of southwestern Taiwan. Mar Geophys Res 39(4):523–535

Hsu SK, Lin SS, Wang SY et al (2018b) Seabed gas emissions and submarine landslides off SW Taiwan. Terr, Atmos Ocean Sci 29(1):7–15

Johansen C, Todd AC, MacDonald IR (2017) Time series video analysis of bubble release processes at natural hydrocarbon seeps in the Northern Gulf of Mexico. Mar Pet Geol 82:21–34

Judd A, Hovland M (2007) Seabed fluid flow: the impact on geology, biology and the marine environment. Cambridge University Press, New York. https://doi.org/10.5860/choice.45-0294

Kvenvolden KA (1993) Gas hydrates-geological perspective and global change. Rev Geophys 31(2):173–187

Leifer I (2015) Seabed bubble flux estimation by calibrated video survey for a large blowout seep in the North Sea. Mar Pet Geol 68:743–752

Li HJ, Song HB, Zhang K et al (2021) A quantitative study on the active cold seep flow field along the Cascadia margin. Chinese J Geophys-Chinese Ed 64(8):2982–2993 (in Chinese with English abstract)

Li J, Roche B, Bull JM et al (2020) Broadband acoustic inversion for gas flux quantification. Appl a methane plume Scanner pockmark, Cent. North Sea. J Geophys Res-Oceans 125 (9):e2020JC016360

Liu B (2017) Gas and gas hydrate distribution around seafloor mound in the Dongsha area, north slope of the South China Sea. Haiyang Xuebao 39(3):68–75 (in Chinese with English abstract)

Liu B, Song H, Guan Y et al (2015) Characteristics and formation mechanism of cold seep system in the northeastern continental slope of South China Sea from sub-bottom profiler data. Chinese J Geophys-Chinese Ed 58(1):247–256 (in Chinese with English abstract)

Liu B, Chen J, Yang L et al (2021) Multi-beam and seismic investigations of the active Haima cold seeps, northwestern South China Sea. Acta Oceanol Sin 40(7):183–197

Luo M, Dale AW, Wallmann K et al (2015) Estimating the time of pockmark formation in the SW Xisha Uplift (South China Sea) using reaction-transport modeling. Mar Geol 364:21–31

Pilcher R, Argent J (2007) Mega-pockmarks and linear pockmark trains on the West African continental margin. Mar Geol 244(1–4):15–32

Rehder G, Brewer PW, Peltzer ET et al (2002) Enhanced lifetime of methane bubble streams within the deep ocean. Geophys Res Lett 29(15):1–4

Römer M, Sahling H, Pape T et al (2012) Quantification of gas bubble emissions from submarine hydrocarbon seeps at the Makran continental margin (offshore Pakistan). J Geophys Res-Oceans 117(C10):1–19

Römer M, Riedel M, Scherwath M et al (2016) Tidally controlled gas bubble emissions: A comprehensive study using long-term monitoring data from the NEPTUNE cabled observatory offshore Vancouver Island. Geochem Geophys Geosyst 17(9):3797–3814

Schroot BM, Klaver GT, Schüttenhelm RTE (2005) Surface and subsurface expressions of gas seepage to the seabed - examples from the Southern North Sea. Mar Pet Geol 22(4):499–515

Sultan N, Bohrmann G, Ruffine L et al (2014) Pockmark formation and evolution in deep water Nigeria: Rapid hydrate growth versus slow hydrate dissolution. J Geophys Res-Solid Earth 119(4):2679–2694

Sun Q, Wu S, Hovland M et al (2011) The morphologies and genesis of mega-pockmarks near the Xisha Uplift. South China Sea. Mar Pet Geol 28(6):1146–1156

Sun Q, Wu S, Cartwright J et al (2012) Shallow gas and focused fluid flow systems in the Pearl River Mouth Basin, northern South China Sea. Mar Geol 315–318:1–14

Sun Q, Wu S, Cartwright J et al (2013) Focused fluid flow systems of the Zhongjiannan Basin and Guangle Uplift South China Sea. Basin Res 25(1):97–111

Talukder AR (2012) Review of submarine cold seep plumbing systems: leakage to seepage and venting. Terr Nova 24(4):255–272

Thielicke W, Stamhuis EJ (2014) PIVlab – towards user-friendly, affordable and accurate digital particle image velocimetry in MATLAB. J Open Res Software 2(e30):1–10

Tryon MD, Brown KM (2004) Fluid and chemical cycling at Bush Hill: implications for gas- and hydrate-rich environments. Geochem Geophys Geosyst 5(12):1–7

Turco F, Ladroit Y, Watson SJ et al (2022) Estimates of methane release from gas seeps at the southern Hikurangi margin New Zealand. Front Earth Sci 10(834047):1–20

Urban P, Köser K, Greinert J (2017) Processing of multibeam water column image data for automated bubble/seep detection and repeated mapping. Limnol Oceanogr Meth 15:1–21

Veloso-Alarcón ME, Urban P, Weiss T et al (2022) Quantitatively monitoring bubble-flow at a seep site offshore Oregon: Field trials and methodological advances for parallel optical and hydroacoustical measurements. Front Earth Sci 10(858992):1–23

von Deimling JS, Rehder G, Greinert J et al (2011) Quantification of seep-related methane gas emissions at Tommeliten North Sea. Cont Shelf Res 31(7–8):867–878

Wan Z, Yao Y, Chen K et al (2019) Characterization of mud volcanoes in the northern Zhongjiannan Basin, western South China Sea. Geol J 54(1):177–189

Wang B, Socolofsky SA, Breier JA et al (2016) Observations of bubbles in natural seep flares at MC 118 and GC 600 using in situ quantitative imaging. J Geophys Res-Oceans 121(4):2203–2230

Wang J, Wu S, Kong X et al (2018a) Subsurface fluid flow at an active cold seep area in the Qiongdongnan Basin, northern South China Sea. J Asian Earth Sci 168:17–26

Wang X, Liu B, Qian J et al (2018b) Geophysical evidence for gas hydrate accumulation related to methane seepage in the Taixinan Basin, South China Sea. J Asian Earth Sci 168:27–37

Wang LJ, Zhu JT, Zhuo HT et al (2019) Seismic characteristics and mechanism of fluid flow structures in the central depression of Qiongdongnan Basin, northern margin of South China Sea. Int Geol Rev 62(7–8):1108–1130

Wang B, Du Z, Luan Z et al (2021) Seabed features associated with cold seep activity at the Formosa Ridge, South China Sea: Integrated application of high-resolution acoustic data and photomosaic images. Deep-Sea Res Part I-Oceanogr Res Pap 177(2021):103622

Wei J, Li J, Wu T et al (2020) Geologically controlled intermittent gas eruption and its impact on bottom water temperature and chemosynthetic communities—A case study in the "HaiMa" cold seeps South China Sea. Geol J 55(9):6066–6078

Westerweel J (2000) Theoretical analysis of the measurement precision in particle image velocimetry. Exp Fluids 29(7):S003-S012

Xu H, Xing T, Wang J et al (2012) Detecting seepage hydrate reservoir using multi-channel seismic reflecting data in Shenhu area. Earth Sci 37(S1):195–202 (in Chinese with English abstract)

Yang L, Liu B, Xu M et al (2018) Characteristics of active cold seepages in Qiongdongnan Sea Area of the northern South China Sea. Chinese J Geophys-Chinese Ed 61(7):2905–2914 (in Chinese with English abstract)

Yu K, Miramontes E, Alves TM et al (2021) Incision of submarine channels over pockmark trains in the South China Sea. Geophys Res Lett 48(e2021GL092861):1–13

Zhang X, Lin J, Jiang H (2018) Time-dependent variations in vertical fluxes of hydrothermal plumes at mid-ocean ridges. Mar Geophys Res 40:245–260

Zhang T, Wu Z, Zhao D et al (2019) The morphologies and genesis of pockmarks in the Reed Basin South China Sea. . Haiyang Xuebao 41(3):106–120 (in Chinese with English abstract)

Zhang K, Guan Y, Song H et al (2020a) A preliminary study on morphology and genesis of giant and mega pockmarks near Andu Seamount, Nansha Region (South China Sea). Mar Geophys Res 41(2):1–12

Zhang K, Song H, Wang H et al (2020b) A preliminary study on the active cold seeps flow field in the Qiongdongnan Sea Area, the northern South China Sea. Chin Sci Bull 65(12):1130–1140 (in Chinese with English abstract)

Zhang W, Liang J, Yang X et al (2020c) The formation mechanism of mud diapirs and gas chimneys and their relationship with natural gas hydrates: insights from the deep-water area of Qiongdongnan Basin, northern South China Sea. Int Geol Rev 62(7–8):789–810

Zhu C, Cheng S, Zhang M et al (2018) Results from Multibeam Survey of the Gas Hydrate Reservoir in the Zhujiang Submarine Canyons. Acta Geol Sin-Engl Ed 92(2):135–138

Zhu S, Li X, Zhang H et al (2020) Types, characteristics, distribution, and genesis of pockmarks in the South China Sea: insights from high-resolution multibeam bathymetric and multichannel seismic data. Int Geol Rev 63(12):1682–1702

Chapter 4
Gas Hydrates at Seeps

Min Luo and Yuncheng Cao

Abstract Gas hydrates have been the focus of intensive research during recent decades due to the recognition of their high relevance to future fossil energy, submarine geohazards, and global carbon and climate changes. Cold seep-related gas hydrate systems have been found in both passive and active margins worldwide. A wealth of data, including seismic imaging, borehole logging, seafloor surveys, and coring, suggest that seep-related gas hydrates are present in the western Taixinan Basin and the Qiongdongnan Basin of the northern South China Sea (SCS). Here, we provide an overview of the current understanding of seep-related gas hydrate systems in the northern SCS and underscore the need for more systematic work to uncover the factors governing the interplay of hydrate dynamics and gas seepage and to quantitatively assess the temporal and spatial variability of gas hydrate and cold seep systems.

4.1 Introduction

Gas hydrates are ice-like structures that are composed of hydrogen-bonded water molecules forming a lattice of cages that encapsulate molecules of natural gases, mainly methane (Sloan 1998). Large quantities of gas hydrates are stored in the submarine continental margin sediments where the conditions required for gas hydrate formation, including high pressure, low temperature, and sufficient supply of methane, are met. Recent decades have evidenced an increasing interest in gas hydrates because they are considered to represent a potential source of fossil energy

M. Luo (✉) · Y. Cao
College of Marine Sciences, Shanghai Ocean University, Shanghai 201306, China
e-mail: mluo@shou.edu.cn

Y. Cao
e-mail: yccao@shou.edu.cn

M. Luo
Laboratory for Marine Geology, Qingdao National Laboratory for Marine Science and Technology, Qingdao 266061, China

© The Author(s) 2023
D. Chen and D. Feng (eds.), *South China Sea Seeps*,
https://doi.org/10.1007/978-981-99-1494-4_4

that could exceed the energy content of all other known fossil fuels combined (Holder et al. 1984; Collett 2002). Additionally, as submarine gas hydrates sequester large amounts of carbon, they are thought to form a great capacitor that could regulate the Earth's climate (Dickens 1999; Archer et al. 2009; Ruppel and Kessler 2017). Furthermore, if large-scale gas hydrate dissociation occurs in response to environmental changes, slope failure could be triggered, and the released methane could aggravate ocean deoxygenation and acidification (Sultan et al. 2004; Biastoch et al. 2011).

Gas hydrate provinces associated with methane seep systems are found in both passive and active margins worldwide (Holbrook et al. 2002; Tryon et al. 2002; Torres et al. 2004; Boswell et al. 2012; Luo et al. 2016). Intensive methane supply in the form of advective methane-bearing fluids or methane bubbles from depth can lead to the formation of massive gas hydrates in the gas hydrate stability zone (GHSZ). As a result, seep-related gas hydrates are usually characterized by shallow burial depths and high saturation. Extensive investigation and exploration of gas hydrates in the South China Sea (SCS) have been taking place since 1999. Seep-related shallow gas hydrate systems were suggested to occur in the western Taixinan Basin and the Qingdongnan Basin, northern SCS, based on geophysical imaging and seafloor observations (Fig. 4.1) (Sha et al. 2015; Wang et al. 2018; Geng et al. 2021). Indeed, seafloor coring and drilling has confirmed their occurrence in both areas (Zhang et al. 2015; Bohrmann et al. 2019; Hu et al. 2019; Wei et al. 2019; Meng et al. 2021; Ren et al. 2022). The western Taixinan Basin is located in the northeastern rifted passive SCS margin and is close to the active continental margin where the SCS oceanic crust subducted eastward beneath the Philippine Sea Plate. Horsts and grabens formed across the margin during the period of rifting (Late Cretaceous–late middle Miocene) and were subsequently buried by marine deposits (Clift and Lin 2001; Zhu et al. 2009). Due to the general extensional tectonic environment, two groups of normal faults striking NW–WNW and NNE–ENE are well developed in the western Taixinan Basin (Zhang et al. 2015). This region is characterized by the widespread presence of NNW–SSE trending topographic ridges due to canyon incision. During the R/V SONNE 177 cruise in 2007, a widely distributed seep carbonate crust called Jiulong Methane Reef was found in the northern part of the western Taixinan Basin, indicating extensive paleo-methane seep events in this area (Suess et al. 2005; Han et al. 2008). The Qiongdongnan Basin, located on the northwestern continental slope of the SCS, contains a Cenozoic sedimentary succession of up to 12 km in thickness and is suggested to have great hydrocarbon and gas hydrate potential (Zhu et al. 2009; Shi et al. 2013; Zhang et al. 2019). A number of mega-pockmarks related to submarine fluid flow together with indicators of shallow gas hydrate occurrence have been reported in the western Qiongdongnan Basin (Sun et al. 2011; Luo et al. 2014). In 2015, "Haima cold seeps" were discovered in the southern Qiongdongnan Basin during ROV surveys launched by the Guangzhou Marine Geological Survey. Living chemosynthesis-based communities (e.g., mussels, tubeworms and clams), gas ebullition, seep carbonates, and massive gas hydrates were observed and recovered from the "Haima cold seeps" (Liang et al. 2017b).

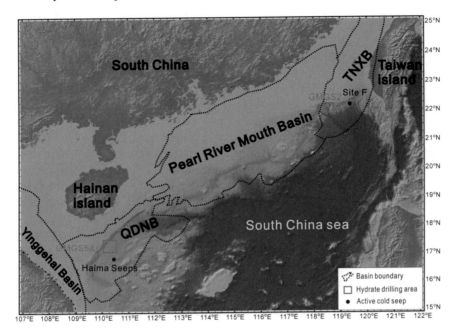

Fig. 4.1 Distribution of cold seep-related gas hydrate systems in the northern South China Sea

The primary goal of this chapter is to provide an overview of the current understanding of shallow gas hydrate systems associated with methane seeps in the western Taixinan Basin and the Qingdongnan Basin, northern SCS by synthesizing the scientific results obtained in the past decade.

4.2 Characteristics of the Shallow Gas Hydrates

4.2.1 Gas Hydrate Textures

Gas hydrate samples in marine sediments can be described in a variety of ways, namely, as massive layers, lenses, veins, nodules, and disseminated grains, based on their macroscopic appearance. In most cases, shallow gas hydrates exhibit different textures from their deeply buried counterparts. Two drilling campaigns have been launched in the western Taixinan Basin: the GMGS2 expedition in 2013 and the SO266 MeBo200 drilling cruise in 2018. During the GMGS2 expedition, 13 sites were drilled, with 10 sites logged while drilling (LWD), along the crest of two seafloor ridges at water depths ranging between 664 and 1420 m (Yang et al. 2014). The primary objective of the GMGS2 expedition was to determine the nature, distribution, and saturation of gas hydrates in this basin. At sites GMGS2-W08, GMGS2-W09, and

GMGS2-W16, shallow gas hydrates with saturations up to ~50% of pore space were observed (Sha et al. 2015). Gas hydrates in the form of veins and nodules were first encountered ~8–10 m below the seafloor (mbsf) at these three sites (Fig. 4.2), where LWD data showed both resistivity and p-wave velocity values (Zhang et al. 2015; Wang et al. 2018). To the east of the drilled ridges during GMGS2, the active seep area, Site F, or Formosa Ridge, was drilled during SO266. Pore water chloride profiles, borehole logging, and IR imaging data all showed clear indications of hydrate-bearing layers in the upper 15–42 mbsf (Bohrmann et al. 2019). The shallow gas hydrates mainly exhibit tubular and nodular appearances. These types of hydrate textures observed during both drilling campaigns are likely indicative of substantial methane supply from the deep subsurface, which is also demonstrated by the observations of seep carbonates accompanied by chemosynthetic bivalve shells both in the retrieved cores and on the seafloor (Chen et al. 2016; Wang et al. 2018; Bohrmann et al. 2019).

Shallow gas hydrates (~4 mbsf) in the Qiongdongnan Basin were recovered for the first time from the "Haima cold seeps" in 2015 (Liang et al. 2017b), although negative pore water Cl and positive $\delta^{18}O$ anomalies, which implied gas hydrate dissociation during core recovery, were previously reported in a pockmark located in the western Qiongdongnan Basin (Luo et al. 2014). The recovered gas hydrate samples appeared in the form of small nodules and thin veins (Liang et al. 2017b; Hu et al. 2019). During the GMGS5 expedition in 2018, four sites (W01, W07, W08 and W09) were drilled in the eastern Qiongdongnan Basin, with the aim of exploring the distribution, abundance, and formation mechanism of the massive gas hydrates associated with seeps. Drilled sites W07, W08, and W09 contained both abundant gas hydrates and authigenic carbonate in the sediments from the upper ~6–23 mbsf. Large hydrate veins with various dip angles and thicknesses, hydrate nodules of different sizes, degassing, and soupy sediments caused by gas hydrate dissociation have been observed in shallow sediments (Liang et al. 2019; Wei et al. 2019). Similar to the drilling results from the western Taixinan Basin, seep carbonate crust and typical seep-related bivalve fragments were also found in the recovered cores and on the seafloor in close proximity to the drilling sites in the eastern Qiongdongnan Basin, which likely indicated the potential linkage of shallow massive hydrates and methane seepage (Ye et al. 2019). However, the shallow sediments of the eastern Qiongdongnan Basin hold greater amounts of gas hydrate than those in the western Taixinan Basin, implying a more sufficient methane supply from the subsurface in the eastern Qiongdongnan Basin. A subsequent gas hydrate drilling expedition (GMGS6) in the western Qiongdongnan Basin was launched in 2019, with the aim of characterizing the distribution and accumulation of gas hydrates in the sandy sediment layers associated with Quaternary channel-levee facies and mass transport deposits (MTDs) (Meng et al. 2021). Site W01 was drilled at a currently active cold seep, where authigenic carbonate crust and massive gas hydrates were observed on the seafloor (Ren et al. 2022). Fracture-filling gas hydrates present as chunks, veins, and nodules first appeared within the MTDs at depths of ~5–28 mbsf at Site W01 (Fig. 4.2). Interestingly, the shallow hydrate-bearing silty sediments were immediately overlain by a thick authigenic carbonate layer as a cap (Meng et al. 2021).

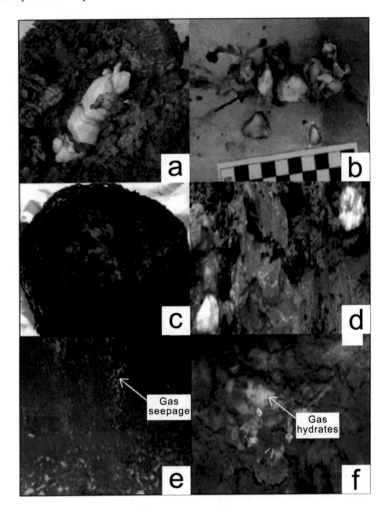

Fig. 4.2 Examples of gas hydrate samples retrieved from shallow sediments in the Qiongdongnan Basin (**a** and **b**) and the western Taixinan Basin (**c** and **d**). Seafloor observations of gas seepages and gas hydrates on the seafloor of the GMGS6 drilling area of the Qiongdongnan Basin are also shown (**e** and **f**) (Modified from Zhang et al. 2015; Wei et al. 2019; Meng et al. 2021 and Ren et al. 2022). Reprinted from Marine and Petroleum Geology, 67, Zhang et al. (2015) Geological features, controlling factors and potential prospects of the gas hydrate occurrence in the east part of the Pearl River Mouth Basin, South China Sea, 356–367, Copyright (2015), with permission from Elsevier; from Marine and Petroleum Geology, 110, Wei et al. (2019) Characteristics and dynamics of gas hydrate systems in the northwestern South China Sea—Results of the fifth gas hydrate drilling expedition, 287–298, Copyright (2019), with permission from Elsevier; from Deep Sea Research Part I: Oceanographic Research Papers, 177, Meng et al. (2021) Quaternary deep-water sedimentary characteristics and their relationship with the gas hydrate accumulations in the Qiongdongnan Basin, Northwest South China Sea, 103,628, Copyright (2021), with permission from Elsevier; and from Journal of Petroleum Science and Engineering, 215, Ren et al. (2022) Sand-rich gas hydrate and shallow gas systems in the Qiongdongnan Basin, northern South China Sea, 110,630, Copyright (2022), with permission from Elsevier

4.2.2 Geochemistry of Hydrate-Bound Hydrocarbons

The sources of light hydrocarbons bound in shallow gas hydrates have been determined for hydrate samples retrieved from both the western Taixinan Basin and the Qiongdongnan Basin. The molecular ratios of hydrocarbon composition (C1/(C2+C3)) and the stable carbon isotopic compositions of methane ($\delta^{13}C$–CH_4) of the shallow gas hydrates retrieved during GMGS2 in the western Taixinan Basin indicate that the hydrate-bound gas is mainly of microbial origin via CO_2 reduction (Fig. 4.3; Liu et al. 2015; Liang et al. 2017a), which is the prevailing methanogenic pathway in marine sediments. This finding is consistent with Raman spectroscopic and X-ray diffraction results that reveal hydrate crystallization in a typical type I cubic lattice structure I (sI), with dominant occupancy of methane in both large and small cages (Liu et al. 2015). Moreover, preliminary onboard analysis of the compositions of void gas also implies a major contribution of biogenic hydrocarbons at Site F (Bohrmann et al. 2019).

In contrast to the biogenic methane that formed the shallow gas hydrates, the hydrate-bound gas in the Qiongdongnan Basin has a mixed biogenic and thermogenic origin based on C1/(C2+ C3) and $\delta^{13}C$–CH_4 (Fig. 4.3) (Lai et al. 2022). Additionally, the detection of C3+hydrocarbons indicates the contribution of thermogenic gas

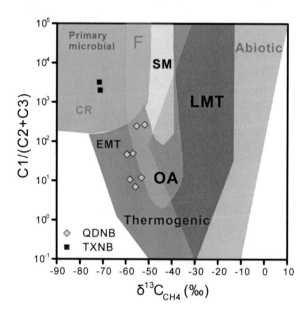

Fig. 4.3 Genetic diagram of $\delta^{13}C_{CH4}$ versus C1/(C2+ C3) according to Milkov and Etiope (2018) that is used to identify the hydrocarbon sources of shallow gas hydrates in the South China Sea. CR–CO_2 reduction; F—methyl-type fermentation; SM—secondary microbial; EMT—early mature thermogenic gas; OA—oil-associated thermogenic gas; LMT—late mature thermogenic gas; QDNB—Qiongdongnan Basin; TXNB—Taixinan Basin. *Data sources* Liu et al. (2015); Liang et al. (2017a); Liang et al. (2019); Lai et al. (2021); Wei et al. (2021); Lai et al. (2022)

produced in deep subsurface sediments to the shallow gas hydrate system (Ye et al. 2019; Lai et al. 2021). Furthermore, both sI and sII types of gas hydrates were found at site W08 by means of Raman spectroscopy and X-ray diffraction (Wei et al. 2021). In-situ temperature measurements at sites W07, W08, and W09 yielded linear geothermal gradients of 102–111 °C km^{-1}, which are substantially higher than those in the background sediments of the Qiongdongnan Basin (65 °C km^{-1}) (Wei et al. 2019). The observed high geothermal gradient was explained as the result of vertical migration of deep-rooted, warm fluids, thereby remobilizing thermogenic hydrocarbons toward the seafloor and contributing to the formation of shallow gas hydrates and cold seeps (Ye et al. 2019).

4.2.3 Relationship Between Cold Seeps and Shallow Gas Hydrates

The drilling area of GMGS2 is located in proximity to the deformation front of SW Taiwan. Several NW–SE trending submarine canyons and seafloor ridges are well developed in the shelf edge down to the lower slope. Faults, gas chimneys, and diapiric structures have been identified from seismic imaging in this area (Wang et al. 2018). The gas hydrates drilled during GMGS2 in the western Taixinan Basin are inferred to have been fed by a continuous biogenic gas supply from the Pliocene and Pleistocene strata into the GHSZ via mud diapirs, gas chimneys, and deep-rooted faults. Although the hydrate-bound gas was inferred to be sourced from microbe-mediated CO_2 reduction, secondary microbial methane derived from the bioconversion of thermogenic organic matter (e.g., oil deposits, coalbeds) also contributed to the biogenic gas supply (Gong et al. 2016). Shallow gas hydrates tended to accumulate in the topographic highs on the ridges, where overpressure caused by free gas generation in the deep subsurface drove the upward migration of gassy fluids. Focused gas flow that contributes to cold seeps and near-surface gas hydrates can either be transported by gas chimneys formed close to the seafloor or by sub-vertical faults connecting the tops and/or flanks of large gas chimneys in subsurface sediments to the seafloor. These gas conduits can be traced to at least as deep as the base of the GHSZ. Although seafloor observations show generally weak methane seepage intensity in the drilling area of GMGS-2, multiple authigenic carbonate-bearing and bioclastic layers retrieved by the drilled cores suggest large-scale, intense methane seepage events in the past (Sha et al. 2015; Wang et al. 2018). In addition, authigenic carbonate and gas hydrates are stratigraphically linked, and analysis of the stable carbon and oxygen isotopic compositions of authigenic carbonate points to vigorous methane seepage likely caused by gas hydrate dissociation (Chen et al. 2016). Similarly, Site F, which was drilled during SO266, also exhibits a clear subsurface plumbing system that connects deep gas reservoirs with the seafloor and closely links shallow gas hydrates and authigenic carbonate (Bohrmann et al. 2019). It is inferred that the accumulation of free gas beneath the base of the GHSZ causes hydraulic fracturing, and the resulting gas venting occurs through the GHSZ to the seafloor (Kunath et al. 2022). Therefore, it is likely that gas hydrate dissociation close to the base of the GHSZ triggered a significant release of methane toward the seafloor, forming authigenic carbonate and gas hydrates in the near-surface sediments of the western Taixinan Basin.

The drilling areas of GMGS5 and GMGS6 are located in the Qiongdongnan Basin, which is one of the major petroliferous basins in the northern SCS. Seismic imaging showed the widespread occurrence of gas chimneys indicated by acoustic blanking and/or chaotic reflections, which were associated with the development of cold seeps in the Qiongdongnan Basin. Large gas chimneys were found to be underlain by pre-Paleogene basement uplift resulting from magmatic intrusion. Faults that were formed during basement uplift could act as effective pathways for transporting deep-seated thermogenic hydrocarbons within the Paleogene source rocks and deeply buried gas reservoirs to shallow sediments in the form of gas chimneys (Geng et al. 2021). In addition, a 3D seismic survey revealed that the widely distributed MTDs and channel systems in the Quaternary strata could facilitate hydrocarbon and gas hydrate accumulation (Cheng et al. 2021; Meng et al. 2022). Geochemical analysis reveals the same source of hydrate-bound gas as natural gas in deep hydrocarbon reservoirs (Lai et al. 2022). These observations suggest a direct coupling of gas hydrates and deep source rocks and/or petroleum reservoirs, which represents a unique hydrate system in the SCS.

Likewise, the authigenic carbonate collected from the seafloor and the drilled cores of the Qiongdongnan Basin also allows the fingerprinting of methane seepage events caused by gas hydrate dissociation (Liang et al. 2017b; Wei et al. 2022). Distinct seep-related seafloor morphologies, e.g., pockmarks and mud volcanoes, are abundant in the western Qiongdongnan Basin and are also postulated to be associated with the waxing and waning of subsurface hydrate reservoirs (Sun et al. 2011; Luo et al. 2015; Yang et al. 2021). The multiple hydrate dissociation-driven methane seepage events imply highly dynamic hydrate systems susceptible to sea-level fluctuation, bottom water temperature change, and sedimentary loading. In addition to vigorous methane bubbling out of the seafloor, some of the seeping methane is currently being converted to gas hydrates in the shallow sediments, as reflected by the observations of positive pore water chloride anomalies within ~10–50 mbsf at Site W08. Therefore, it is likely that the deep petroleum system and gas hydrate reservoirs controlled the development of cold seeps and the accumulation of shallow high-saturation gas hydrates in the Qiongdongnan Basin (Fig. 4.4).

4.3 Summary and Perspectives

Shallow gas hydrates associated with cold seeps have been discovered in the western Taixinan Basin and the Qiongdongnan Basin of the SCS. Logging data and seafloor drilling confirmed the presence of gas hydrates in the upper ~5–30 mbsf. Additionally, seafloor observations showed bare gas hydrates on the seafloor at the active seep sites in the Qiongdongnan Basin. Shallow gas hydrates are generally fracture fillings,

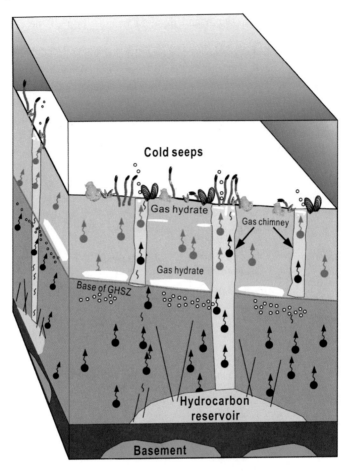

Fig. 4.4 Schematic diagram showing the linkage between the hydrates and seeps at the surface and hydrate/hydrocarbon reservoir at depth

with various fabrics, including massive layers, veins, and nodules, which are typical structures of seep-related hydrates. Light hydrocarbons, originating predominantly from microbe-mediated CO_2 reduction, are trapped in gas hydrates recovered from the western Taixinan Basin. In contrast, light hydrocarbons bound in gas hydrates of the Qiongdongnan Basin are sourced from a mixture of biogenic and thermogenic gas. The upward migration of free gas through the GHSZ is facilitated by faults that connect the deep petroleum reservoir and gas-charged sediment with the seafloor. Periodic release of free gas leads to the precipitation of gas hydrates and the formation of authigenic carbonate in shallow sediments, featuring a seep-related gas hydrate system.

Although significant progress has been made in understanding gas hydrate systems in the SCS, the multi-stage evolution of cold seeps and hydrate dynamics remain poorly constrained. More in-depth and detailed examination of the relationship between deep-seated petroleum systems and gas hydrate reservoirs and the coupling of deep and shallow hydrate systems are needed. The SCS is bounded by the convergent margin to the east, rifted margin to the north, and transform margin to the west, thereby offering an ideal opportunity to explore and compare seep-related hydrate systems in various tectonic settings. Moreover, as the seep-related hydrate systems in the SCS exist in a unique setting characterized by multiple tectonic regimes, rapid fluctuations in sediment delivery toward the northern continental margin with glacial–interglacial cycles, and variations in both the temperature and strength of bottom currents, the underlying mechanisms of the spatial and temporal variability of hydrate dynamics and gas seepage in the SCS generally remain elusive. The most important key to addressing the abovementioned open questions is to combine long-term monitoring and numerical modeling approaches to quantitatively assess the temporal and spatial variability of gas hydrate and cold seep systems.

Acknowledgements The authors are grateful to Prof. Joris Gieskes for his constructive reviews. Tingcang Hu and Weiding Li are thanked for their help with figures.

References

Archer D, Buffett B, Brovkin V (2009) Ocean methane hydrates as a slow tipping point in the global carbon cycle. Proc Natl Acad Sci USA 106(49):20596–20601

Biastoch A, Treude T, Rüpke LH et al (2011) Rising Arctic Ocean temperatures cause gas hydrate destabilization and ocean acidification. Geophys Res Lett 38(8):L08602

Bohrmann G, Ahrlich F, Bergenthal M et al (2019) MeBo200 Methane Hydrate Drillings Southwest of Taiwan, TaiDrill, Cruise No. SO266/1, 15 October–18 November 2018, Kaohsiung (Taiwan) Kaohsiung (Taiwan). https://doi.org/10.48433/cr_so266_1

Boswell R, Collett TS, Frye M et al (2012) Subsurface gas hydrates in the northern Gulf of Mexico. Mar Pet Geol 34(1):4–30

Chen F, Hu Y, Feng D et al (2016) Evidence of intense methane seepages from molybdenum enrichments in gas hydrate-bearing sediments of the northern South China Sea. Chem Geol 443:173–181

Cheng C, Jiang T, Kuang Z et al (2021) Seismic characteristics and distributions of Quaternary mass transport deposits in the Qiongdongnan Basin, northern South China Sea. Mar Pet Geol 129:105118

Clift P, Lin J (2001) Preferential mantle lithospheric extension under the South China margin. Mar Pet Geol 18(8):929–945

Collett TS (2002) Energy resource potential of natural gas hydrates. AAPG Bull 86(11):1971–1992

Dickens GR (1999) The blast in the past. Nature 401(6755):752–755

Geng M, Zhang R, Yang S et al (2021) Focused Fluid Flow, Shallow Gas Hydrate, and Cold Seep in the Qiongdongnan Basin, Northwestern South China Sea. Geofluids 2021:5594980

Gong J, Xu L, Lu H (2016) Contribution of thermogenic organic matter to the formation of biogenic gas hydrate: Evidence from geochemical and microbial characteristics of hydrate-containing sediments in the Taixinan Basin, South China Sea. Mar Pet Geol 80:432–449

Han X, Suess E, Huang Y et al (2008) Jiulong methane reef: microbial mediation of seep carbonates in the South China Sea. Mar Geol 249(3–4):243–256

Holbrook WS, Lizarralde D, Pecher IA et al (2002) Escape of methane gas through sediment waves in a large methane hydrate province. Geology 30(5):467–470

Holder GD, Kamath VA, Godbole SP (1984) The potential of natural gas hydrates as an energy resource. Ann Rev Energy 9:427–445

Hu Y, Luo M, Liang Q et al (2019) Pore fluid compositions and inferred fluid flow patterns at the Haima cold seeps of the South China Sea. Mar Pet Geol 103:29–40

Kunath P, Crutchley G, Chi WC et al (2022) Episodic venting of a submarine gas seep on geological time scales: Formosa Ridge, Northern South China Sea. J Geophys Res-Solid Earth 127(9):e2022JB024668

Lai H, Qiu H, Liang J et al (2022) Geochemical characteristics and gas-to-gas correlation of two leakage-type gas hydrate accumulations in the Western Qiongdongnan Basin, South China Sea. Acta Geol Sin-Engl Ed 96(2):680–690

Lai H, Fang Y, Kuang Z et al (2021) Geochemistry, origin and accumulation of natural gas hydrates in the Qiongdongnan Basin, South China Sea: implications from site GMGS5-W08. Mar Pet Geol 123:104774

Liang J, Zhang W, Lu J et al (2019) Geological occurrence and accumulation mechanism of natural gas hydrates in the eastern Qiongdongnan Basin of the South China Sea: Insights from site GMGS5-W9-2018. Mar Geol 418:106042

Liang J, Fu S, Chen F et al (2017a) Characteristics of methane seepage and gas hydrate reservoir in the Northeastern Slope of South China Sea. Nat Gas Geosci 28(5):761–770 (In Chinese with English abstract)

Liang Q, Hu Y, Feng D et al (2017b) Authigenic carbonates from newly discovered active cold seeps on the northwestern slope of the South China Sea: constraints on fluid sources, formation environments, and seepage dynamics. Deep-Sea Res Part I-Oceanogr Res Pap 124:31–41

Liu C, Meng Q, He X et al (2015) Characterization of natural gas hydrate recovered from Pearl River Mouth basin in South China Sea. Mar Pet Geol 61:14–21

Luo M, Chen L, Tong H et al (2014) Gas Hydrate Occurrence Inferred from Dissolved Cl$^-$ Concentrations and $\delta^{18}O$ Values of Pore Water and Dissolved Sulfate in the Shallow Sediments of the Pockmark Field in Southwestern Xisha Uplift, Northern South China Sea. Energies 7(6):3886–3899

Luo M, Dale AW, Wallmann K et al (2015) Estimating the time of pockmark formation in the SW Xisha Uplift (South China Sea) using reaction-transport modeling. Mar Geol 364:21–31

Luo M, Dale AW, Haffert L et al (2016) A quantitative assessment of methane cycling in Hikurangi Margin sediments (New Zealand) using geophysical imaging and biogeochemical modeling. Geochem Geophys Geosyst 17(12):4817–4835

Meng M, Liang J, Kuang Z et al (2022) Distribution characteristics of quaternary channel systems and their controlling factors in the Qiongdongnan Basin, South China Sea. Front Earth Sci 10:902517

Meng M, Liang J, Lu J et al (2021) Quaternary deep-water sedimentary characteristics and their relationship with the gas hydrate accumulations in the Qiongdongnan Basin, Northwest South China Sea. Deep-Sea Res Part I-OceanogrRes Pap 177:103628

Milkov AV, Etiope G (2018) Revised genetic diagrams for natural gases based on a global dataset of >20,000 samples. Org Geochem 125:109–120

Ren J, Cheng C, Xiong P et al (2022) Sand-rich gas hydrate and shallow gas systems in the Qiongdongnan Basin, northern South China Sea. J Pet Sci Eng 215,Part B:110630

Ruppel CD, Kessler JD (2017) The interaction of climate change and methane hydrates. Rev Geophys 55(1):126–168

Sha Z, Liang J, Zhang G et al (2015) A seepage gas hydrate system in northern South China Sea: seismic and well log interpretations. Mar Geol 366:69–78

Shi W, Xie Y, Wang Z et al (2013) Characteristics of overpressure distribution and its implication for hydrocarbon exploration in the Qiongdongnan Basin. J Asian Earth Sci 66:150–165

Sloan ED (1998) Gas hydrates: review of physical/chemical properties. Energy Fuels 12(2):191–196

Sultan N, Cochonat P, Foucher JP et al (2004) Effect of gas hydrates melting on seafloor slope instability. Mar Geol 213(1–4):379–401

Suess E, Huang Y, Wu N et al (2005) Cruise report SO-177, Sino-German cooperative project, South China Sea. In: Distribution, Formation and Effect of Methane & Gas Hydrate on the Environment. IFM-GEOMAR Reports. http://store.pangaea.de/documentation/Reports/SO177.pdf

Sun Q, Wu S, Hovland M et al (2011) The morphologies and genesis of mega-pockmarks near the Xisha Uplift, South China Sea. Mar Pet Geol 28(6):1146–1156

Torres ME, Wallmann K, Trehu AM et al (2004) Gas hydrate growth, methane transport, and chloride enrichment at the southern summit of Hydrate Ridge, Cascadia margin off Oregon. Earth Planet Sci Lett 226(1–2):225–241

Tryon MD, Brown KM, Torres ME (2002) Fluid and chemical flux in and out of sediments hosting methane hydrate deposits on Hydrate Ridge, OR, II: Hydrological processes. Earth Planet Sci Lett 201(3–4):541–557

Wang X, Liu B, Qian J et al (2018) Geophysical evidence for gas hydrate accumulation related to methane seepage in the Taixinan Basin, South China Sea. J Asian Earth Sci 168:27–37

Wei J, Wu T, Miao X et al (2022) Massive natural gas hydrate dissociation during the Penultimate Deglaciation (~130 ka) in the South China Sea. Front Mar Sci 9:875374

Wei J, Liang J, Lu J et al (2019) Characteristics and dynamics of gas hydrate systems in the northwestern South China Sea—results of the fifth gas hydrate drilling expedition. Mar Pet Geol 110:287–298

Wei J, Wu T, Zhu L et al (2021) Mixed gas sources induced co-existence of sI and sII gas hydrates in the Qiongdongnan Basin, South China Sea. Mar Pet Geol 128:105024

Yang J, Lu M, Yao Z et al (2021) A geophysical review of the Seabed Methane Seepage features and their relationship with gas hydrate systems. Geofluids 2021:9953026

Yang S, Zhang G, Zhang M et al (2014) Complex gas hydrate system in the Dongsha area, South China Sea: results from drilling expedition GMGS2. In: Proceedings of the 8th International Conference on Gas Hydrate (ICGH8e2014), p 12

Ye J, Wei J, Liang J et al (2019) Complex gas hydrate system in a gas chimney, South China Sea. Mar Pet Geol 104:29–39

Zhang G, Liang J, Lu J et al (2015) Geological features, controlling factors and potential prospects of the gas hydrate occurrence in the east part of the Pearl River Mouth Basin, South China Sea. Mar Pet Geol 67:356–367

Zhang W, Liang J, Su P et al (2019) Distribution and characteristics of mud diapirs, gas chimneys, and bottom simulating reflectors associated with hydrocarbon migration and gas hydrate accumulation in the Qiongdongnan Basin, Northern slope of the South China Sea. Geol J 54(6):3556–3573

Zhu W, Huang B, Mi L et al (2009) Geochemistry, origin, and deep-water exploration potential of natural gases in the Pearl River Mouth and Qiongdongnan Basins, South China Sea. AAPG Bull 93(6):741–761

Chapter 5
Cold Seep Macrofauna

Yi-Xuan Li, Yanan Sun, Yi-Tao Lin, Ting Xu, Jack Chi Ho Ip, and Jian-Wen Qiu

Abstract In deep-sea chemosynthetic ecosystems, macrofaunal diversity and distribution are determined by geochemical environments generated by fluid seepage. The South China Sea is located in the northwestern Pacific Ocean with a passive continental shelf, containing over 40 seep sites. In this chapter, we provide a summary of the macrofaunal diversity and distribution at two active hydrocarbon seeps, Haima cold seep and Site F, with updated information based on samples collected from recent cruises. There are at least 81 macrofaunal species from eight phyla, 14 classes, and 34 orders, highlighting their high diversity of the South China Sea. The two active seep regions share ten species, but their communities present different structures represented by mussel beds, clam beds, and clusters of two siboglinid tubeworms. The four community types all occur at Haima cold seep. The seep community at Site F, characterized by the co-dominance of the bathymodioline mussel *Gigantidas platifrons* and the squat lobster *Shinkaia crosnieri*, resembles the vent communities in the Okinawa Trough.

Y.-X. Li · Y.-T. Lin · J. C. H. Ip · J.-W. Qiu (✉)
Department of Biology, Hong Kong Baptist University, Hong Kong, China
e-mail: qiujw@hkbu.edu.hk

Y.-X. Li
e-mail: liyi-xuan@life.hkbu.hk.edu

Y.-T. Lin
e-mail: linyitao0@outlook.com

J. C. H. Ip
e-mail: jackip@hkbu.edu.hk

Y.-X. Li · Y. Sun · Y.-T. Lin · T. Xu · J.-W. Qiu
Southern Marine Science and Engineering Guangdong Laboratory (Guangzhou), Guangzhou, China
e-mail: sunyanan8733@gmail.com

T. Xu
e-mail: tingxu@ust.hk

Y. Sun · T. Xu
Department of Ocean Science, The Hong Kong University of Science and Technology, Hong Kong, China

© The Author(s) 2023
D. Chen and D. Feng (eds.), *South China Sea Seeps*,
https://doi.org/10.1007/978-981-99-1494-4_5

5.1 Background

5.1.1 Cold Seeps in the South China Sea

Since the first report of hydrocarbon seeps in the South China Sea (SCS) in 2005, more than 40 seep sites have been discovered, mostly distributed in the northern SCS (Suess et al. 2005; Chen et al. 2006; Niu and Feng 2017; Wang et al. 2022). Among them are three active seep areas—two located off southwest Taiwan (Yam seep and Site F), and the other off Hainan (Haima cold seep) (Fig. 5.1). The Yam seep (Tseng et al. 2023), located on the Four Way Closure Ridge with a depth around 1500–1700 m, is characterized by low seepage intensity. Little is known about the epifaunal community of the Yam seep, except for the presence of the bathymodiolines *Gigantidas platifrons* and *Bathymodiolus securiformis* (Klaucke et al. 2016; Kuo et al. 2019). Site F, located at the summit of Formosa Ridge—a small area of about 180 m × 180 m with well-developed authigenic carbonates at water depths around 1120 to1150 m, has been extensively surveyed in recent years (Lin et al. 2007; Feng and Chen 2015). The Haima cold seep is the largest active seep region in SCS, covering an area of around 350 km^2 at water depths between 1350 to 1525 m. It is located in the Qiongdongnan Basin at the northwest continental margin of the SCS. Several epifaunal surveys have been conducted at Site F and Haima during the past decade, which have shed light on their epifaunal compositions and potential trophic relationships (Feng et al. 2015, 2018a; Zhao et al. 2020; Ke et al. 2022).

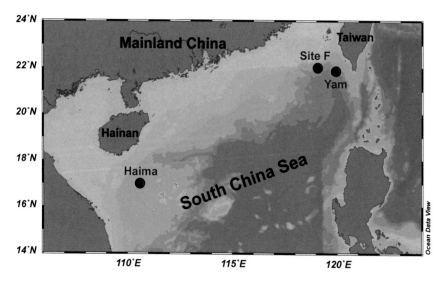

Fig. 5.1 Distribution of active deep-sea hydrocarbon seep fields in the South China Sea

5.1.2   Seep Macrofauna of the South China Sea

Suess et al. (2005) was the first to systematically survey the SCS cold seeps in 2004, but in this cruise only inactive cold seeps were found in the northeastern slopes. He reported evidence of past seep activities, as shown by the extensive development of authigenic carbonate rocks at depths from 500–800 m. He also reported empty shells of bivalves that are supposed to host endosymbiotic chemosynthetic bacteria, including shells from a mud bottom with high concentrations of methane at 3000 m depth in Haiyang 4 area that were later identified as *Archivessica nanshaensis* (Li et al. 2023), and some vesicomyid clam and shells, and coral skeletons on authigenic carbonate rocks at the Jiulong methane reef.

Since the first discovery of an active cold seep in the SCS in 2007 (Lin et al. 2007; Machiyama et al. 2007), several studies of the cold seep epifauna have been conducted in this region, including Mollusca (Chen et al. 2018; Xu et al. 2019; Lin et al. 2022a), Annelida (Zhang et al. 2018; Xu et al. 2022), Crustaceans (Dong and Li 2015; Li 2015), Echinodermata (Li et al. 2021; Nethupul et al. 2022) and other taxa (Gong et al. 2015). Six studies have attempted to describe the epifaunal communities inhabiting the Site F and Haima cold seep (Feng et al. 2018a; Zhao et al. 2020; Xu et al. 2020; Dong et al. 2021; Ke et al. 2022; Wang et al. 2022). Feng et al. (2018a) presented a list of 30 species from six phyla, including eight species of Mollusca, 12 species of Arthropoda, three species of Annelida, two species of Porifera, one species of Cnidaria and four species of Chordata. They found only three common species (*Gigantidas platifrons*, *Bathyacmaea lactea* and *Munidopsis verrilli*) in two seep areas. Zhao et al. (2020) expanded the list of epibenthic animals at Site F to 28 species, including 10 new records. In the Haima region, Xu et al. (2020) reported six seep sites with various methane seepage strengths, and different macrofaunal communities, including HM-1 with clusters of *Archivesica marissinica* and empty *Archivesica* shells along with mussels and holothuroids, HM-2 with a dense *G. haimaensis* mussel bed and scattered *Paraescarpia* tubeworms at the edge, HM-3 with massive mussel beds and *Phymorhynchus buccinoides* snails, HM-4 with authigenic carbonates with empty *Archivesica* shells and a few holothuroids, HM-5 with carbonate mounds as a deep-sea fish habitat, and HM-6 with several patches of mussels. They reported species from 11 families and seven classes observed by video or photographic records. Dong et al. (2021) reported a total of 34 species, including 12 species of Mollusca, seven species of Arthropoda, two Annelida and one species of Echinodermata. More recently, Ke et al. (2022) reported a total of 30 species, including 13 species of Mollusca, six species of Arthropoda, seven species of Annelida and three other taxa. Among these, six were new records from the Haima area.

Besides reporting the biodiversity of cold-seep macrofauna, several studies have determined their stable isotopes (i.e., $\delta^{13}C$, $\delta^{15}N$, $\delta^{34}S$) (Feng et al. 2018b; Lin et al. 2020; Zhao et al. 2020; Ke et al. 2022; Wang et al. 2022), which contributed our understanding of the trophic modes of seep specialists and habitat generalists. For

instance, Zhao et al. (2020), using a two-source stable isotope mixing model, estimated that methane contributed from 30% in the king crab *Lithodes longispina* to
91% in the mussel *Gigantidas platifrons* to the carbon pool of the common animals
at Site F. Lin et al. (2020) constructed trophic models on seep and non-seep fauna
off Taiwan. They found that the seep megafauna had specialized diet niches, whilst
the seep-associated *L. longispina* was the top predator. Moreover, they also indicated
that this king crab utilized energy from the neighboring deep-sea ecosystems.

Studies have also been conducted to unveil the genetic divergence and connectivity
of seep macrofauna within the SCS (Yao et al. 2022), and between the SCS and nearby
vent and seep ecosystems in Northwest Pacific (Shen et al. 2016; Yang et al. 2016;
Xu et al. 2017, 2018, 2019, 2021; Xiao et al. 2020). These species have different
life-history characteristics that may determine their dispersal routes (i.e., via bottom,
mid-water or surface currents), dispersal distances, and genetic differentiation.

Evolutionary studies have also been conducted to reveal the genetic adaptations
in cold seep macrofauna, including the evolution of the mitochondrial gene orders of
deep-sea polynoids (Zhang et al. 2018) and bathymodiolines (Liu et al. 2018; Zhang
et al. 2022), as well as the genome- or transcriptome-level metabolic complementarity
between bathymodiolines (Sun et al. 2017; Lin et al. 2022b), vesicomyid clams (Lan
et al. 2019; Ip et al. 2021), siboglinid tubeworms (Yang et al. 2019; Sun et al. 2021)
and their symbiotic bacteria.

In this chapter, we summarize the previous research findings of seep macrofauna
in the SCS and our new findings from cruises taken in 2021 and 2022. We focus on
species diversity at the Haima and Site F seeps, and compare them with those vent
fields and seep areas in the Northwest Pacific.

5.2 Faunal Species Diversity

We compiled a total of 81 seep-associated macrofauna from SCS, including 13
species that have not been reported in previous studies (Table 5.1). These species
belong to eight phyla, 14 classes, and 34 orders, where 42 macrobenthos were identified to species level, 31 species were identified only to the genus level and the rest
were assigned to high taxonomic ranks. They include 35 species reported from Site
F and 58 species from Haima, and 10 species from both cold seep areas, including
four molluscs (*G. platifrons*, "*B.*" *aduloides*, *Phymorhynchus buccinoides*, *Bathyacmaea lactea*), two polychaetes (*P. echinospica* and *Branchipolynoe pettiboneae*),
three decapods (*A. longirostris*, *M. lauensis* and *M. verrilli* and one ophiuroid (*H.
haimaensis*). Previous studies have pointed out the substantial differences in seep
macrofaunal abundance and species compositions between Site F and Haima (Zhao
et al. 2020; Dong et al. 2021; Ke et al. 2022). For instance, the mussel *G. haimaensis*
and the limpet *Bathyacmaea lactea* were dominant at Haima, while the mussel *G.
platifrons* and the limpet *Bathyacmaea nipponica* were dominant at Site F. The crus-
tacean *S. crosnieri* at Site F, but it was not found at Haima. The sea anemone *Actinernus* sp., the tubeworm *Sclerolinum annulatum* and the clam *A. marissinica* were

endemic to Haima cold seep. These observations imply that geographic settings along with other unknown abiotic and biotic factors may influence the spatial distribution and associated macrobenthic communities in these two seep areas.

Site F and Haima host a total of 30, 21 and 11 species of molluscs, crustaceans and annelids, respectively. A previous study summarizing 42 vent and seep areas in the Northwest Pacific revealed 2 − 38 species of molluscs, crustaceans and annelids (Nakajima et al. 2014), therefore, the faunal diversity of the two seep areas in the SCS is relatively high.

5.2.1 Mollusca

Mollusca is the most diverse phylum of seep macrofauna in the SCS, with 30 species belonging to four classes, including 10 species of Bivalvia, 16 species of Gastropoda, three species of Polyplacophora, and one species of Cephalopoda. A total of eight species of molluscans have been recorded from Site F, and 25 from Haima. Among them, only *G. platifrons*, *"B." aduloides*, *Phymorhynchus buccinoides* and *Bathyacmaea lactea* have been recorded from both cold seep areas. The bathymodioline mussels *G. haimaensis*, and *Nypamodiolus samadiae*, and the vesicomyid clam *A. marissinica* appear to be endemic to the SCS (Fig. 5.2a, b; Chen et al. 2018; Xu et al. 2019; Lin et al. 2022a).

Most non-symbiotic molluscan species live on mussel beds. At Site F, six of the reported molluscs are non-symbiotic, i.e., *Enigmaticolus nipponensis* (previously named as *E. inflatus*, see Chen et al. 2020b), *Provanna* snails and *Bathyacmaea* limpets were commonly found on mussel beds of *G. platifrons*. The octopus *Benthocopus* sp. might not be a seep-endemic species; it might take advantage of the large amount of bathymodiolines as potential food resources, but this hypothesis needs to be tested with further stable isotope and gut content analyses. Likewise, most non-symbiotic molluscs at Haima are mainly observed on the *G. haimaensis* mussel beds (Fig. 5.2e). A specimen of the grey limpet *Puncturella* sp. was collected during a 2022 cruise, but it is not as common as *Bathyacmaea lactea*. The large gastropod *P. buccinoides* sometimes form as dense aggregates on the edge of mussel beds and clam beds, which often move along the same direction. *Yoldiella* sp., a small bivalve (often less than 1 cm) previously identified as *Malletia*, was commonly found in sediment of the mussel beds, clam beds and tubeworm clusters. The glass scallops *Catillopecten* sp., previously identified as *Propeamussium* sp. (Dong et al. 2021; Ke et al. 2022), formed a small aggregate on the empty clam shells (Fig. 5.2d), and have been found swimming into the water column when disturbed.

At Site F, *G. platifrons* is the dominant species, which harbors symbiotic methane-oxidizing bacteria (MOB; Gammaproteobacteria) inside its gill epithelial cells (Sun et al. 2017), and sulfur-oxidizing bacteria (SOB; Campylobacteria) on the gill surface (Sun et al. 2022; Lin et al. 2022b). *"Bythymodiolus" aduloides*, which is rare at both Site F and Haima, also harbors symbiotic SOB in its gill epithelial cells (Lorion et al. 2009). While, at Haima, seven reported bivalves, i.e., four from Bathymodiolinae,

Table 5.1 Species list of seep fauna in the South China Sea

Phylum	Class	Order	Species	Haima	Site F	Yam	Report Reference
Porifera		Hexactinellida	Semperella jiaolongae		+		[9]
			Euplectellidae gen. et sp.		+		[9]
Platyhelminthes	Platyhelminthes	Polycladida	Discocelidae gen. et sp.	+*			This study
Cnidaria	Anthozoa	Actiniaria	Actinernus sp.	+			[16, 18]
			Hormathiidae gen. et sp.	+*			This study
		Scleralcyonacea	Anthomastus sp.		+		[14]
			Chrysogorgia sp.		+		[3, 14]
Mollusca	Bivalvia	Solemyida	Acharax sp. (previously identified as Solemya sp.)	+			[16]
		Nuculanida	Yoldiella sp.	+			[16, 18]
		Mytilida	Gigantidas haimaensis	+			[11, 18]
			Gigantidas platifrons*	+	+	+	[7, 12, 14, 18]
			Gigantidas securiformis			+	[12]
			"Bathymodiolus" aduloides*	+	+		[14, 18]
			Nypamodiolus samadiae	+			[17]
		Lucinida	Thyasira sp.	+			[16]; This study
		Pectenida	Catillopecten sp.	+			[18]
		Venerida	Archivesica marissinica	+			[8]
	Gastropoda	Caenogastropoda	Provanna glabra	+			[16, 18]
			Provanna fenestrata	+			[18]
			Provanna subglabra		+		[14]
		Lepetellida	Puncturella sp.	+*			This study

(continued)

Table 5.1 (continued)

Phylum	Class	Order	Species	Haima	Site F	Yam	Report Reference
		Neogastropoda	Aulacofusus sp.	+			[18]
			Enigmaticolus nipponensis		+		[14]
			Tractolira sp.	+			[18]
			Plicifusus sp.	+			[16]
			Scabrotrophon scitulus	+			[16, 18]
			Phymorhynchus buccinoides*	+	+		[9, 16, 18]
			Phymorhynchus sp.	+			[18]
			Phaenomenella samadiae	+*			This study
		Patellogastropoda	Bathyacmaea lactea*	+	+		[9, 18]
			Bathyacmaea nipponica		+		[14]
			Neolepetopsidae gen. et sp.	+			[9]
		Trochida	Margarites sp.	+			[16]
	Polyplacophora	Lepidopleurida	Leptochiton sp.	+			[18]
			Leptochiton tenuidontus	+			[16]
		Chitonida	Thermochiton sp.	+*			This study
	Cephalopoda	Octopoda	Benthoctopus sp.		+		[9]
Annelida	Polychaeta	Sabellida	Paraescarpia echinospica*	+	+		[7, 14, 18]
			Sclerolinum annulatum	+			[18, 19]
		Phyllodocida	Branchipolynoe pettiboneae*	+	+		[10, 14, 18]
			Branchinotogluma cf. japonicus		+		[14]
			Lepidonotopodium cf. okinawae		+		[10, 14]

(continued)

Table 5.1 (continued)

Phylum	Class	Order	Species	Haima	Site F	Yam	Report Reference
			Glycera sp.	+			[18]
			Nereis sp.	+			[18]
			Sirsoe sp.	+			[18]
		Terebellida	Amphisamytha sp.		+		[14]
		Scolecida (infraclass)	Nicomache sp.	+*			This study
			Capitella sp.	+			[18]
Arthropoda	Malacostraca	Amphipoda	Eurythenes maldoror	+			[16]
			Paralicella sp.	+			[18]
		Isopoda	Bathynomus jamesi	+			[16]
		Decapoda	Alvinocaris longirostris *	+	+		[16, 18]
			Alvinocaris kexue	+*	+		This study
			Munidopsis lauensis *	+	+		[12, 16]
			Munidopsis tuberosa		+		[14]
			Munidopsis verrilli*	+	+		[9, 14]
			Munidopsis pilosa	+			[16]
			Nematocarcinus sp.	+			[16]
			Uroptychus jiaolongae		+		[9]
			Uroptychus spinulosus		+		[9, 16]
			Acanthephyra faxoni		+		[9]
			Globospongicola spinulatus		+		[2, 9]
			Lithodes longispina		+		[4, 9]

(continued)

Table 5.1 (continued)

Phylum	Class	Order	Species	Haima	Site F	Yam	Report Reference
			Shinkaia crosnieri		+		[5]
			Glyphocrangon sp.		+		[1]
			Neolithodes brodiei	+			[18]
			Paralomis sp.	+			[16]
			Chaceon manningi		+		[14]
			Parapaguridae gen. et sp.	+ *			This study
Echinodermata	Holothuroidea	Apodida	*Chiridota heheva*	+			[13, 18]
		Aspidochirotida	*Paelopatides* sp.	+ *			This study
	Echinoidea	Spatangoida	*Paleopneustina* sp.	+ *			This study
	Asteroidea	Velatida	*Pteraster* sp.		+		[14]
	Ophiuroidea	Amphilepidida	*Histampica haimaensis**	+	+		[15, 16]
		Amphilepidida	*Amphiuridae* gen. et sp.	+			[16]
		Ophiacanthida	*Ophiophthalmus serratus*	+			[16, 20]
Chordata	Actinopteri	Perciformes	*Psychrolutes inermis*	+	+		This study; [14]
			Coryphaenoides sp.	+ *	+		This study; [9]
			Synaphobranchus sp.	+ *	+		This study; [9]
			Bathypterois sp.		+		[9]
	Chondrichthyes	Chimaeriformes	*Hydrolagus* sp.	+ *			This study
In total: 8	14	34	81	58	35	2	

Species with asterisks are shared between the Haima cold seep and Site F. Species with asterisks by side of " + " indicate the newly recorded species from this study. This table is updated from the following references: [1] Lin et al. 2013; [2] Tan et al. 2015; [3] Li 2017; [4] Wang et al. 2016; [5] Shen et al. 2016; [6] Wang and Sha 2017; [7] Sun et al. 2017; [8] Chen et al. 2018; [9] Feng et al. 2018a; [10] Zhang et al. 2018; [11] Xu et al. 2019; [12] Kuo et al. 2019; [13] Thomas et al. 2020; [14] Zhao et al. 2020; [15] Li et al. 2021; [16] Dong et al. 2021; [17] Lin et al. 2022a; [18] Ke et al. 2022; [19] Xu et al. 2022; [20] Nethupul et al. 2022

Fig. 5.2 Common seep macrofauna in the Haima cold seep region. **a** mussel bed (dominated by *Gigantidas haimaensis*); **b** clam bed (*Archivesica marissinica*); **c** siboglinid tubeworms (*Sclerolinum annulatum*; Xu et al. 2022); **d** glass scallops (*Catillopecten* sp.) on empty clam shells; (**e**) epibenthos on mussel beds, including *Actinernus* sp., *Puncturella* sp., *Provanna* spp., *Munidopsis lauensis*, *Histampica haimaensis*; **f** swimming shrimp around mussel beds (*Nematocarcinus* sp.); **g** *Alvinocaris longirostris* on the mussel bed; **h** two hermit crabs on mud surface near a clam bed and one carried an orange sea anemone (Hormathiidae gen. et sp.); **i–m** some swimming deep-sea species, including Chordata, jellyfish and sea cucumber (*Paelopatides* sp.). Water depth of above species, ~1360–1430 m. Photographed by ROV *"Pioneer"*

one from Solemyidae, one from Lucinidae and one from Pliocardiinae, are known for their association with symbiotic chemosynthetic bacteria (Sun et al. 2017; Ip et al. 2021). *Gigantidas haimaensis* dominates the seepage area, forming mussel beds and providing shelters for many other macrobenthos (Fig. 5.2a; Xu et al. 2019). The other three bathymodiolines are rare: a few *G. platifrons* and *N. samadiae* were observed within mussel beds of *G. haimaensis* while "*B.*" *aduloides* individuals were found on carbonate rocks surrounding a *P. echinospica* cluster. The different community structures, habitat preferences and abundances reflect the availability of methane and hydrogen sulfide in their habitats. These characteristics are reflected in their stable isotope compositions. The $\delta^{13}C$ values of *G. platifrons* (–66‰ in Site F) and *G. haimaensis* (~–50‰ in Haima) are low, indicating methanotrophic nutrition, whereas the other two bathymodiolines have $\delta^{13}C$ values of ~–35‰, inferring their thiotrophic nutrition (Feng et al. 2018b; Ke et al. 2022). In fact, *G. platifrons* and *G. haimaensis* harbor not only endosymbiotic gammaproteobacterial MOB, but also epibiotic campylobacterial SOB on gill surface, although the function of the epibionts remains unclear (Lin et al. 2022b). In addition, "*B.*" *aduloides* and *N. samadiae* symbiosis with gammaproteobacterial SOB in their gill cells, and these symbionts are closely related to those of other bathymodioline endosymbionts (Lorion et al. 2009; Lin et al. 2022a). Moreover, the gill epithelial cells of *N. samadiae* also harbor Spirochaetes, but their function in the host's nutrition and other aspects of biology such as immunity is unknown (Lin et al. 2022a). While the bathymodiolines are found on the surfaces of sediment or authigenic carbonate rocks, the other three bivalves are semi-infaunal or infaunal. The vesicomyid clam *Archivesica marissinica* lives at the edge of mussel assemblages or away from mussels with half of their shells buried into the sediment (Fig. 5.2b). Like other vesicomyids, they extend their foot into the sediment to obtain hydrogen sulfide from the pore water and extend their inhalant siphon into the water column to pump in oxygen- and carbon dioxide-rich water. Its gill epithelial cells harbor chemosynthetic SOB (Lan et al. 2019; Ip et al. 2021), as indicated by $\delta^{13}C$ value of ~ –35‰ (Ke et al. 2022). Different from the bathymodiolines that are inferred to obtain their symbionts from the ambient environment via horizontal transmission during the larval development in each generation, these vesicomyids transfer their symbionts from parent to offspring via maternal transmission (Funkhouser and Bordenstein 2013). As long-term obligate intracellular lifestyle has promoted genetic drift, the endosymbionts' genome size has been greatly reduced, missing many genes required for free living, including cellular envelope, signal transport, environmental sensing, motility, cell cycle control, and recombination (Ip et al. 2021). *Acharax* sp. (identified as *Solemya* sp. in Dong et al. 2021) and *Thyasira* sp., also bearing chemosynthetic symbionts (Giudice and Rizzo 2022; Taylor et al. 2022), were discovered from sediment samples collected in the margins of mussel beds and clam beds. Both species are rare at Haima, and their detailed symbiotic relationships have not been analyzed, although they are expected to be associated with SOB.

5.2.2 Arthropoda

Arthropoda is the second largest phylum among the reported seep macrobenthos, consisting of 21 seep specialists and non-seep obligate species, among them *Uroptychus jiaolongae* and *U. spinulosus* have been reported only from Site F. Among the 21 species, 12 were from Haima and Site F, respectively, and two of them were found in both seep areas. *Shinkaia crosnieri*, although very abundant at Site F (Zhao et al. 2020), is absent at Haima. *Alvinocaris* and *Munidopsis* are common on mussel beds of both seep areas (Fig. 5.2e, g). *Alvinocaris* shrimps are frequently sheltered among mussels while *Munidopsis* sp. are distributed on mussels. Three species of *Munidopsis* have been reported from the SCS seeps, but it is impossible to distinguish them from videos or photographs taken on-site. *Alvinocaris longirostris* is commonly found on mussel beds. Having been reported from vents in the Manus Basin (Wang and Sha 2017) and Okinawa Trough (Watanabe and Kojima 2015), and a seep in Sagami Bay (Fujikura et al. 1996), *A. longirostris* is perhaps the most widely distributed species among the animals inhabiting chemosynthetic ecosystems in western Pacific. *Alvinocaris kexueae*, reported herein for the first time from the SCS seeps, has been reported from the Manus Basin only (Wang and Sha 2017). These shrimps rely on a consortium of chemosynthetic bacteria comprising mainly SOB on their gill surface for nutrition, which is reflected by their stable isotope signature (δ^{13}C value: ~ –42‰ to –40‰; Ke et al. 2022). *Shinkaia crosnieri* obtains nutrition by feeding directly on episymbiotic bacteria grown on the chaetae that are densely distributed on their ventral side (Watsuji et al. 2014), but its dietary composition may shift during ontogenetic development from a dominance of SOB to a dominance of MOB, indicated by ~ –54‰ to –42‰ δ^{13}C value for individuals of different sizes (Zhao et al. 2020).

At Haima, opportunistic scavenging amphipods *Eurythenes maldoror* and *Parallcella* sp. and the isopod *Bathynomus jamesi* have been captured using trap nets. Large crabs *Neolithodes brodiei* and *Paralomis* sp. have been captured using either scoop nets or directly using the manipulators of the ROV. Although no stable isotope data are available for some crabs, they should be opportunistic predators not specific to seepages like *Chaceon manningi*, *Lithodes longispina* and *Glyphocrangon* sp. found at Site F (Zhao et al. 2020; Ke et al. 2022). We also collected two hermit crabs from Haima in 2022 (Fig. 5.2h), one in the family Parapaguridae associated with a sea anemone. Currently, no detailed taxonomic information is available for these hermit crabs.

5.2.3 Annelida

A total of 11 species of annelids are known in the SCS seeps, including two species of Sabellida, six species of Phyllodocida, one species of Terebellida and two species of Scolecida. Among them, eight species have been recorded from Haima and five from Site F, with only two of them recorded from both cold seep areas (Table 5.1).

As important modifiers of local habitats, Siboglinid tubeworms are among the conspicuous fauna in cold seeps. Both siboglinid tubeworms (*Sclerolinum annulatum* and *Paraescarpia echinospica*) in the SCS form dense tube bushes, with their roots extending deep into the sediment. *Paraescarpia echinospica* forms dense populations at Haima, but it is rare at Site F (Zhao et al. 2020). While *Sclerolinum annulatum* (Fig. 5.2c) forms dense patches at Haima only (Xu et al. 2022). Both tubeworm species harbor thiotrophic gammaproteobacterial SOB as the electron donor for energy production and nutrition (Yang et al. 2019; Sun et al. 2021; Xu et al. 2022).

Three species of scale worms, including *Branchipolynoe pettiboneae*, *Branchinotogluma* cf. *japonicus*, and *Lepidonotopodium* cf. *okinawae*, have been reported from Site F, while only *B. pettiboneae* has been recorded from Haima. *Branchipolynoe pettiboneae* is a parasite of the bathymodiolines *G. haimaensis* and *G. platifrons* in the SCS, commonly found in their mantle cavity. The other two species are free-living, having been found crawling on bathymodioline shells or authigenic carbonate rocks.

Amphisamytha sp., a deposit feeder living inside their self-secreted membrane tubes, has been found only at Site F, with their tubes attached to the carapace and claws of *S. crosnieri*. Two infaunal species (*Capitella* sp. and *Nemache* sp.) and three free-living species (*Glycera* sp., *Nereis* sp. and *Sisoe* sp.) have been recorded at Haima seep, although these species have yet to be identified at species level.

5.2.4 Other Taxa

Seven species of Echinodermata have been recorded from the SCS seeps. Among them, the holothuroid *Chiridota heheva* and the ophiuroid *Histamipica haimaensis* are abundant at Haima. *Histampica haimaensis* in the SCS has not been reported from elsewhere, but *C. heheva* is widely distributed from seeps, vents and wood-falls in Gulf of Guinea, Southwest Indian Ridge and Cayman Rise (Thomas et al. 2020). *Histampica haimaensis* is common on bathymodioline shells, and *P. echinospica* and *S. annulatum* tubes. *C. heheva* often lies on mussel beds and clam beds, as well as bare sediment surfaces. In addition to *H. haimaensis,* two other morphologically distinct ophiuroids have been discovered in this region. *Ophiophthalmus serratus* (Nethupul et al. 2022) is big, with a disc length of ~ 15 mm and the disc having granules on the dorsal side and no plates and exposed radial shield. It has a similar distribution to *H. haimaensis,* but its abundance is much lower. Amphiuridae sp., with a disc length of ~ 10 mm, is rare in this region. It is easily distinguished from the other two species by having a basal-linked radial shield.

In 2022, we found two individuals of brown-color sediment-burying echinoids at Haima belonging to the suborder Paleopneustina. This is the first echinoid species discovered in SCS seeps, but its species identity remains unknown. A purple-color holothuroid (*Paelopatides* sp.) was found swimming around the seep site. Members of this genus of sea cucumbers are common in the Pacific Ocean (Fig. 4. in Martinez et al. 2019). At Site F, the starfish *Pteraster* sp. (Zhao et al. 2020) and the brittle star *H. haimaensis* (Li et al. 2021) are the only reported echinoderms. Like many annelids, most seep echinoderms in the SCS have not been identified at the species level.

Several species of minor taxa, including Platyhelminthes, Porifera and Cnidaria have also been reported from the SCS seeps. A species of Platyhelminthes (Discocelidae gen. et sp.) was first found on mussel beds of the Haima in 2022, but its species identity is unknown. Two species of Porifera (*Semperella jiaolongae* and Euplectellidae gen. et sp.) and two species of soft corals (*Anthomastus* sp. and *Chrysogorgia* sp.) were reported from Site F (Gong et al. 2015; Li 2017; Zhao et al. 2020). Two species of sea anemones were found only at Haima, including the purple *Actinernus* sp., which commonly attached to shells or on the sediment surface, and the orange Hormathiidae gen. et sp., which was rare, with only a few individuals observed on the sediment surface close to mussel beds or clam beds, as well as on the shell of a hermit crab (Fig. 5.2h).

Several species were captured at the seep areas by videos during previous research cruises but were not captured physically, such as an unidentified jellyfish, a swimming holothuroid (*Paelopatides* sp.) and fishes (Fig. 5.2i–m). The raindrop fish *Psychrolutes inermis* has been found in both seep areas, often staying on mussel beds (Xu et al. 2020; Zhao et al. 2020). Given its low mobility and its eggs were found on carbonate rocks (Xu et al. 2020; but it was mistakenly identified as *Spectrunculus grandis*), this fish species is likely a seep specialist, potentially depending on other seep animals for nutrition. Species of *Coryphaenoides*, *Synaphobranchus*, *Bathypterois* and *Hydrolagus* fishes were commonly found swimming the above seep areas. Given their strong mobility, they are like opportunity predators seeking food in the seep areas but not specific to these chemosynthetic habitats.

5.3 Connectivity with Other Northwestern Chemosynthetic Ecosystems

Quite a few chemosymbiotic species and non-chemosymbiotic taxa discovered in the SCS seep communities have also been reported from other vent and seep fields in the Northwest Pacific (Watanabe et al. 2010; Xu et al. 2020; Zhao et al. 2020; Dong et al. 2021). Dong et al. (2021) suggested that the community structure of Haima cold seep is similar to that of chemosynthesis-based communities in Sagami Bay (Fujikura et al. 2012). That needs more evidence, considering *Paraescarpia echinospica* also distributed in Site F and some vents and some species are endemic to Haima (*G. haimaensis*, *A. marissinica*, *N. samadiae* etc.). In contrast, the seep community at Site F is similar to that of the hydrothermal vents in the Okinawa Trough owing to the "*G. platifrons*– *S. crosnieri*" assemblage and other non-symbiont species (Watanabe and Kojima 2015; Feng et al. 2018a; Zhang et al. 2018; Dong et al. 2021). They share several symbiont-hosting species (i.e., *G. platifrons*, "*B.*" *aduloides*, *S. crosnieri*, *A.*

longirostris and *P. echinospica*), and quite a few non-symbiont hosting taxa. This high faunal similarity in the two regions indicates that the seep and vent animals may share some common environmental requirements, although that are supposed to be quite different (Levin et al. 2016).

Recent rapid advances in molecular technologies have enabled us to decipher the connectivity and biogeography of deep-sea organisms from a genomic perspective. To date, population genomic studies based upon genome- or transcriptome-wide single-nucleotide polymorphisms (SNPs) have been performed on three dominant macrofauna species in the Northwest Pacific, including the bathymodioline mussel *G. platifrons* (Xu et al. 2018), the munidopsid squat lobster *S. crosnieri* (Chen et al. 2020a; Xiao et al. 2020), and the patellogastropod limpet *B. nipponica* (Xu et al. 2021). Among them, *G. platifrons* and *B. nipponica* both spawn eggs into the water column for fertilization, with larvae of *G. platifrons* possibly being planktotrophic while those of *B. nipponica* being lecithotrophic (Chen et al. 2019; Ponder et al. 2020). By contrast, females of *S. crosnieri* brood oil-rich eggs attached to the pleopod and were inferred to produce lecithotrophic larvae (Miyake et al. 2010).

With the application of genome-wide SNP markers, two semi-isolated lineages of *G. platifrons* were discovered in the Northwest Pacific (Xu et al. 2018). One was in the seep area at Site F and the other spanned vent fields in the Okinawa Trough and the Off Hatsushima seep in Sagami Bay (Xu et al. 2018). Additionally, a minor genetic divergence was revealed between its local populations in the the southern Okinawa Trough and those across the middle Okinawa Trough and the Sagami Bay (Xu et al. 2018). Similarly, with the application of either genome-wide SNPs or transcriptome-wide SNP markers, two genetic groups of *S. crosnieri* have been detected in the Northwest Pacific, including one at Site F and the other at the vent fields in the Okinawa Trough (Chen et al. 2020b; Xiao et al. 2020). In comparison, four habitat-linked genetic groups were characterized for *B. nipponica* in the Northwest Pacific by using genome-wide SNP markers (Xu et al. 2021). Among them, three seep genetic groups distributed at Site F in the SCS, the Kuroshima Knoll seep in the Ryukyu Arc, and the Off Hatsushima seep in Sagami Bay, respectively, while one vent genetic group dwelled in hydrothermal vents across the southern to middle Okinawa Trough (Xu et al. 2021).

The genetic differentiation patterns of the three deep-sea macrofaunal species collectively highlight the barrier effect of the Luzon Strait: potentially due to the relatively small amount of water involved, the Luzon Strait may represent a genetic barrier or serve as a dispersal barrier to facilitate the genetic divergence of deep-sea species that inhabit its two sides (Xu et al. 2018, 2021). Furthermore, the discordances in genetic divergence of these three deep-sea macrofauna imply the differences in life-history traits and/or environmental adaptabilities or habitat types may have played vital roles in shaping their population connectivity and biogeographic patterns (e.g., Xiao et al. 2020; Xu et al. 2021). Therefore, more information on the reproductive strategies, larval biology and life history of the seep- and/or vent-endemic species, coupled with better deep-sea hydrodynamic models, is crucial to assess the population connectivity of deep-sea macrofauna in the Northwest Pacific.

5.4 Perspectives

In this chapter, we summarized and updated the profiles of cold-seep macrofauna in the SCS. Integrative morphological and molecular analyses have resulted in the description of nine new species and plenty of new records from these areas over the last decade. However, there are still some knowledge gaps in our understanding of the biodiversity of SCS seeps: (1) as new species have been discovered with more sampling efforts, there should be still undescribed species waiting our discovery, especially those hiding under the sediment surface; (2) among the collected species, quite a few have not been identified to the species level, which hinders our understanding of their phylogenetic position and divergence history; (3) many of the collected species have not been subjected to stable isotope analysis, therefore their trophic levels remain unclear; (3) among the symbiont-hosting species, only a few (i.e. *G. haimaensis*, *G. platifrons*, *A. marissinica* and *P. echinospica*) have been subjected to detailed studies of symbiotic relationships, therefore more such studies should be conducted to reveal the diversity of symbiosis in these seep-dwelling animals, and understand how they may have exploited the ecological niches; (4) although quite a few species found on the SCS seeps are widely distributed in the Northwest Pacific, only three species (i.e., *G. platifrons*, *S. crosnieri* and *B. nipponica*) have been subjected to population genetic studies in order to understand the divergence and gene flow among their different populations. For the remaining species, population genetic studies should be conducted to define the biogeographic regions of deep-sea chemosynthetic ecosystems, and devise conservation plans for these ecosystems that are under increasing threats from human activities, especially mining for metal-rich sulfur deposits from vent fields and extraction of methane hydrate from cold seeps (Levin et al. 2016).

Acknowledgements Funding was provided by RGC/GRF of Hong Kong SAR Government (Grants: 12101320, 12102222). We thank the crews of R/V *"Xiangyanghong 01"* and ROV *"Pioneer"* for assistance in sampling and photography.

References

Chen Z, Yan W, Chen M et al (2006) Discovery of seep carbonate nodules as new evidence for gas venting on the northern continental slope of South China Sea. Chin Sci Bull 51(10):1228–1237

Chen C, Okutani T, Liang Q et al (2018) A noteworthy new species of the family Vesicomyidae from the South China Sea (Bivalvia: Glossoidea). Venus (j Malacol Soc Jpn) 76(1–4):29–37

Chen C, Watanabe HK, Nagai Y et al (2019) Complex factors shape phenotypic variation in deep-sea limpets. Biol Lett 15(10):20190504

Chen J, Hui M, Li YL et al (2020a) Genomic evidence of population genetic differentiation in deep-sea squat lobster *Shinkaia crosnieri* (crustacea: Decapoda: Anomura) from Northwestern Pacific hydrothermal vent and cold seep. Deep-Sea Res Part I-Oceanogr Res Pap 156:103188

Chen C, Xu T, Fraussen K et al (2020b) Integrative taxonomy of enigmatic deep-sea true whelks in the sister-genera *Enigmaticolus* and *Thermosipho* (Gastropoda: Buccinidae). Zool J Linn Soc 193(1):230–240

Dong D, Li XZ (2015) Galatheid and chirostylid crustaceans (Decapoda: Anomura) from a cold seep environment in the northeastern South China Sea. Zootaxa 4057(1):91–105

Dong D, Li XZ, Yang M et al (2021) Report of epibenthic macrofauna found from Haima cold seeps and adjacent deep-sea habitats, South China Sea. Mar Life Sci Tech 3(1):1–12

Feng D, Chen D (2015) Authigenic carbonates from an active cold seep of the northern South China Sea: New insights into fluid sources and past seepage activity. Deep-Sea Res Part II-Top Stud Oceanogr 122:74–83

Feng D, Cheng M, Kiel S et al (2015) Using *Bathymodiolus* tissue stable carbon, nitrogen and sulfur isotopes to infer biogeochemical process at a cold seep in the South China Sea. Deep-Sea Res Part I-Oceanogr Res Pap 104:52–59

Feng D, Qiu JW, Hu Y et al (2018a) Cold seep systems in the South China Sea: an overview. J Asian Earth Sci 168:3–16

Feng D, Peckmann J, Li N et al (2018b) The stable isotope fingerprint of chemosynthesis in the shell organic matrix of seep-dwelling bivalves. Chem Geol 479:241–250

Funkhouser LJ, Bordenstein SR (2013) Mom knows best: the universality of maternal microbial transmission. PLoS Biol 11(8):e1001631

Fujikura K, Hashimoto J, Fujiwara Y et al (1996) Community ecology of the chemosynthetic community at Off Hatsushima site, Sagami Bay, Japan-II: comparisons of faunal similarity. JMSTC J Deep Sea Res 12:133–153 (in Japanese with English summary)

Fujikura K, Lindsay DJ, Kitazato H et al (2012) Marine biodiversity in Japanese waters. PLoS ONE 5(8):e11836

Gong L, Li XZ, Qiu J-W (2015) Two new species of Hexactinellida (Porifera) from the South China Sea. Zootaxa 4034(1):182–192

Giudice AL, Rizzo C (2022) Bacteria associated with benthic invertebrates from extreme marine environments: promising but underexplored sources of biotechnologically relevant molecules. Mar Drugs 20(10):617

Ip JC-H, Xu T, Sun J et al (2021) Host-endosymbiont genome integration in a deep-sea chemosymbiotic clam. Mol Biol Evol 38(2):502–518

Kuo M-Y, Kang D-R, Chang C-H et al (2019) New records of three deep-sea *Bathymodiolus* mussels (Bivalvia: Mytilida: Mytilidae) from hydrothermal vent and cold seeps in Taiwan. J Mar Sci Technol-Taiwan 27(4):352–358

Ke Z, Li R, Chen Y et al (2022) A preliminary study of macrofaunal communities and their carbon and nitrogen stable isotopes in the Haima cold seeps, South China Sea. Deep-Sea Res Part I-Oceanogr Res Pap 184:103774

Klaucke I, Berndt C, Crutchley G et al (2016) Fluid venting and seepage at accretionary ridges: the four way closure ridge offshore SW Taiwan. Geo-Mar Lett 36(3):165–174

Làn Y, Sun J, Zhang WP et al (2019) Host-symbiont interactions in deep-sea chemosymbiotic vesicomyid clams: Insights from transcriptome sequencing. Front Mar Sci 6:680

Levin LA, Baco AR, Bowden DA et al (2016) Hydrothermal vents and methane seeps: rethinking the sphere of influence. Front Mar Sci 3:72

Li XZ (2015) Report on two deep-water caridean shrimp species (Crustacea: Decapoda: Caridea: Alvinocarididae, Acanthephyridae) from the northeastern South China Sea. Zootaxa 3911(1):130–138

Li XZ (2017) Taxonomic research on deep-sea macrofauna in the South China Sea using the Chinese deep-sea submersible Jiaolong. Integr Zool 12(4):270–282

Li QH, Li Y-X, Na JY et al (2021) Description of a new species of *Histampica* (Ophiuroidea: Ophiothamnidae) from cold seeps in the South China Sea and analysis of its mitochondrial genome. Deep-Sea Res Part I-Oceanogr Res Pap 178:103658

Li Y-X, Zhang YJ, Ip JC-H et al (2023) Phylogenetic context of a deep-sea clam (Bivalvia: Vesicomyidae) revealed by DNA from 1,500-years-old shells. Zool Res 44(2):353–356

Liu J, Liu HL, Zhang HB (2018) Phylogeny and evolutionary radiation of the marine mussels (Bivalvia: Mytilidae) based on mitochondrial and nuclear genes. Mol Phylogenet Evol 126:233–240

Lin S, Lim YS, Liu CS et al (2007) Formosa Ridge, a cold seep with densely populated chemosynthetic community in the passive margin, southwest of Taiwan. Geochim Cosmochim Acta 71:A582–A582

Lin C-W, Tsuchida S, Lin SW et al (2013) Munidopsis lauensis Baba & de Saint Laurent, 1992 (Decapoda, Anomura, Munidopsidae), a newly recorded squat lobster from a cold seep in Taiwan. Zootaxa 3737(1):92–96

Lin Y-T, Kiel S, Xu T et al (2022a) Phylogenetic placement, morphology and gill-associated bacteria of a new genus and species of deep-sea mussel (Mytilidae: Bathymodiolinae) from the South China Sea. Deep-Sea Res Part I-Oceanogr Res Pap 190:103894

Lin Y-T, Xu T, Ip JC-H et al (2022b) Interactions among deep-sea mussels and their epibiotic and endosymbiotic chemoautotrophic bacteria: Insights from multi-omics analysis. Zool Res 44(1):106–125

Lin ZY, Chen HW, Lin H-J (2020) Trophic model of a deep-sea ecosystem with methane seeps in the South China Sea. Deep-Sea Res Part I-Oceanogr Res Pap 159:103251

Lorion J, Duperron S, Gros O et al (2009) Several deep-sea mussels and their associated symbionts are able to live both on wood and on whale falls. Proc Royal Soc B-Biol Sci 276(1654):177–185

Machiyama H, Lin S, Fujikura K et al (2007) Discovery of "hydrothermal" chemosynthetic community in a cold seep environment, Formosa Ridge: seafloor observation results from first ROV cruise, off southwestern Taiwan. EOS Trans AGU 88(52):OS23A-1041

Martinez MI, Solís-Marín FA, Penchaszadeh PE (2019) First report of Paelopatides (Synallactida, Synallactidae) for the SW Atlantic, with description of a new species from the deep-sea off Argentina. Zool Anz 278:21–27

Miyake H, Kitada M, Itoh T et al (2010) Larvae of deep-sea chemosynthetic ecosystem animals in captivity. Cah Biol Mar 51(4):441–450

Nakajima R, Yamakita T, Watanabe H et al (2014) Species richness and community structure of benthic macrofauna and megafauna in the deep-sea chemosynthetic ecosystems around the Japanese archipelago: an attempt to identify priority areas for conservation. Divers Distrib 20(10):1160–1172

Nethupul H, Stöhr S, Zhang HB (2022) Review of Ophioplinthaca Verrill, 1899 (Echinodermata, Ophiuroidea, Ophiacanthidae), description of new species in Ophioplinthaca and Ophiophthalmus, and new records from the Northwest Pacific and the South China Sea. Zookeys 1099:155–202

Niu MY, Feng D (2017) Ecosystems of cold seeps in the South China Sea. In: Kallmeyer J (Ed) Life at vents and seeps. Walter de Gruyer GmbH, Berlin/Boston, pp 139–160

Ponder WF, Lindberg DR, Ponder JM (2020) Biology and evolution of the Mollusca. CRC Press, Boca Raton (FL)

Suess E, Huang Y, Wu N et al (2005) South China Sea: distribution, formation and effect of methane & gas hydrate on the environment. RV SONNE cruise report SO 177, Sino-German Cooperative Project, vol IFM-GEOMAR report no. 4. IFM-GEO-MAR, Kiel. https://oceanrep.geomar.de/id/eprint/1989

Sun J, Zhang Y, Xu T et al (2017) Adaptation to deep-sea chemosynthetic environments as revealed by mussel genomes. Nat Ecol Evol 1(5):0121

Sun YN, Sun J, Yang Y et al (2021) Genomic signatures supporting the symbiosis and formation of chitinous tube in the deep-sea tubeworm Paraescarpia echinospica. Mol Biol Evol 38(10):4116–4134

Sun Y, Wang MX, Zhong ZS et al (2022) Adaption to hydrogen sulfide-rich environments: strategies for active detoxification in deep-sea symbiotic mussels Gigantidas Platifrons. Sci Total Environ 804:150054

Shen Y, Kou Q, Chen W et al (2016) Comparative population structure of two dominant species, *Shinkaia crosnieri* (Munidopsidae: *Shinkaia*) and *Bathymodiolus platifrons* (Mytilidae: *Bathymodiolus*), inhabiting both deep-sea vent and cold seep inferred from mitochondrial multi-genes. Ecol Evol 6(11):3571–3582

Taylor JD, Glover EA, Yuen B et al (2022) Closing the gap: a new phylogeny and classification of the chemosymbiotic bivalve family Lucinidae with molecular evidence for 73% of living genera. J Molluscan Stud 88(4):eyac025

Tan TW, Tomoyuki K, Chen CL et al (2015) *Globospongicola jiaolongi* Jiang K, Li a junior subjective synonym of *G. spinulatus* Komai & Saito, 2006 (Crustacea: Decapoda: Stenopodidea: Spongicolidae). Zootaxa 4072(5):579–584

Thomas EA, Liu RY, Amon D et al (2020) *Chiridota heheva*—the cosmopolitan holothurian. Mar Biodivers 50(6):109–110

Tseng Y, Römer M, Lin S et al (2023) Yam Seep at four-way closure ridge–a prominent active gas seep system at the accretionary wedge SW offshore Taiwan. Int J Earth Sci in press

Watanabe H, Fujikura K, Kojima S et al (2010) Japan: Vents and seeps in close proximity. In: Kiel S (ed) The vent and seep biota: aspects from microbes to ecosystems. Springer, Dordrecht, pp 279–401

Watanabe H, Kojima S (2015) Vent Fauna in the Okinawa trough. In: Ishibashi JI et al (Eds) Subseafloor biosphere linked to hydrothermal systems: TAIGA Concept. Springer Tokyo, pp 449–459

Watsuji T-O, Yamamoto A, Motoki K et al (2014) Molecular evidence of digestion and absorption of epibiotic bacterial community by deep-sea crab *Shinkaia crosnieri*. ISME J 9(4):821–831

Wang T-W, Ahyong S, Chan T-Y (2016) First records of *Lithodes longispina* Sakai, 1971 (Crustacea: Decapoda: Anomura: Lithodidae) from southwestern Taiwan, including a site in the vicinity of a cold seep. Zootaxa 4066(2):173–176

Wang YR, Sha ZL (2017) A new species of the genus *Alvinocaris* Williams and Chace, 1982 (Crustacea: Decapoda: Caridea: Alvinocarididae) from the Manus Basin hydrothermal vents, Southwest Pacific. Zootaxa 4226(1):126–136

Wang XD, Guan HX, Qiu J-W et al (2022) Macro-ecology of cold seeps in the South China Sea. Geosyst Geoenviron 1(3):100081

Xiao Y, Xu T, Sun J et al (2020) Population genetic structure and gene expression plasticity of the deep-sea vent and seep squat lobster *Shinkaia crosnieri*. Front Mar Sci 7:587686

Xu T, Sun J, Lv J et al (2017) Genome-wide discovery of single nucleotide polymorphisms (SNPs) and single nucleotide variants (SNVs) in deep-sea mussels: potential use in population genomics and cross-species application. Deep-Sea Res Part II-Top Stud Oceanogr 137:318–326

Xu T, Sun J, Watanabe H et al (2018) Population genetic structure of the deep-sea mussel *Bathymodiolus platifrons* (Bivalvia: Mytilidae) in the Northwest pacific. Evol Appl 11(10):1915–1930

Xu T, Feng D, Tao J et al (2019) A new species of deep-sea mussel (Bivalvia: Mytilidae: *Gigantidas*) from the South China Sea: morphology, phylogenetic position, and gill-associated microbes. Deep-Sea Res Part I-Oceanogr Res Pap 146:79–90

Xu HC, Du MR, Li JT et al (2020) Spatial distribution of seepages and associated biological communities within Haima cold seep field South China Sea. J Sea Res 165:101957

Xu T, Wang Y, Sun J et al (2021) Hidden historical habitat-linked population divergence and contemporary gene flow of a deep-sea patellogastropod limpet. Mol Biol Evol 38(12):5640–5654

Xu T, Sun Y, Wang Z et al (2022) The morphology, mitogenome, phylogenetic position, and symbiotic bacteria of a new species of *Sclerolinum* (Annelida: Siboglinidae) in the South China Sea. Front Mar Sci 8(8):793645

Yao GY, Zhang H, Xiong PP et al (2022) Community characteristics and genetic diversity of macrobenthos in haima cold seep. Front Mar Sci 9:920327

Yang CH, Tsuchida S, Fujikura K et al (2016) Connectivity of the squat lobsters *Shinkaia crosnieri* (Crustacea: Decapoda: galatheidae) between cold seep and hydrothermal vent habitats. Bull Mar Sci 92(1):17–31

Yang Y, Sun J, Kwan YH et al (2019) Genomic, transcriptomic, and proteomic insights into the symbiosis of deep-sea tubeworm holobionts. ISME J 14(1):135–150

Zhao Y, Xu T, Law YS et al (2020) Ecological characterization of cold-seep epifauna in the South China Sea. Deep-Sea Res Part I-Oceanogr Res Pap 163:103361

Zhang YJ, Sun J, Rouse GW et al (2018) Phylogeny, evolution and mitochondrial gene order rearrangement in scale worms (Aphroditiformia, Annelida). Mol Phylogenet Evol 125:220–231

Zhang K, Sun J, Xu T et al (2022) Phylogenetic relationships and adaptation in deep-sea mussels: insights from mitochondrial genomes. Int J Mol Sci 22(4):1900

Chapter 6
Symbioses from Cold Seeps

Chaolun Li, Minxiao Wang, Hao Wang, Li Zhou, Zhaoshan Zhong, Hao Chen, and Yan Sun

Abstract Establishing symbiosis between bacteria and invertebrates can significantly enhance energy transfer efficiency between them, which may aid in shaping the flourishing community in deep-sea chemosynthetic ecosystems, including cold seeps, hydrothermal vents, and organic falls. The symbionts utilize the chemical energy from reductive materials to fix carbon, and the hosts absorb the nutrients for growth through farming, milking, or both. Moreover, symbiosis can enhance the sustainability of both participants to survive in harsh conditions. However, the exact process and the regulatory network of symbiosis are still unknown. The cold seeps in

C. Li (✉) · M. Wang · H. Wang · L. Zhou · Z. Zhong · H. Chen · Y. Sun
Center of Deep-Sea Research, and Key Laboratory of Marine Ecology and Environmental
Sciences, Institute of Oceanology, Chinese Academy of Sciences, Qingdao 266071, China
e-mail: lcl@scsio.ac.cn

M. Wang
e-mail: wangminxiao@qdio.ac.cn

H. Wang
e-mail: haowang@qdio.ac.cn

L. Zhou
e-mail: lizhou@qdio.ac.cn

Z. Zhong
e-mail: zhongzhaoshan@qdio.ac.cn

H. Chen
e-mail: chenhao@qdio.ac.cn

Y. Sun
e-mail: ysun@qdio.ac.cn

C. Li
South China Sea Institute of Oceanology, Chinese Academy of Sciences, Guangzhou 510301,
China

University of Chinese Academy of Sciences, Beijing 10049, China

M. Wang · H. Wang · L. Zhou · Z. Zhong · H. Chen · Y. Sun
Laboratory for Marine Ecology and Environmental Science, Qingdao National Laboratory for
Marine Science and Technology, Qingdao 266071, China

D. Chen and D. Feng (eds.), *South China Sea Seeps*,
https://doi.org/10.1007/978-981-99-1494-4_6

the South China Sea offer natural laboratories to study the composition, ecological functions, and regulatory mechanisms of deep-sea symbioses. In this chapter, we focused on two dominant species, a deep-sea mussel *Gigantidas platifrons* and a squat lobster *Shinkaia crosnieri,* which represent endosymbiosis and episymbiosis, respectively, at Site F to summarize our understanding of deep-sea chemosymbiosis. We also discussed some promising avenues for future studies, such as deep-sea in situ experiments to show the exact responses of deep-sea organisms, culture-dependent experiments with genetic operations to validate the functions of critical genes, and microscale omics to elucidate the possible interactions at subcellular levels.

6.1 Introduction

Gravitational and tectonic forces can create leakage pathways, termed cold seeps, that allow subsurface methane to rise from the sea floor (Fig. 6.1; Boetius and Wenzhöfer 2013). Cold seeps are often located on geologically activated and passive continental margins. Leakage fluids are rich in reduced materials, such as methane, short-chain alkanes, and hydrogen sulfide, from deep to shallow sediments or even seawater (Ceramicola et al. 2018). Compared with traditional photosynthetic ecosystems, the particularity of cold seeps is their dependence on chemosynthetic bacteria for primary productivity. The discoveries of such lightless ecosystems have revolutionized our understanding of the origin and energy sources of life on Earth. Since their discovery in the late 1970s and early 1980s (Lonsdale 1979; Paull et al. 1984), an increasing number of cold seeps have been explored with accumulating knowledge of their specialized functions in extreme ecosystems. The most amazing discovery in these ecosystems is the organisms surviving independently of sunlight and form the marvelous landscape of a dense invertebrate community (Fig. 6.2). The macrobenthos fauna is apparently nourished by reduced chemicals seeping out of the sea floor (Kennicutt et al. 1985). To facilitate energy and matter transfer between invertebrates and microbes, they have established close relationships by symbiosis (Husnik et al. 2021). The giant gutless tubeworm *Riftia pachyptila* is the first discovered deep-sea invertebrate with endosymbionts. They used reduced sulfur compounds to fix carbon dioxide and transported the organic carbon to the hosts (Dubilier et al. 2008). The discovery of *R. pachyptila* symbiosis has triggered a series of searches for symbiosis in vents and cold seeps in the deep sea.

Symbiotic relationships with chemosynthetic bacteria have been widely discovered in diverse groups of macrofauna, including intracellular symbiosis (endosymbiosis) represented by bathymodiolin mussels, thyasirids, and vesicomyids in the phylum Mollusca; tubeworms in the phylum Annelida (Dubilier et al. 2008); and extracellular symbiosis (episymbiosis) represented by galatheid crabs (Watsuji et al. 2010) and alvinocaridid shrimp (Polz and Cavanaugh 1995) in the phylum Arthropoda. The host animal must be able to interact with the symbiont at the tissue, cell, and molecular levels to maintain this high-efficiency chemosymbiosis (Hinzke et al. 2019). The host is able to provide the symbiont with necessary gases, such as sulfide,

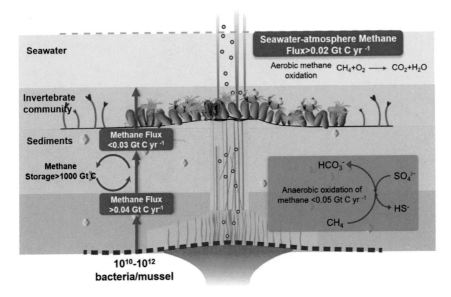

Fig. 6.1 Schematic view of the cold-seep ecosystem with methane fluxes at continental slopes. Reprinted from Nature Geoscience, 6, Boetius and Wenzhöfer (2013) Seafloor oxygen consumption fuelled by methane from cold seeps, 725–734, Copyright (2022), with permission from Springer Nature

methane, oxygen, and carbon dioxide, in large quantities to support the fast growth of the symbiont (Hinzke et al. 2021). Moreover, the host also needs to control the symbiont population by selectively digesting a portion of the symbiont and then extracting and transporting the nutrient for its own growth (Gomez-Mendikute et al. 2005; Hirokawa and Noda 2008; Renoz et al. 2015). To date, this unique symbiosis (between invertebrates and chemosynthetic bacteria) has become a model for understanding animal-microorganism interactions and has been intensely studied. Here, we provide a brief review of studies on the composition, function, and regulatory mechanism of symbiosis in cold seeps in the South China Sea (SCS), focusing on two dominant species, *G. platifrons* and *S. crosnieri,* at Site F in the SCS.

6.2 Adaptive Characteristics of a Symbiotic Lifestyle

Concerning the location where the symbionts dwell, the symbiotic microbiome can form associations with invertebrates, mainly in two forms, endosymbiosis and episymbiosis. Episymbionts occur on the surface of the host, whereas endosymbionts occur internally, either within a body cavity but outside the cells (extracellular) or within cells (intracellular). Bathymodiolin mussels are the research model

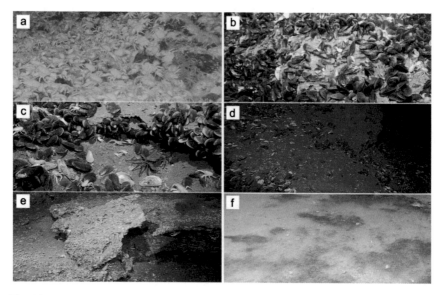

Fig. 6.2 Photos showing the characterized habitat at Site F of the South China Sea. **a** the assemblage of the squat lobster; **b, c** mussel beds harboring nails, limpets and Alvin shrimps; **d** shell debris with tiny sponges; **e** authigenic carbonate covered with sponges and corals; **f** bacterial mats. Reprinted from Journal of Marine Systems, 218, Cao et al. (2021) In situ detection of the fine scale heterogeneity of active cold seep environment of the Formosa Ridge, the South China Sea, 103,530, Copyright (2022), with permission from Elsevier

of endosymbiosis. These mussels host SOX and/or MOX in specialized gill epithelial cells, namely, bacteriocytes (Halary et al. 2008). To obtain reduced chemicals in deep-sea cold seeps, they have evolved some special characteristics, such as enlarged gills and hypertrophic bacteriocytes (Duperron 2010). The exchange surfaces of the bathymodiolin mussels are approximately 20-fold higher than those found in similar-sized coastal mussels (Duperron et al. 2016). With long-term symbiotic adaptation, the digestive systems of deep-sea mussels have degraded, such as intestinal and lip degradation or disappearance (Fisher and Childress 1986).

These mussels have developed refined structures to sustain the bacteriocytes. The gills of *G. platifrons* have a typical filibranch bivalve structure involving cilia ventilation that provides symbiotic bacteria with necessary gases, such as methane and oxygen, from the seep fluid (Fig. 6.3). Each gill filament's lateral face is covered with bacteriocytes, with the exception of the heavily ciliated filament tip. Each bacteriocyte hosts a dense population of a single species of methane-oxidizing symbiont. Symbiont-free intercalary cells also appear between the bacteriocytes. Scanning electron microscopy (SEM) images showed that each bacteriocyte's apex protrudes slightly above the surrounding intercalary cells (Fig. 6.4). Although the surface of the bacteriocytes is smooth, the intercalary cells have microvilli and cilia on their surfaces, which may help slow down water flow and improve gas exchange (Fig. 6.4b′). A transmission electron microscope (TEM) image showed

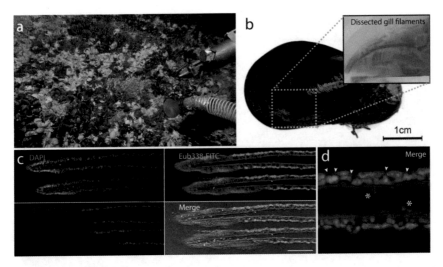

Fig. 6.3 The symbiosis of the deep-sea mussel *G. platifrons*. **a** Sampling of the deep-sea mussel *G. platifrons* at Site F. **b** A healthy *G. platifrons* specimen; the inset shows the dissected gill filaments. **c** Double-FISH with FITC-labeled eubacteria probe EU338, Cy3-labeled *G. platifrons* symbiont probe. **d** Magnified merged image showing the bacteriocytes that host symbionts, nonsymbiotic intercalary cells (white arrowhead), and amoebocytes (white asterisk). Scale bar of Panel (**c**): 50 μm. Reprinted from Deep-Sea Research Part I, 151, Wang et al. (2019) Comparative transcriptomic analysis 624 illuminates the host-symbiont interactions in the deep-sea mussel *Bathymodiolus platifrons*, 103,082, Copyright (2022), with permission from Elsevier

that the bacteriocytes have a compartmentalized internal ultrastructure (Fig. 6.4c). The methane-oxidizing symbionts (Fig. 6.4d and d′), which have the characteristic intracytoplasmic membrane structure of type I methanotroph bacteria, are engulfed in intracellular vacuoles and aggregated at the apex of the bacteriocytes (Fig. 6.3c and d). Primary and secondary lysosomes are also located in the lower part of the bacteriocytes (Fig. 6.4c).

Shinkaia crosnieri is another dominant species at site F and serves as the model for episymbiont studies. *S. crosnieri* raises complex microbial communities, rather than certain symbiotic species, that mainly attach to the setae of appendages (Fig. 6.5; Xu et al. 2022). Rod-shaped Methylococcales attach to the setae surfaces or via the formation of filamentous aggregations with diameters between 0.8 and 1.5 μm. Both Thiotrichales and Campylobacteriales form filament-like aggregations but differ in diameter from 2 to 3 μm and 0.4 to 0.6 μm, respectively. Flavobacteriales were observed as small, rod-shaped cells attached to other filamentous bacteria, indicating that they may utilize the organic carbon generated by other chemosynthetic bacteria. These associations are consistent with observations in biofilms growing on the surfaces of a black smoker chimney in the Loki's Castle vent field (Stokke et al. 2015). Episymbiont biofilms tend to develop overlapping colonies with other bacteria that occur in close proximity, thereby facilitating metabolite sharing among symbionts. *S. crosnieri* can comb out its ventral setae using the third maxilliped to

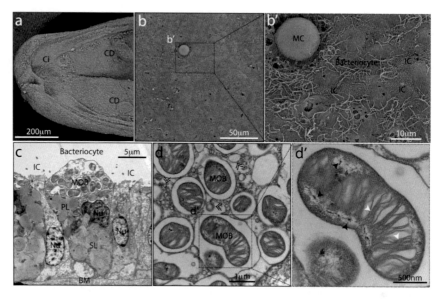

Fig. 6.4 Electron microscopy analysis of the gill of *G. platifrons* **a** An SEM image of the heavily ciliated tip of the *G. platifrons* gill tip. Ci: ciliary, CD: ciliary disk. **b** An SEM image of the lateral face of the *G. platifrons* gill, showing the bacteriocytes and intercalary cells. **b′** A magnified region in **b** showing the detailed surface structures of the bacteriocyte and intercalary cells. The surfaces of bacteriocytes protrude slightly higher than the surrounding intercalary cells. There are cilia and microvillae on the surface of intercalary cells. MC: mucus cell, IC: intercalary cell. **c** A TEM image of the longitudinal section of the bacteriocytes and adjacent intercalary cells. **d** TEM image of methane-oxidizing symbiotic bacteria covered in vacuoles. **d′** A magnified image showing the detailed structure of the symbiont. White arrowheads: intracytoplasmic membrane structure, which is a characteristic feature of type I methanotroph bacteria. Black arrowheads: DNA of the symbiotic bacteria. Reprinted from iScience, 24, Wang et al. (2021) Molecular analyses of the gill symbiosis of the bathymodiolin mussel *Gigantidas platifrons*,101,894, Copyright (2022), with permission from Elsevier

consume the episymbionts. Special ethological features that enhance primary productivity were also recorded for this species. The lobster waves its arm in the reducing fluid to increase the productivity of its epibionts by removing boundary layers that may otherwise limit carbon fixation (Thurber et al. 2011). In contrast, no apparent setae can be observed in another sympatric deep-sea squat lobster, *Munidopsis verrilli* (Dong and Li 2015).

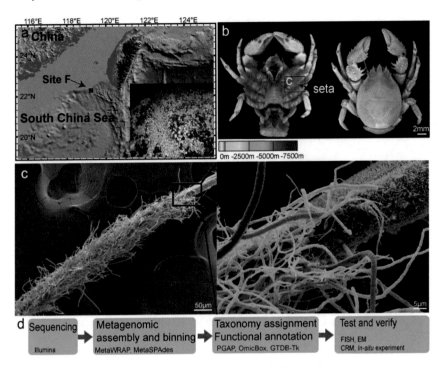

Fig. 6.5 *S. crosnieri* setae symbiont sampling information. **a** The geographic location of the cold seep site studied here (i.e., Site F) and image of the *S. crosnieri* community adjacent to a methane seepage taken by the ROV *Faxian*. **b** Ventral and dorsal views of *S. crosnieri*, with the setae on the appendages highlighted. **c** SEM image showing the distribution of key episymbiont taxa on the setae based on FISH analyses. Coloring of taxa is as follows: blue: Methylococcales; red: Thiotrichales; yellow: Campylobacterales; green: Methylococcales or Thiotrichales. Reprinted from mSystems, 7, Xu et al. (2022) Metabolism Interactions Promote the Overall Functioning of the Episymbiotic Chemosynthetic Community of *Shinkaia crosnieri* of Cold Seeps, e00320-22, Copyright (2022), with permission from ASM journals

6.3 Taxonomic Composition and Carbon Fixation Pathway of the Symbionts

Although 6 proteobacteria could be recovered from the gills of the *G. platifrons,* only the methanotrophic bacteria *Methyloprofundas spp.* were found inside the bacteriocytes with high abundance (up to 95%). Based on the metatranscriptomic dataset, the methane-oxidizing process, which generates formaldehyde, is the most prominent metabolic process in the symbiont (Fig. 6.6; Wang et al. 2021). The three particulate methane monooxygenases that conduct the first step of the reaction by converting methane to methanol were the most highly expressed protein-coding genes. Furthermore, two downstream pathways, the tetrahydromethanopterin (H_4MPT) and ribulose monophosphate (RuMP) pathways, which detoxify formaldehyde and generate energy and biomass, were highly active in the symbiont.

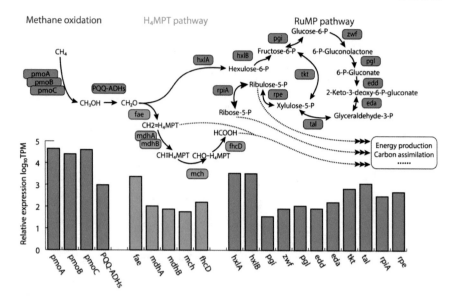

Fig. 6.6 An overview of the symbiont methane oxidation and downstream H4MPT and RuMP pathways. The histogram at the bottom shows the relative gene expression levels ($\log_{10}$TPM) of enzymes of these three pathways. Transcripts encoding enzymes in these three metabolic processes dominate the transcriptome of the *G. platifrons* symbiont. Reprinted from iScience, 24, Wang et al. (2021) Molecular analyses of the gill symbiosis of the bathymodiolin mussel *Gigantidas platifrons*,101, 894, Copyright (2022), with permission from Elsevier

Complex episymbionts have been observed in the setae of *S. crosnieri*. Methylococcales (42.3%) was the most abundant order, followed by Thiotrichales (14.8%), Campylobacteriales (4.9%), Flavobacteriales (4.4%), Chitinophagales (3.8%), and Nitrosococcales (2.2%). Metagenomic binning was carried out to further elucidate the taxonomic composition and functions of the symbionts. The MAGs clustered into five monophyletic clades in the orders Thiotrichales, Nitrosococcales, Methylococcales, Chitinophagales, and Flavobacteriales (Fig. 6.7a), which represented all of the dominant taxa recovered in the phyloFlash 16S ribosomal RNA (rRNA) gene-based analysis. Genome Taxonomy Database Toolkit (GTDB-Tk)-based taxonomic assessments indicated that many MAGs were divergent from known reference genomes, suggesting that they may represent unknown genera and even unknown families. The complex episymbiotic community can supply hosts with fixed organic carbons. One-carbon metabolic pathways, including methane oxidation, methanol oxidation, and the RuMP pathway, were the most notable functional categories encoded by the Methylococcales and Nitrosococcales MAGs. Methane oxidation was particularly prominent in the Methylococcales MAGs. Thiotrichales and Campylobacteriales can fix carbon dioxide via the reverse tricarboxylic acid (rTCA) and Calvin-Benson-Bassham (CBB) pathways, respectively, using electrons obtained by the oxidation of reduced sulfur.

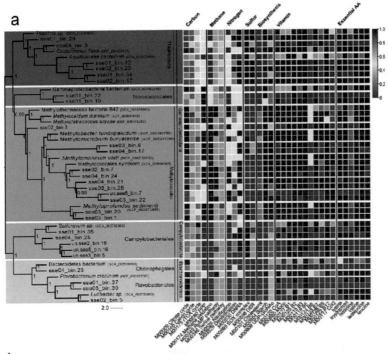

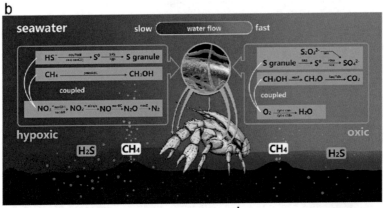

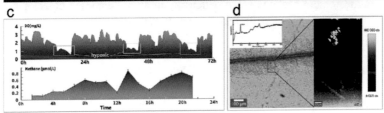

◄**Fig. 6.7** An overview of the symbiont's composition and metabolic interactions with *S. crosnieri*. **a** Phylogenomic assignment, relative abundance, and metabolic potential of the dominant MAGs of the setae along with reference genomes. Phylogeny of 28 high-quality MAGs recovered from the *S. crosnieri* setae. The colors of the tree represent order-level taxonomic groups for MAGs. Heatmap colors represent the completeness of KEGG modules; **b** Schematic of the predicted overall community strategy. During the hypoxic phase, the symbionts oxidize methane and sulfide to methanol and sulfur globules by coupling to nitrate respiration. The stable sulfur and methane intermediates can then be stored as inclusions or released to the environment and taken up by other symbionts. During the oxic phase, episymbionts can use oxygen to oxidize sulfur and methanol to conserve additional energy to support cellular growth; **c** Temporal measurements of DO and methane concentrations inside the *S. crosnieri* community; **d** Confocal Raman microscopy imaging showing elemental sulfur in the episymbiont communities. Reprinted from mSystems, 7, Xu et al. (2022) Metabolism Interactions Promote the Overall Functioning of the Episymbiotic Chemosynthetic Community of *Shinkaia crosnieri* of Cold Seeps, e00320-22, Copyright (2022), with permission from ASM journals

6.4   Nutrient Transfer Between Host and Symbionts

Formations of chemosymbiotic associations, especially endosymbiosis, in the deep sea are mainly nutrient driven. Deep-sea chemosymbiotic invertebrates are suitable for studying metabolic interaction mechanisms due to their exclusive and robust association with chemosynthetic bacteria (Dubilier et al. 2008). With the application of new approaches such as single-cell spatial omics and subcellular imaging, the metabolic interaction mechanisms of these symbiotic systems in cold seeps and hydrothermal vents were revealed (Ponnudurai et al. 2017; Geier et al. 2020). Recently, a comprehensive picture of metabolic cooperation in *G. platifrons* symbiosis was reported with state-of-the-art single-cell transcriptome and spatial transcriptome sequencing (Chen et al. 2021). It has been demonstrated that deep-sea mussels have reshaped bacteriocyte metabolism remarkably to maximize symbiotic profits for both partners. For example, bacteriocytes encode dozens of genes involved in the biosynthesis and transport of carbohydrates, lipids, amino acids, and vitamins to improve the acquisition of nutrition from symbionts. Nevertheless, the exact metabolites transferred between the host and endosymbionts are less known. It is now widely recognized that methanotrophic endosymbionts can provide sterol intermediates to the mussel host, while endosymbiotic sulfur oxidizers may compensate for the host's putative deficiency in amino acid and cofactor biosynthesis (Ponnudurai et al. 2017; Takishita et al. 2017). In turn, the host replenishes essential biosynthetic TCA cycle intermediates for their endosymbiont. Recently, it was demonstrated that bacteriocytes could retrieve fructose-6-phosphate (F6P) directly from methanotrophic symbionts by sugar phosphate exchanger genes and convert it into glucose-6-phosphate (G6P), fructose-1,6-bisphosphate (F-1,6-BP) and glyceraldehyde-3-phosphate (G6P) before redistributing them back to both the host and symbionts as supplements of gluconeogenesis and the tricarboxylic acid cycle (TCA cycle). A direct supply of ammonia from the host to symbionts via ammonium transporters was also reported (Chen et al. 2022). The bioconversion and exchange of sugar phosphate and ammonia between the host and symbionts provided an efficient and direct way to coordinate the metabolism of both partners, which greatly promoted the efficiency and profits of symbiosis.

In addition, metatranscriptome sequencing and gene coexpression network analysis of *G. platifrons* under laboratory maintenance with a gradual loss of symbionts also provided evidence of possible nutrient interactions between the two parts (Sun et al. 2022a). One-day short-term maintenance triggered global transcriptional perturbation in symbionts but little gene expression changes in mussel hosts; the genes that changed were mainly involved in responses to environmental changes (Fig. 6.8a). Long-term maintenance with depleted symbionts induced a metabolic shift in the mussel host. The most notable changes were the suppression of sterol biosynthesis and the complementary activation of terpenoid backbone synthesis in response to the reduction in bacteria-derived terpenoid sources. In addition, gene expression of the host proteasomes responsible for amino acid deprivation caused by symbiont depletion was significantly upregulated. Additionally, a significant correlation between host microtubule motor activity and symbiont abundance was revealed, suggesting the possible function of microtubule-based intracellular trafficking in the nutritional interaction of symbiosis (Fig. 6.8b). Overall, the dynamic transcriptomic changes of *G. platifrons* during the loss of symbionts highlighted the nutritional dependence of the host on terpenoid compounds and essential amino acids from their endosymbionts and proposed the possible important function of microtubule-based intracellular trafficking in symbiosis.

In *S. crosnieri*, the metabolic potentials of the dominant taxa have been reconstructed based on KEGG module completeness visualizations in Fig. 6.7a (Xu et al. 2022). The episymbionts exhibited the potential to provide most nutrients to their hosts, including hydrocarbons, vitamins, and essential amino acids. In return, invertebrate hosts can supply vitamin B6 to episymbionts, which cannot synthesize it on their own (Lan et al. 2021). The nutrient contributions to the host are also likely complementary. For example, the essential amino acid histidine could only be provided by some Campylobacteriales and Thiotrichales, and the steroid precursor lanosterol could only be synthesized by Methylococcales. The complex functionally complemented symbiotic community may be adaptive and selected by the host. The setae and symbionts can be observed in the intestinal tracts of *S. crosnieri*, and the symbionts can be digested by intestinal digestive enzymes (Watsuji et al. 2015). The results indicate that *S. crosnieri* ingests the epibionts using maxillipeds and assimilates them via its digestive organs as a nutrient source.

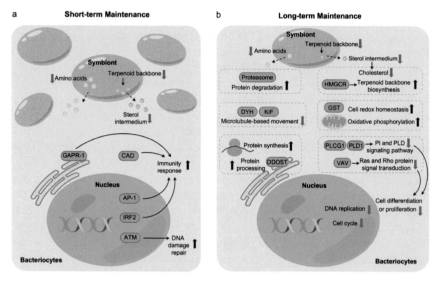

Fig. 6.8 Schematic diagram summarizing the overall metabolic responses of *G. platifrons* holobionts to laboratory maintenance. **a** Transcriptional responses to one-day short-term maintenance. The biosynthesis of essential amino acids and sterol intermediates was downregulated in the endosymbionts after short-term maintenance. The DNA damage repair-related genes and immune-related genes of the mussel host were upregulated. CAD: Cis-aconitate decarboxylase, AP-1: Transcription factor AP-1, IRF2: Interferon regulatory Factor 2, ATM: Serine-protein kinase ATM, GAPR-1: Golgi-associated plant pathogenesis-related protein; **b** Transcriptional responses to long-term maintenance for 25 and 35 days. The biosynthesis of essential amino acids and sterol intermediates was continuously downregulated in symbionts, and nutrients were deficient due to symbiont depletion. Sterol biosynthesis in the mussel host was suppressed, and the biosynthesis of the terpenoid backbone and proteasomes was complementarily upregulated in response to the reduction in bacteria-derived nutrition sources. The microtubule-based movement and regulatory pathways for DNA replication and cell differentiation or proliferation were also downregulated. In contrast, genes related to protein synthesis and processing, mitochondrial oxidative phosphorylation, and cell redox homeostasis were upregulated. DYH: Dynein heavy chain, KIF: Kinesin-like protein, OST48: Dolichyl-diphosphooligosaccharide-protein glycosyltransferase 48 kDa subunit, HMGCR: 3-hydroxy-3-methylglutaryl-coenzyme a reductase, GST: Glutathione S-transferase, PLCG1: 1-phosphatidylinositol 4,5-bisphosphate phosphodiesterase gamma-1, PLD1: Phospholipase D1, VAV: Proto-oncogene vav. Green arrows indicate the downregulated genes or pathways. Red arrows indicate the upregulated genes or pathways. Reprinted from Molecular Ecology, Sun et al. (2022a, b) Insights into symbiotic interactions from metatranscriptome analysis of deep-sea mussel *Gigantidas platifrons* under long term laboratory maintenance, https://doi.org/10.1111/mec.16765, Copyright (2022), with permission from John Wiley and Sons

6.5 The Molecular Basis for Establishing and Maintaining Symbiosis

The establishment of endosymbiosis is undoubtedly the most important step in deep-sea adaptation (Stewart et al. 2005). Bathymodiolin species rely on symbiotic chemoautotrophic bacteria in their specialized gill epithelial cells, which are called bacteriocytes, for all or part of their nutrition as well as environmental adaptations. Bathymodiolin mussel symbiosis is a good model to study host-symbiont interactions, especially how symbionts are actively selected and maintained.

In deep-sea mussels, 1314 positively selected genes were mainly involved in cellular homeostasis, immune response, and transporters. The genes upregulated in the deep-sea mussels were mainly involved in the maintenance of cellular homeostasis, transport, and immune reactions. Based on these results, a conceptual scheme of the interactions of hosts and symbiosis is given (Fig. 6.9). One of the most extraordinary traits of bathymodiolin mussels is their endosymbiosis. Pattern recognition receptors (PRRs) appear to be different in both organisms, and their expression levels also differ between deep-sea and shallow-water mussels. This polymorphism might somehow generate different immune functions and pathogen resistance to support individual adaptation. In summary, chemosynthetic deep-sea mussels might have evolved with reduced PRR patterns and some specific PRRs (e.g., Toll-like receptors, TLR and Complement C1q domain containing, C1qDC proteins) to allow symbiont entry and to successfully maintain them in bacteriocytes. Lineage-specific and positively selected TLRs and highly expressed C1qDC proteins were identified in the gills of the bathymodiolins, suggesting their possible functions in symbiont recognition. However, the PRRs of the bathymodiolins were globally reduced, facilitating the invasion and maintenance of the symbionts obtained by either endocytosis or phagocytosis. Additionally, various transporters were positively selected or more highly expressed in the deep-sea mussels, indicating a means by which necessary materials could be provided for the symbionts. Key genes supporting lysosomal activity were also positively selected or more highly expressed in the deep-sea mussels, suggesting that nutrients fixed by the symbionts can be absorbed in a "farming" way when the symbionts are digested by lysosomes (Zheng et al. 2017). Regulation of key physiological processes, including lysosome activity, apoptosis, and immune reactions, is needed to maintain a stable host-symbiont relationship, but the mechanisms are still unclear.

Transcriptomic dynamic analysis during desymbiosis and symbiosis reestablishment was adopted to further depict the possible processes by which the host selects and maintains the specified symbionts. Healthy *G. platifrons* with dense endogenous symbionts were collected from a cold seep in the SCS and then "desymbionted" by maintaining specimens in an atmospheric pressure lab cultivation system. Furthermore, these conditioned mussels were challenged by either symbionts or environmental bacteria to determine the different host-bacteria recognition mechanisms induced by environmental bacteria or symbionts (Wang et al. 2019). As shown in Fig. 6.10a, the analysis of DEGs between untreated and symbiont-loss samples revealed that metabolic and cellular organizational changes in the gill were associated with increased ribosomal activities, ubiquitin–proteasome systems, and autophagy. Additionally, the differentially expressed immune genes suggest that the host recognizes and interacts with endosymbionts through PRRs, especially Toll-like receptors. In addition, the DEGs between symbiont-treated and environmental bacteria-treated samples shed light on the mechanism of symbiont recognition (Fig. 6.10b). Symbiont treatment not only partially reversed the transcriptomic changes caused by symbiont

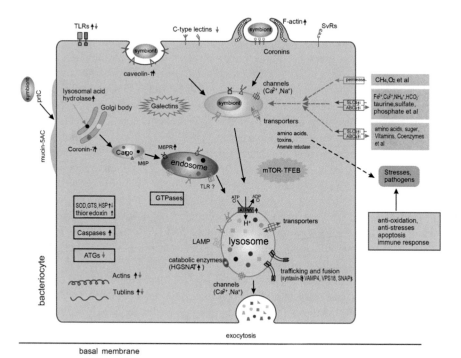

Fig. 6.9 A conceptual scheme of the host–symbiont interactions in the bacteriocyte. The symbionts are located inside vacuoles (green) surrounded by the bacteriocyte cytosol (blue). An overview of the major processes discussed in this article is shown. The underlined bold indicates the orthologs that are positively selected in deep-sea mussels; the red up arrow indicates the orthologs that are expressed consistently more highly in deep-sea mussels; the blue down arrow indicates the orthologs that are expressed consistently more highly in shallow-sea mussels; the dots inside lysosomes indicate lysosomal acid hydrolase, some of which are positively selected and possess higher expression levels in deep-sea mussels (e.g., cathepsins). Key genes for the regulation of lysosome activities (mTOR and TFEB) are neither positively selected nor highly expressed in deep-sea mussels and are colored gray. TLRs: Toll-like receptors; SvRs: scavenger receptors; SLCs: solute carrier family; ABCs: ATP-binding cassette transporters; SOD: manganese superoxide dismutase; GST: glutathione S-transferase; CYPs: cytochrome P450s; ATGs: autophagy-related proteins; M6P: mannose-6-phosphate; M6PR: cation-independent mannose-6-phosphate receptor-like; ATPeV: vacuolar H + -ATPase; VPSs: vacuolar protein sorting proteins; VAMPs: vesicle-associated membrane proteins; SNAPs: synaptosome-associated proteins; HGSNAT: heparan-α glucosaminide N-acetyltransferase; mTOR: mammalian target of rapamycin; TFEB: transcription factor enriched bacteriocytes. Reprinted from Molecular Ecology, 26, Zheng et al. (2017) Insights into deep-sea adaptations and host symbiont interactions: A comparative transcriptome study on Bathymodiolus mussels and their coastal relatives, 5133–5148, Copyright (2022), with permission from John Willey and Sons

loss but also suppressed the host immune system, which might facilitate symbiont entrance into and survival within host bacteriocytes. Taken together, these results suggest that the host-symbiont system in *G. platifrons* is tightly regulated but also has plasticity to fit environmental fluctuations.

In the environmental bacteria-treated samples, the PRRs were upregulated. The environmental bacteria were first recognized by PRRs and then engulfed by phagocytosis. The internalized bacteria were then eliminated by antibacterial proteins such

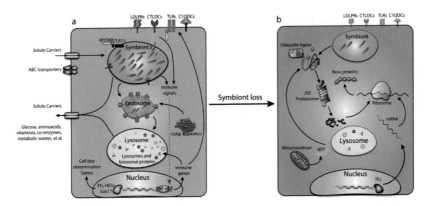

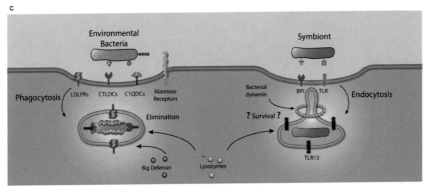

Fig. 6.10 A conceptual scheme of the host-symbiont interaction. **a** comparison of a, regular bacteriocytes (0Day samples); and in **b** the symbiont-loss bacteriocytes (integration of 7Day and 12Day_Blank samples). **c** A proposed model for the different host-bacteria recognition mechanisms induced by environmental bacteria or symbionts. In the environmental bacteria-treated samples, the PRRs were upregulated. The environmental bacteria were first recognized by PRRs and then engulfed by phagocytosis. The internalized bacteria were then eliminated by antibacterial proteins such as C-type lectins, lysozymes, and big defensin. In the symbiont-treated samples, the expression of phagocytosis-related PRRs, such as low density lipoprotein receptors (LDLPR), c-type lectin-like containing (CTLDC), C1qDCs, and mannose receptors, was suppressed. Bacterial permeabilization inducing protein (BPI) and TLRs bind to microbe-associated molecular patterns (MAMP) on symbionts and then induce endocytosis. The internalized symbionts were recognized by TLR13. Reprinted from Deep-Sea Research Part I, 151, Wang et al. (2019) Comparative transcriptomic analysis 624 illuminates the host-symbiont interactions in the deep-sea mussel *Bathymodiolus platifrons*, 103,082, Copyright (2022), with permission from Elsevier

as C-type lectins. In the symbiont-treated samples, the expression of phagocytosis-related PRRs, such as low-density lipoprotein receptor (LDLPR) and mannose receptors, was suppressed. Bacterial permeabilization inducing (BPI) protein and TLRs bind to microbe-associated molecular patterns (MAMPs) on symbionts and then induce endocytosis (Fig. 6.10c).

PRRs in the host play irreplaceable roles in the immune recognition of symbionts. To further decode the immune recognition of symbiosis, several PRR proteins that may interact with methanotrophic symbionts were studied to understand their molecular characteristics and expression patterns (Chen et al. 2019; Li et al. 2020, 2021). Six candidate PRRs, including leucine-rich-repeat-domain-containing protein, TLR13, TLR2, integrin, vacuolar sorting protein and matrix metalloproteinase 1, were selected by methane-oxidizing bacteria or lipopolysaccharide pull-down assays, suggesting their key role in symbiont recognition. The key roles of the PRRs were validated by FISH and immunoblotting. Combined with their expression patterns in bacteriocytes, six PRR proteins were shown to be involved in the recognition of symbionts during the establishment of symbiosis between *G. platifrons* mussels and methane-oxidizing bacteria.

Bacteriocytes are the core units for symbiosis in *G. platifrons*. However, the homeostasis of the system depends on the cooperation of various cells of the symbiotic tissue, the gill. The gill of the bathymodiolin mussel is a complicated tissue that executes other basic physiological functions, such as respiration, transportation, growth, and immune defense (Neumann and Kappes 2003; Nguyen et al. 2019). The bathymodiolin gill includes a variety of cell types. Beyond its bacteriocytes, the bathymodiolin gill filament has several nonsymbiotic components: the tip and edge of each gill filament are covered with cilia cells and mucous cells, and on the lateral face of the gill filament, numerous intercalary cells fill the intercellular spaces among the bacteriocytes (Fiala-Médioni et al. 1986; Fujiwara et al. 2000; Barry et al. 2002). These accumulating data demonstrate that spatial expression information for genes is crucial for understanding symbiosis. However, because transcriptomic studies use whole-gill tissue homogenate as research material, valuable spatial information is lost. To overcome this drawback, *G. platifrons* bacteriocytes were enriched for RNA-seq studies (Wang et al. 2021). In comparative analyses of genes enriched in the bacteriocytes and whole gill and in metatranscriptomic analyses of the symbiont, an integrated view of the molecular mechanisms that directly govern symbiosis in *G. platifrons* was proposed. This breakdown of the complex symbiotic tissue allowed us to characterize the host-symbiont interactions further. GO analyses of the DEGs suggest the presence of transport activities between the bacteriocytes and nonsymbiotic cells. Contrary to expectations, 6 of the top 10 enhanced GO terms in the enriched bacteriocyte samples were involved in microtubule activities. Many of the DEGs in these GO terms encode a variety of tubulins. These results suggest that microtubule activities are significantly enhanced in bacteriocytes. The whole gill samples with enriched GO terms align well with the nonsymbiotic part of the cellular activities of the gill. The genes enriched in "Small molecule transport" and "Transporter activity" indicate that the gill's nonsymbiotic components actively shuffle or exchange nutrients and metabolites with the bacteriocytes. In addition, several GO terms that were related to DNA replication initiation, including genes involved in controlling cell proliferation, such as DNA replication licensing factors, proliferating cell nuclear antigen, and eukaryotic translation initiation factor, were enriched in whole gill samples. Finally, two terms related to the extracellular matrix, "Extracellular space" and "Extracellular region," were significantly enriched in whole gill samples, which correlate with the gill's strong secretion activities. All of the above data showed that the gill's nonsymbiotic parts play crucial roles in maintaining and protecting the symbiosis; the bacteriocytes supply the symbiont with metabolites, control the symbiont population, and shelter the symbiont from phage infection, and the symbiont contributes products from methane oxidation and energy production.

6.6 Adaptation of Deep-Sea Symbiotic Invertebrates to the Harsh Environment

Due to underground fluid seepage, chemosynthetic ecosystems, including cold seeps and hydrothermal vents, are sometimes hostile to megafauna, as the concentrations of hydrogen sulfide or heavy metals are usually high (Zhou et al. 2020, 2021). Among the deep-sea symbiotic invertebrates, deep-sea bathymodiolin mussels are ubiquitous in most cold seeps and hydrothermal fields, where they can accumulate toxic materials, including metals, and thus could serve as an ideal model to investigate the toxicological responses of deep-sea organisms to metal exposure.

Similar to coastal mussels, deep-sea mussels have adopted similar antioxidant defense mechanisms to detoxify metals. A field investigation of metal concentrations and antioxidant enzymes in deep-sea bathymodiolin mussels from four different deep-sea geochemical settings showed that deep-sea mussel gills generally exhibited higher metal enrichment than the mantle. Mussels from hydrothermal vents usually had higher metal concentrations (Fe, Cr, Cd, and Pb) than those from cold seeps, which could be related to their higher contents in fluids or sediments. However, despite quite different metal loads among the geochemical environment settings, Mn, Zn, and Cu concentrations varied over a smaller range across the sampling sites, implying biological regulation of these elements by deep-sea mussels. Although the vent ecosystem is harsher than the cold seep ecosystem, antioxidant enzyme concentrations in deep-sea mussels were not so different, suggesting that some adaptive or compensatory mechanisms may occur in chronically polluted deep-sea mussels. Principal component analysis allowed for distinguishing different deep-sea settings, indicating that bathymodiolin mussels are robust indicators of their living environments (Fig. 6.11a). To further understand the molecular mechanisms of deep-sea mussels in response to metal exposure, indoor metal exposure experimental studies with deep-sea mussels were conducted. The results showed that metal was able to induce general cellular injury, oxidative stress, and disturbances in the metabolism of amino acids, carbohydrates, and lipids. In addition, different metals activated different molecular responses in deep-sea mussels (Fig. 6.11b). Specifically, the monosaccharide D-allose, which is involved in suppressing mitochondrial reactive oxygen species production, was uniquely downregulated in deep-sea mussels under Cd exposure. Alterations in dopamine, as well as dopamine-related and serotonin-related metabolites, in deep-sea mussels under Cu exposure were found, pointing to perturbation of neurotransmission (Fig. 6.11c).

Hydrogen sulfide (H_2S) is another highly toxic molecule for animals living in deep-sea environments because it can bind tightly to cytochrome c oxidase (COX) and block mitochondrial oxidative phosphorylation (Kelley et al. 2016). In *G. platifrons*, the sulfide detoxification and adaptative strategy were detected in both the symbiotic bacterial communities and the host mussels through the comparison of *G. platifrons* inhabiting hydrothermal vent and methane seep environments with distinct sulfide concentrations. First, symbiotic bacteria, including epibionts and endosymbionts, play a role in sulfide detoxification. As indicated in the analysis of 16S rRNA locus sequencing and metatranscriptome sequencing of *G. platifrons* collected from a Site F cold methane seep in the SCS and a hydrothermal vent at Iheya North Knoll in the Mid-Okinawa Trough, the composition of bacterial communities in the gills of the vent and seep mussels was different (Sun et al. 2022b). Methane-oxidizing bacteria, endosymbionts belonging to *Methyloprofundus* (Hirayama et al. 2022), were the most abundant species across all samples. The second most abundant taxon belonged to Campylobacteria, a common sulfur-oxidizing episymbiont bacteria in bathymodiolin mussels (Assié et al. 2016, 2020). Moreover, the relative abundance of Campylobacterota episymbionts was higher in hydrothermal vent mussels (with an average of 39.2%) than in methane seep mussels (with an average of 18.5%). Metatranscriptome sequencing was conducted for each of the three replicates of hydrothermal vent and methane seep mussels, and functional enrichment analysis of the differentially expressed genes (DEGs) in bacterial symbionts revealed that the most significantly enriched pathway was the sulfur metabolism pathway (ko00920, adjusted P value = 1.99E − 10). Notably, a key sulfide-oxidizing gene, sulfide:quinone oxidoreductase (sqr), derived from the methanotrophic endosymbiont, was significantly upregulated in vent mussels, indicating the oxidization of intracellular sulfide by the endosymbiont. In the mussel host, genes participating in mitochondrial oxidative phosphorylation and energy generation, as well as sulfur metabolism, were upregulated in the vent mussels. In summary, the possible adaptations of *G. platifrons* holobionts for sulfide tolerance and protection were hypothesized (Fig. 6.12): (1) exclusion of environmental sulfide by the presence of abundant sulfur-oxidizing epibiotic bacteria colonizing the surfaces of gill filaments with direct contact with the external reduced environment and forming the front line against the potentially toxic environment; (2) oxidization of the intracellular sulfide in methanotrophic endosymbionts aggregated at the apex of the gill bacteriocytes using the sqr gene, processing additional electron and energy sources to their mussel host, and improved fitness to environmental adaptation; and (3) rapid detoxification through the mitochondrial sulfide oxidation pathway of the mussel host, which oxidizes sulfide to less toxic sulfur compounds and simultaneously provides electrons to mitochondrial oxidative phosphorylation and drives ATP synthesis.

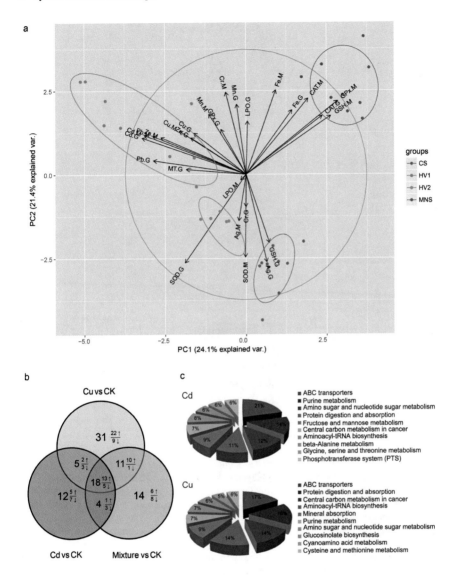

Fig. 6.11 **a** Principal component analysis (PCA) ordination plots of metal accumulation and metal biomarker levels in B. mussels from four different deep-sea geothermal sites sampled (CS: one cold seep from the South China Sea; HV1: hydrothermal vent 1 from the Okinawa Trough; HV2: hydrothermal vent 1 from the Okinawa Trough; MN: one hydrothermal vent from Manus). Reprinted from Science of The Total Environment, 707, Zhou et al. (2020) Metal adaptation strategies of deep-sea Bathymodiolus mussels from a cold seep and three hydrothermal vents in the West Pacific,136,046, Copyright (2022), with permission from Elsevier; **b** Venn diagrams of differentially accumulated metabolites, revealing those commonly or uniquely regulated metabolites occurring in G. platifrons under 7 days of exposure to Cu, Cd, or their mixture; **c** The top ten metabolic pathways that were significantly affected by exposure of *G. platifrons* to Cu, Cd. Figure b and c was reprinted from Aquatic Toxicology, 236, Zhou et al. (2021) Biochemical and metabolic responses of the deep sea mussel *Bathymodiolus platifrons* to cadmium and copper exposure, 105,845, Copyright (2022), with permission from Elsevier

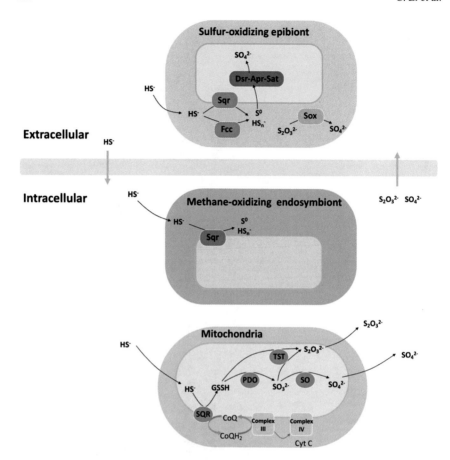

Fig. 6.12 Schematic adaptative strategies of sulfide detoxification hypothesized in the *G. platifrons* holobiont. The epibionts, endosymbionts, and mussel hosts collaborate on sulfide detoxification from the extracellular to the intracellular space. The extracellular epibiotic bacteria oxidize the environmental sulfide through Sqr and Fcc while simultaneously oxidizing other reduced sulfur compounds (S^0 and $S_2O_3^{2-}$) by the Dsr-Apr-Sat and Sox enzyme system. Methanotrophic endosymbionts are densely aggregated at the apical top of gill bacteriocytes and protect cells by oxidization of intracellular sulfide through the *sqr* gene. Intracellular sulfide is also oxidized by the mitochondrial sulfide oxidation pathway in the mussel host. Symbols represent bacterial genes involved in the oxidization of reduced sulfide compounds: Sqr: sulfide:quinone reductase; Fcc: sulfide dehydrogenase (flavocytochrome c); Dsr: dissimilatory sulfite reductase; Apr: adenosine phosphosulfate reductase; Sat: sulfate adenylyltransferase; Sox: sox multiple-enzyme system soxABCXYZ. Symbols represent genes in the mitochondrial sulfide oxidation pathway: SQR: sulfide quinone oxidoreductase; PDO: persulfide dioxygenase; TST: thiosulfate sulfurtransferase; SO: sulfite oxidase. Note: In the diagram, the sulfide oxidization process in the epibiont represents genes and pathways found in the sulfur-oxidizing Campylobacterota and Gammaproteobacteria communities instead of a single species of bacteria. Reprinted from Science of the Total Environment, 804, Sun et al. (2022a, b) Adaption to hydrogen sulfide-rich environments: Strategies for active detoxification in deep-sea symbiotic mussels, *Gigantidas platifrons*, 150,054, Copyright (2022), with permission from Elsevier

The mobile squat lobster *S. crosnieri* may adopt different strategies to cope with abiotic environmental stress. Environmental conditions are unstable in cold seeps due to temporal variation in reduced compounds that are released from bottom currents (Fig. 6.7c) that could further shape regional chemical spatial distributions in the subseafloor (Girard et al. 2020). Methane concentrations were negatively correlated with oxygen concentrations (Cao et al. 2021). However, lobster symbionts require oxygen and reduced substrates such as methane to efficiently fix carbon, which may be simultaneously unavailable in their natural habitats. Thus, different strategies could be used by chemosynthetic symbionts to solve this paradox and achieve more efficient energy conservation. For example, episymbionts can conserve energy through the oxidation of diverse electron donors (sulfide, thiosulfate, sulfur, methane, ammonia, and hydrogen) coupled with oxygen or nitrate respiration, thereby allowing the cells to utilize adaptations to both oxic and hypoxic environments (Nakagawa and Takai 2008). Sulfur globules were observed in our samples (Fig. 6.7d). Although methanol was not directly quantified in these samples, it is reasonable to speculate that methanol accumulation occurred in the MOBs since pmoABC was expressed at 20-fold higher levels than lanthanide-containing methanol dehydrogenases (xoxF) in hypoxic conditions. In addition, globin-like proteins containing heme domains that may sense O_2 were encoded by many of the symbiont MAGs and could facilitate their dynamic responses to fluctuating oxygen environments. Here, a conceptual interactive model was proposed (Fig. 6.7b). During the hypoxic phase, the symbionts oxidize methane and sulfide to methanol and sulfur globules by coupling to nitrate respiration. The stable sulfur and methane intermediates could then be stored as inclusions (Zbinden et al. 2008) or released to the environment and taken up by other symbionts (van Grinsven et al. 2021). During the oxic phase, episymbionts can use oxygen to oxidize sulfur and methanol to conserve additional energy to support cellular growth. Hence, the stable intermediates likely play essential roles in metabolic interactions and adaptations to fluctuating environments that commonly occur at vents and seeps. The cooperation of episymbiotic communities could lead to greater adaptability to chemosynthetic ecosystems, thereby providing greater levels of energy supplies to invertebrate hosts.

6.7 Summary and Perspectives

Deep-sea chemosynthetic symbiotic animals represent a unique model for understanding the interaction between multicellular metazoan and unicellular microorganisms (Dubilier et al. 2008; Moya et al. 2008; Sogin et al. 2021). Several model deep-sea chemosynthetic symbiotic animals, such as squat lobsters, vestimentiferan tubeworms and bathymodiolin mussels, have been intensively studied in the past few decades. Through detailed morphological characterization, ultrastructural imagery analyses, and molecular cloning and characterization, in recent years, with the rapid development of large-scale "OMICs" platforms and computational power, including next-generation direct RNA-seq transcriptomic analysis, large-scale proteomics profiling, and whole-genome sequencing and assembly techniques, the detailed molecular mechanisms governing host-symbiont interactions have been gradually revealed. We now understand that host animals have evolved a sophisticated molecular cascade to communicate and interact with the symbiont. The host also utilizes cellular mechanisms, such as its immune pathways and lysosomes, to recognize, digest and harvest symbionts to harvest nutrients for its growth.

Nevertheless, several fundamental questions in deep-sea symbiosis still need to be answered. For example, it is still largely unknown how host animals recognize symbionts and establish intracellular symbiosis at early developmental stages. How adult host animals produce new bacteriocytes and recruit symbionts is also unclear. Furthermore, at the subcellular scale, how the host senses and regulates the symbiont population, what molecules are used for host-symbiont communication, and what types of metabolites are shared between the symbiont and host still need further investigation. Hopefully, with emerging new technologies, such as deep-sea sampling, single-cell transcriptomics and proteomics sequencing, and high-resolution mass spectrometry, these intriguing scientific questions will be answered in the future.

Symbiosis with chemoautotrophic bacteria was discovered over 40 years ago, and it is now recognized as an important milestone event that shaped the diversity and evolution of deep-sea invertebrates such as sponges, nematodes, crustaceans, annelids, and mollusks. The cold seeps in the SCS provide us with a natural laboratory to understand the sophisticated processes between invertebrates and symbionts. Future studies should focus on discovery-driven expeditions to reveal the full biodiversity of chemosymbiotic associations and their ecological, as well as mechanistic, research to uncover the biochemical, physiological, and molecular mechanisms that govern interactions between symbiotic partners. Revolutionary changes are needed to understand deep-sea organisms. Deep-sea in situ research is undoubtedly the future direction. Moreover, the emerging microscale omics methodology offers us new research opportunities. More works based on experimental verification, such as in vitro culture of symbionts, gene function verification and manipulation of gene expression, should be attempted to better reveal these metabolic interaction mechanisms and confirm similarities and differences between various symbiotic systems.

Acknowledgements Funding was provided by the NSFC (Grants: 42,030,407 and 42,076,091). We are grateful to the captain and crew of the research vessel (R/V) *Kexue*, as well as the ROV *Faxian* for assistance with sample collection. Li Mengna and Ye Ziyun are thanked for her assistance with paper reviewing, which has greatly improved the quality of the chapter.

References

Assié A, Borowski C, van der Heijden K et al (2016) A specific and widespread association between deep-sea *Bathymodiolus* mussels and a novel family of Epsilonproteobacteria. Environ Microbiol Rep 8(5):805–813

Assié A, Leisch N, Meier DV et al (2020) Horizontal acquisition of a patchwork Calvin cycle by symbiotic and free-living Campylobacterota (formerly Epsilonproteobacteria). ISME J 14(1):104–122

Barry JP, Buck KR, Kochevar RK et al (2002) Methane-based symbiosis in a mussel, *Bathymodiolus platifrons*, from cold seeps in Sagami Bay. Japan. Invertebr Biol 121(1):47–54

Boetius A, Wenzhöfer F (2013) Seafloor oxygen consumption fuelled by methane from cold seeps. Nat Geosci 6(9):725–734

Cao L, Lian C, Zhang X et al (2021) In situ detection of the fine scale heterogeneity of active cold seep environment of the Formosa Ridge, the South China Sea. J Mar Syst 218:103530

Ceramicola S, Dupré S, Somoza L et al (2018) Cold seep systems. In: Micallef A, Krastel S, Savini A (eds) Submarine Geomorphology. Springer International Publishing, Chem, pp 367–387

Chen H, Wang M, Zhang H et al (2019) An LRR-domain containing protein identified in *Bathymodiolus platifrons* serves as intracellular recognition receptor for the endosymbiotic methane-oxidation bacteria. Fish Shellfish Immunol 93:354–360

Chen H, Wang M, Zhang H et al (2021) microRNAs facilitate comprehensive responses of Bathymodiolinae mussel against symbiotic and nonsymbiotic bacteria stimulation. Fish Shellfish Immunol 93:420–431

Chen H, Li M, Wang M et al (2022) Single-cell perspectives on the function and development of deep-sea mussel bacteriocytes. bioRxiv https://doi.org/10.1101/2022.05.28.493830

Dong D, Li X (2015) Galatheid and chirostylid crustaceans (Decapoda: Anomura) from a cold seep environment in the northeastern South China Sea. Zootaxa 4057(1):91–105

Dubilier N, Bergin C, Lott C (2008) Symbiotic diversity in marine animals: the art of harnessing chemosynthesis. Nat Rev Microbiol 6(10):725–740

Duperron S (2010) The diversity of deep-sea mussels and their bacterial symbioses. In: Kiel S (ed) The vent and seep biota: aspects from microbes to ecosystems. Springer, Netherlands, Dordrecht, pp 137–167

Duperron S, Quiles A, Szafranski KM et al (2016) Estimating Symbiont Abundances and Gill Surface Areas in Specimens of the Hydrothermal Vent Mussel *Bathymodiolus puteoserpentis* Maintained in Pressure Vessels. Front Mar Sci 3:1–12

Fiala-Médioni A, Métivier C, Herry A et al (1986) Ultrastructure of the gill of the hydrothermal-vent mytilid *Bathymodiolus* sp. Mar Biol 92:65–72

Fisher CR, Childress JJ (1986) Translocation of fixed carbon from symbiotic bacteria to host tissues in the gutless bivalve *Solemya reidi*. Mar Biol 93:59–68

Fujiwara Y, Takai K, Uematsu K et al (2000) Phylogenetic characterization of endosymbionts in three hydrothermal vent mussels: influence on host distributions. Mar Ecol-Prog Ser 208:147–155

Geier B, Sogin EM, Michellod D et al (2020) Spatial metabolomics of in situ host-microbe interactions at the micrometre scale. Nat Microbiol 5(3):498–510

Girard F, Sarrazin J, Olu K (2020) Impacts of an eruption on cold-seep microbial and Faunal Dynamics at a Mud Volcano. Front Mar Sci 7:241. https://doi.org/10.3389/fmars.2020.00241

Gomez-Mendikute A, Elizondo M, Venier P et al (2005) Characterization of mussel gill cells in vivo and in vitro. Cell Tissue Res 321(1):131–140

Halary S, Riou V, Gaill F et al (2008) 3D FISH for the quantification of methane- and sulphur-oxidizing endosymbionts in bacteriocytes of the hydrothermal vent mussel *Bathymodiolus azoricus*. ISME J 2(3):284–292

Hinzke T, Kleiner M, Meister M et al (2021) Bacterial symbiont subpopulations have different roles in a deep-sea symbiosis. Elife 10:e58371

Hinzke T, Kleiner M, Breusing C et al (2019). Host-microbe interactions in the Chemosynthetic *Riftia pachyptila* Symbiosis. mBio 10(6):e02243–19

Hirayama H, Takaki Y, Abe M et al (2022) Multispecies populations of methanotrophic methylopro-fundus and cultivation of a likely dominant species from the Iheya north deep-sea hydrothermal field. Appl Environ Microbiol 88(2):e0075821

Hirokawa N, Noda Y (2008) Intracellular transport and kinesin superfamily proteins, KIFs: structure, function, and dynamics. Physiol Rev 88(3):1089–1118

Husnik F, Tashyreva D, Boscaro V et al (2021) Bacterial and archaeal symbioses with protists. Curr Biol 31(13):R862–R877

Kelley JL, Arias-Rodriguez L, Martin DP et al (2016) Mechanisms underlying adaptation to life in hydrogen sulfide-rich environments. Mol Biol Evol 33(6):1419–1434

Kennicutt MC, Brooks JM, Bidigare RR et al (1985) Vent-type taxa in a hydrocarbon seep region on the Louisiana slope. Nature 317:351–353

Lan Y, Sun J, Chen C et al (2021) Hologenome analysis reveals dual symbiosis in the deep-sea hydrothermal vent snail *Gigantopelta aegis*. Nat Commun 12(1):1165

Li M, Chen H, Wang M et al (2020) Identification and characterization of endosymbiosis-related immune genes in deep-sea mussels *Gigantidas platifrons*. J Oceanol Limnol 38(4):1292–1303

Li M, Chen H, Wang M et al (2021) A Toll-like receptor identified in *Gigantidas platifrons* and its potential role in the immune recognition of endosymbiotic methane oxidation bacteria. PeerJ 9:e11282

Lonsdale P (1979) A deep-sea hydrothermal site on a strike-slip fault. Nature 281:531–534

Moya A, Peretó J, Gil R et al (2008) Learning how to live together: genomic insights into prokaryote–animal symbioses. Nat Rev Genet 9(3):218–229

Nakagawa S, Takai K (2008) Deep-sea vent chemoautotrophs: diversity, biochemistry and ecological significance. FEMS Microbiol Ecol 65(1):1–14

Neumann D, Kappes H (2003) On the growth of bivalve gills initiated from a Lobule-Producing Budding Zone. Biol Bull 205(1):73–82

Nguyen TTK, Tran HV, Vu TT et al (2019) Peptide-modified electrolyte-gated organic field effect transistor: application to Cu^{2+} detection. Biosens Bioelectron 127:118–125

Paull CK, Hecker B, Commeau R et al (1984) Biological communities at the florida Escarpment Resemble Hydrothermal Vent Taxa. Science 226(4677):965–967

Polz MF, Cavanaugh CM (1995) Dominance of one bacterial phylotype at a Mid-Atlantic Ridge hydrothermal vent site. Proc Natl Acad Sci U S A 92(16):7232–7236

Ponnudurai R, Kleiner M, Sayavedra L et al (2017) Metabolic and physiological interdependencies in the *Bathymodiolus azoricus* symbiosis. ISME J 11(2):463–477

Renoz F, Noel C, Errachid A et al (2015) Infection dynamic of symbiotic bacteria in the pea aphid *Acyrthosiphon pisum* gut and host immune response at the early steps in the infection process. PLoS ONE 10(3):e0122099

Sogin EM, Kleiner M, Borowski C et al (2021) Life in the dark: phylogenetic and physiological diversity of chemosynthetic symbioses. Annu Rev Microbiol 75:695–718

Stewart FJ, Newton ILG, Cavanaugh CM (2005) Chemosynthetic endosymbioses: adaptations to oxic–anoxic interfaces. Trends Microbiol 13(9):439–448

Stokke R, Dahle H, Roalkvam I et al (2015) Functional interactions among filamentous Epsilon-proteobacteria and Bacteroidetes in a deep-sea hydrothermal vent biofilm. Environ Microbiol 17(10):4063–4077

Sun Y, Wang M, Chen H et al (2022a) Insights into symbiotic interactions from metatranscriptome analysis of deep-sea mussel *Gigantidas platifrons* under long-term laboratory maintenance. Mol Ecol. https://doi.org/10.1111/mec.16765

Sun Y, Wang M, Zhong Z et al (2022b) Adaption to hydrogen sulfide-rich environments: Strategies for active detoxification in deep-sea symbiotic mussels, *Gigantidas platifrons*. Sci Total Environ 804:150054

Takishita K, Takaki Y, Chikaraishi Y et al (2017) Genomic Evidence that Methanotrophic Endosymbionts Likely Provide Deep-Sea *Bathymodiolus* Mussels with a Sterol Intermediate in Cholesterol Biosynthesis. Genome Biol Evol 9(5):1148–1160

Thurber AR, Jones WJ, Schnabel K (2011) Dancing for food in the deep sea: bacterial farming by a new species of yeti crab. PLoS ONE 6(11):e26243

van Grinsven S, Damste JSS, Harrison J et al (2021) Nitrate promotes the transfer of methane-derived carbon from the methanotroph *Methylobacter* sp. to the methylotroph *Methylotenera* sp. in eutrophic lake water. Limnol Oceanogr 66(3):878–891

Wang H, Zhang H, Wang M et al (2019) Comparative transcriptomic analysis illuminates the host-symbiont interactions in the deep-sea mussel *Bathymodiolus platifrons*. Deep-Sea Res Part I-Oceanogr Res Pap 151:103082

Wang H, Zhang H, Zhong Z et al (2021) Molecular analyses of the gill symbiosis of the bathymodiolin mussel *Gigantidas platifrons*. iScience 24(1):101894

Watsuji TO, Nakagawa S, Tsuchida S et al (2010) Diversity and function of epibiotic microbial communities on the galatheid crab *Shinkaia Crosnieri*. Microbes Environ 25(4):288–294

Watsuji T-o, Yamamoto A, Motoki K et al (2015) Molecular evidence of digestion and absorption of epibiotic bacterial community by deep-sea crab *Shinkaia crosnieri*. ISME J 9(4):821–831

Xu Z, Wang M, Zhang H et al (2022) Metabolism Interactions Promote the Overall Functioning of the Episymbiotic Chemosynthetic Community of *Shinkaia crosnieri* of Cold Seeps. mSystems 7(4): e00320–22

Zbinden M, Shillito B, Le Bris N et al (2008) New insigths on the metabolic diversity among the epibiotic microbial communitiy of the hydrothermal shrimp Rimicaris exoculata. J Exp Mar Biol Ecol 359(2):131–140

Zheng P, Wang M, Li C et al (2017) Insights into deep-sea adaptations and host-symbiont interactions: A comparative transcriptome study on *Bathymodiolus* mussels and their coastal relatives. Mol Ecol 26(19):5133–5148

Zhou L, Cao L, Wang X et al (2020) Metal adaptation strategies of deep-sea *Bathymodiolus* mussels from a cold seep and three hydrothermal vents in the West Pacific. Sci Total Environ 707:136046

Zhou L, Li M, Zhong Z et al (2021) Biochemical and metabolic responses of the deep-sea mussel *Bathymodiolus platifrons* to cadmium and copper exposure. Aquat Toxicol 236:105845

Chapter 7
New Biogeochemical Proxies in Seep Bivalves

Xudong Wang, Steffen Kiel, and Dong Feng

Abstract Reduced compounds dissolved in seeping fluids, such as methane and hydrogen sulfide, are the main energy sources in submarine cold seep systems, where they nourish the unique chemosynthesis-based ecosystems. Chemosymbiotic bivalves are the dominant macrofauna in many of these ecosystems and have been extensively studied due to their large biomass (hundreds of individuals per square meter), their symbiotic relationships with chemotrophic bacteria (methanotrophic bivalves: methane-oxidizing bacteria; thiotrophic bivalves: sulfur-oxidizing bacteria), and because they are unique archives of biogeochemical processes. In this chapter, we briefly introduce the advancements in seep bivalve research worldwide and then summarize the trophic modes and geographic distribution of seep bivalves in the South China Sea. Thereafter, the biogeochemical processes, such as the enzymatic strategy and energy transfer of seep bivalves, are generalized by integrating the trace elements and stable isotope data of the soft tissues and their corresponding calcareous shells of seep bivalves. Overall, we highlight the past contributions and current knowledge in this field and outline opportunities and future directions to expand this area of research.

7.1 Introduction

Since the discovery of hydrothermal vents and cold seeps in the 1980s (Corliss et al. 1979; Paull et al. 1984), the widespread chemosynthetic ecosystems on the seabed have revolutionized our understanding of life on Earth (Fisher 1995; Dubilier

X. Wang (✉) · D. Feng
College of Marine Sciences, Shanghai Ocean University, Shanghai 201306, China
e-mail: xd-wang@shou.edu.cn

D. Feng
e-mail: dfeng@shou.edu.cn

S. Kiel
Department of Palaeobiology, Swedish Museum of Natural History, Stockholm, Sweden
e-mail: Steffen.Kiel@nrm.se

© The Author(s) 2023
D. Chen and D. Feng (eds.), *South China Sea Seeps*,
https://doi.org/10.1007/978-981-99-1494-4_7

et al. 2008). Within chemosynthetic environments, the high abundance of reduced compounds, such as hydrogen sulfide (H_2S) and methane (CH_4), serve as energy sources for carbon-fixing microbes. These microbes, in turn, form the base of the food web of dense biological communities and make significant contributions to primary production in the deep sea, deep-marine biodiversity, and marine geochemical cycles (Baker et al. 2010).

Deep-sea cold seeps are chemosynthetic ecosystems that host highly productive communities and thus enhance regional diversity on continental margins (Levin et al. 2016). Due to their ubiquity and dominance in cold seep ecosystems, chemosymbiotic bivalves—in particular bathymodiolin mussels and vesicomyid clams—have attracted the attention of researchers since the discoveries of these ecosystems. They were first investigated in the Gulf of Mexico with respect to their biogeochemical behavior (Paull et al. 1985; Childress et al. 1986; Cordes et al. 2009) and were subsequently found and studied at virtually all active cold seeps worldwide (Nankai accretionary wedge: Fiala-Médioni et al. 1993; Håkon Mosby mud volcano: Gebruk et al. 2003; pockmarks on the West African margin: Olu-Le Roy et al. 2007; Concepción methane seep area off Chile: Sellanes et al. 2008; New Zealand: Thurber et al. 2010; Marmara Sea: Ritt et al. 2012; North Atlantic and Mediterranean Sea: Duperron et al. 2013; Gulf of Cadiz: Rodrigues et al. 2013; Pacific coast of Costa Rica: Levin et al. 2015; Guaymas Basin: Portail et al. 2015; Arctic: Åström et al. 2019; Krishna-Godavari Basin: Peketi et al. 2022). After nearly 40 years of research, the understanding of the unique symbiotic relationships between seep-inhabiting bivalves and chemotrophic bacteria and the related biogeochemical processes has grown considerably (Fisher 1995; Dubilier et al. 2008; Petersen and Dubilier 2009; Lorion et al. 2013; Decker et al. 2014; Petersen et al. 2016).

Encouraging progress has also been made in discovering and understanding seep-inhabiting bivalves in the South China Sea (SCS) in the last 20 years through a large amount of investment in capital and technology. To date, three active cold seep regions (Site F, Haima and Yam) with seabed manifestations of significant gas seepage and active biological activity have been identified in the SCS, and bivalves represent the dominant macrofauna at these sites. The species, their trophic modes, and their distribution in the SCS are shown in Table 7.1 and Fig. 7.1. Among the chemosymbiotic bivalves, there are five mussel species (*Bathymodiolus aduloides* and *Gigantidas platifrons*, Hashimoto and Okutani 1994; *Gigantidas securiformis*, Okutani et al. 2003; *Gigantidas haimaensis*, Xu et al. 2019; *Nypamodiolus samadiae*, Lin et al. 2022), and one clam species (*Archivesica marissinica*, Chen et al. 2018) has been found. In addition, seven bivalve species with heterotrophic or unknown feeding strategies were found in (or near) site F and Haima seep (Han et al. 2008; Jiang et al. 2019; Ke et al. 2022).

Anatomically, chemosymbiotic bivalves are composed of soft tissue (gill, mantle, foot, siphons, etc.) and a calcareous shell, making them a unique archive for a diversity of research questions (Fig. 7.2). Soft tissue preserves information of the microenvironment that the bivalve inhabits and allows the reconstruction of biogeochemical processes, but it only provides an "instantaneous snapshot" of the brief time immediately before sampling. In contrast, the calcareous shell can provide insights

Table 7.1 Bivalve species from the cold seep ecosystems in the South China Sea

Species	Trophic modes	Distribution	References
Gigantidas platifrons	Methanotroph	Site F, Haima, Yam	Hashimoto and Okutani (1994)
Gigantidas haimaensis		Haima	Xu et al. (2019)
Gigantidas securiformis		Yam	Okutani et al. (2003); Kuo et al. (2019)
"Bathymodiolus" aduloides	Thiotroph	Site F, Haima	Hashimoto and Okutani (1994)
Nypamodiolus samadiae		Haima	Lin et al. (2022)
Archivesica marissinica			Chen et al. (2018)
Vesicomya rhombica	Unknown		Jiang et al. (2019)
Spinosipella xui			
Malletia sp.			Ke et al. (2022)
Propeamussium sp.			
Acharax sp.		Near Site F	Han et al. (2008)
Conchocele sp.			
Lucinoma sp.			Personal observation

into various environmental conditions during the entire lifetime of the bivalve and can potentially preserve this information for a long time and thus provide insights into evolutionary processes over geological timescales. Therefore, chemosymbiotic bivalves are excellent model organisms—their soft tissues and calcareous shell can be used to explore current and past biogeochemical processes, respectively (Wang et al. 2022a, b).

In this chapter, we compile the latest elemental and isotopic data from soft tissues and calcareous shells of diverse seep bivalves in the SCS. The trace element contents and stable isotope compositions of bivalves are outstanding indicators for evaluating their living environment, physiological functions, pathways of energy, and food sources. Together with the understanding of important biogeochemical processes of deep-sea cold seeps, we discuss the variations in geochemical composition in seep-inhabiting bivalves and provide new insights based on recent findings.

7.2 Tissue Element Compositions

Only limited studies have reported trace elements in the gill, mantle and shell of the bathymodiolin mussel *G. platifrons* (Wang et al. 2017; Zhou et al. 2020), systematic comparative studies (inter- and intraspecific comparison and comparison between soft tissues and shells) are still sparse. Generally, trace element contents are ranked in decreasing order, $Zn > Fe > Cu > Sr > Ag > As > Al > Mn > Ba > Pb \approx Cd > Ni$

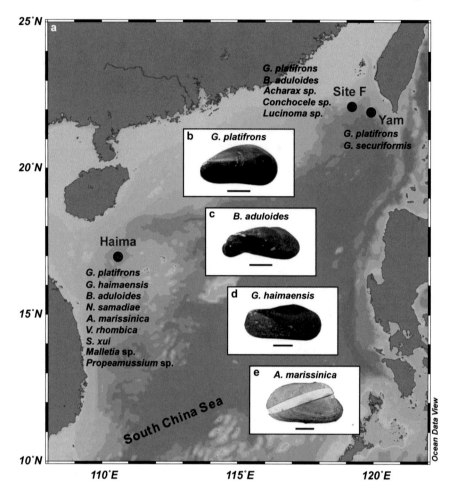

Fig. 7.1 Distribution of known bivalves in cold seep ecosystems in the South China Sea thus far. **a** The active seep sites—Site F, Haima and Yam—are highlighted by red circles. Representative seep bivalves are shown in **b** *Gigantidas platifrons*; **c** *Bathymodiolus aduloides* (modified from Feng et al. 2015); **d** *Gigantidas haimaensis*; and **e** *Archivesica marissinica* (modified from Wang et al. 2022a, b, c). Scale bars in **b–e** are 3 cm. See Table 7.1 for the abbreviations

> Co, and it seems that the contents of transition metal elements (e.g., Zn, Fe, Cu) in soft tissues are much higher than those in corresponding shells, indicating that bivalves may have strong physiological requirements for these elements (Fig. 7.3). Undoubtedly, through more measurements over a larger geographic area, longer timespan, and more specimens than species, the interactions among trace elements, the local environments and the physiology of bivalves may be further verified.

In contrast, the contents of rare earth elements (REEs) in seep bivalve soft tissues of the SCS have been well studied (Wang et al. 2020). The aerobic oxidation of methane is thought to be largely dependent upon the use of REEs, but to date, the

Fig. 7.2 Anatomy of *Gigantidas haimaensis.* Typical features of bivalves include the calcareous shell and soft tissues, such as the gill, mantle and foot

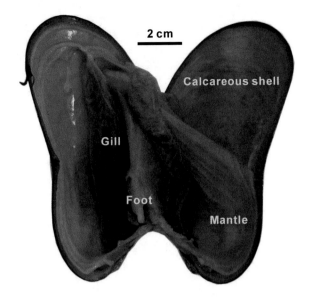

effects of this process on their abundance in bacteria or organisms living in symbiosis with methanotrophs remain to be evaluated. Wang et al. (2020) selected *G. platifrons* (methanotroph) and *B. aduloides* (thiotroph) from Site F and *G. haimaensis* (methanotroph) and *A. marissinica* (thiotroph) from the Haima seep site to compare the effects of different regions and different trophic modes on the contents and patterns of REEs in seep bivalves. They showed that the soft tissues of all methanotrophic bivalves had significant lanthanum (La) enrichment, whereas such a pattern was not observed in thiotrophic bivalves (Fig. 7.4). These results demonstrate that methanotrophic bivalves prospering at methane seeps display distinctive La enrichments linked to the enzymatic activities of their symbionts and that methanotrophic bacteria efficiently fractionates REE distributions in organisms and possibly in the environment.

7.3 Tissue Isotope Compositions

It is well recognized that measuring the carbon, nitrogen, and sulfur isotope compositions of the macrofauna inhabiting cold seeps is one of the most powerful approaches to identifying their food and energy sources, as well as the determination of complex trophic interactions, including symbiosis and heterotrophy (Paull et al. 1985; Childress et al. 1986; Brooks et al. 1987; Carlier et al. 2010; Decker and Olu 2012; Demopoulos et al. 2019; Ke et al. 2022). Indeed, it is possible to distinguish consumers that assimilate carbon through chemosynthesis (more [13]C- and ^{15}N-depleted) from those that rely on photosynthetic primary production (Paull et al. 1985; Brooks et al. 1987). This is because methane is much more ^{13}C-depleted (usually $\delta^{13}C_{CH4} < -40‰$; Whiticar 1999) than dissolved inorganic carbon (DIC)

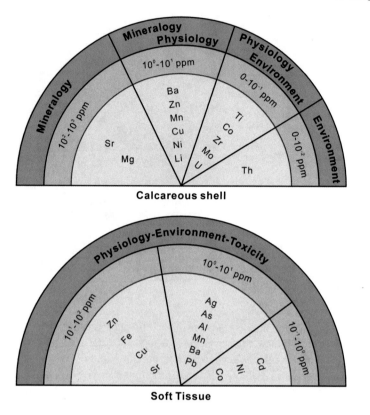

Fig. 7.3 General overview of trace elements analyzed in the shell and soft tissue of seep bivalves, their approximate concentrations, and the factors potentially controlling their behavior (summarized based on data from Wang et al. 2022a)

in the water column ($\delta^{13}C_{DIC} \approx 0‰$), and methane is assimilated by bacteria with only a small carbon isotopic fractionation (Alperin et al. 1988). Within cold seep areas, methane is mainly consumed in the surface sediment by anaerobic oxidation of methane (AOM) coupled with sulfate reduction and produces DIC and sulfide (Boetius et al. 2000), which generates a high microbial biomass (usually a consortium of anaerobic methanotrophic archaea and sulfate-reducing bacteria) and thus can provide a significant supply of methane-derived carbon to heterotrophic bacteria and higher-level consumers (Boetius et al. 2000). The $\delta^{13}C$ signatures of symbiotic and heterotrophic seep-related metazoans indicate whether they preferentially rely on methanotrophically or thiotrophically derived carbon (Childress et al. 1986; Brooks et al. 1987) because the net $\delta^{13}C$ fractionation between a consumer and its diet is small (typically < 1‰; Vander Zanden and Rasmussen 2001; McCutchan et al. 2003). Methanotrophic bacteria are usually more ^{13}C-depleted than thiotrophic bacteria that fix DIC from the water column (Paull et al. 1985). However, $\delta^{13}C$ values of symbiotic species that exclusively rely on thiotrophic bacteria may also be strongly

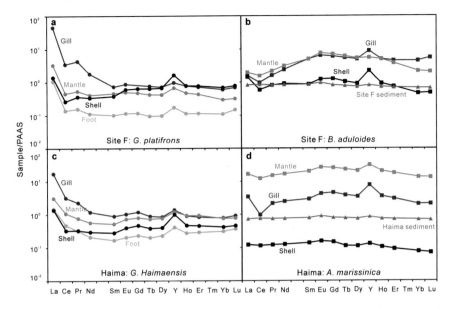

Fig. 7.4 REE + Y patterns normalized to the Post-Archean Australian Shale (PAAS; Pourmand et al. 2012) for seep bivalves in the South China Sea (data from Wang et al. 2020). Abbreviations: (1) *G. platifrons* = *Gigantidas platifrons*; (2) *B. aduloides* = *Bathymodiolus aduloides*; (3) *G. haimaensis* = *Gigantidas haimaensis*; (4) *A. marissinica* = *Archivesica marissinica*

modified when the DIC originates from a diversity of sources, including byproducts of methane oxidation (Lösekann et al. 2008).

Sulfur stable isotopes are useful in discriminating organic matter (OM) from the water column (by phytoplankton) or from reduced sediments (by chemosynthetic microorganisms). In general, the $\delta^{34}S$ values of seawater sulfate are quite homogeneous at ~+21‰. Carbon fixation by phytoplankton results in only a small negative fractionation of sulfur isotopes, resulting in a narrow range in the sulfur isotope composition of oceanic particulate OM (+17‰ < $\delta^{34}S$ < +21‰; Peterson and Fry 1987). Therefore, benthic organisms that depend exclusively on phytoplanktonic production show $\delta^{34}S$ values with similar signatures because sulfur isotopes do not significantly fractionate between trophic levels (McCutchan et al. 2003). In contrast, dissimilatory sulfate reduction by bacteria within the sediment results in strong fractionation (−25‰ < $\delta^{34}S$ < +5‰; Carlier et al. 2010). Consequently, organisms that assimilate these reduced compounds exhibit low $\delta^{34}S$ values as well. To some extent, additional $\delta^{34}S$ analyses may allow differentiation between the input of photosynthetic and chemosynthetic material for seep-related organisms (Brooks et al. 1987; MacAvoy et al. 2005).

The carbon, nitrogen, and sulfur isotopes of *G. platifrons*, *B. aduloides* and *A. marissinica* from cold seeps in the SCS have been investigated (Fig. 7.5; Feng et al. 2015; Zhao et al. 2020; Ke et al. 2022). According to the observations of the available data, some general patterns emerge:

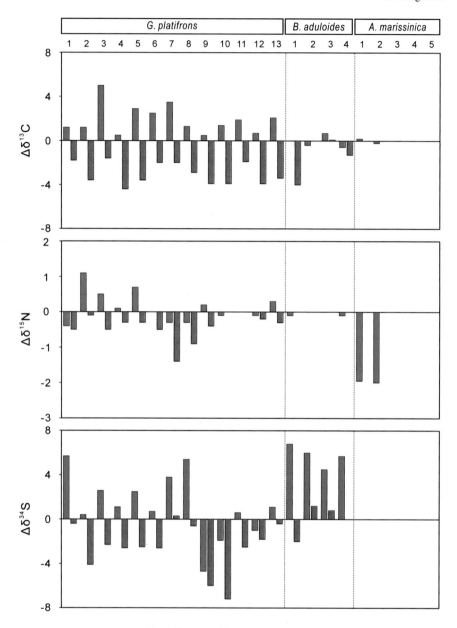

Fig. 7.5 Differences (Δ) in $\delta^{13}C$, $\delta^{15}N$, and $\delta^{34}S$ values among the gill, mantle and foot of seep bivalves (data from Feng et al. 2015, 2018 and unpublished). Blue rectangle: $\delta^{xx}C/N/S_{\text{gill-mantle}}$; orange rectangle: $\delta^{xx}C/N/S_{\text{gill-foot}}$. Abbreviations: *G. platifrons* = *Gigantidas platifrons*; *B. aduloides* = *Bathymodiolus aduloides*

(1) Carbon and sulfur isotopic compositions of seep bivalve soft tissues are related to their respective trophic modes. The carbon isotope values of methanotrophic bivalves ($-77.6‰$ to $-51.2‰$) are obviously lower than those of thiotrophic bivalves ($-37.7‰$ to $-33.0‰$), while the sulfur isotope values show the opposite trend (methanotrophic bivalves: $+1.4‰$ to $+19.3‰$; thiotrophic bivalves: $-12.9‰$ to $+14.6‰$).

(2) The characteristics of the carbon isotope signatures of different soft tissue parts of methanotrophic bivalves are consistent ($\delta^{13}C_{Mantle} < \delta^{13}C_{Gill} < \delta^{13}C_{Foot}$), which has implications for the carbon transmission process in *G. platifrons*.

(3) It is suggested that different nitrogen sources are used by bivalves, in which the main nitrogen source of *B. aduloides* is organic nitrogen from the sediments, while *G. platifrons* mainly uses NH_4^+ (Sun et al. 2017; Wang et al. 2022a, b, c). However, it remains unclear whether different nitrogen sources are responsible for the rather stable nitrogen isotope signature of *B. aduloides* ($+1.1‰$ to $+1.2‰$) and the lower and highly variable nitrogen isotope signature of *G. platifrons* ($-0.9‰$ to $+1.3‰$). Obviously, the nitrogen isotope signature archived by seep bivalves should be further explored.

7.4 Element and Isotope Compositions of Shells

By comparison, prospecting and investigation of the calcareous shells of these bivalves lags behind that of their soft tissue. Generally, due to some inherent methodological and physiological limitations related to the shell (the record of the whole life process – years to decades of accumulation, the high degree of mixing with seawater, bulk shell sampling methods, etc.), it seems that the shell reflects some insignificant geochemical signals that inevitably blend the information of seawater and carbonate. For example, trace element concentrations ranked in decreasing order for calcareous shells of *G. haimaensis* and *A. marissinica* are Sr > Mg > Ba > Zn ≈ Mn > Cu > Ni > Li ≈ Ti > Co ≈ Zr ≈ Mo ≈ U > Th, obviously because carbonate is rich in trace elements, such as Sr, Mg and Ba (Fig. 7.3; Wang et al. 2022a). However, few studies have demonstrated that shells, similar to soft tissue, can potentially be used to reconstruct biogeochemical processes (Feng et al. 2018; Wang et al. 2020). Although the REE contents of the shells are lower than those of the corresponding soft tissues, characteristic of La enrichment is also observed in the methanotrophy-related shells (Fig. 7.4). In addition, generally consistent characteristics between the carbon, nitrogen and sulfur isotopic information contained in the shell organic matter (SOM) and its corresponding soft tissue mantle are present (Fig. 7.6). In contrast to the quickly degrading soft tissue, the bivalve shell has a high preservation potential, including its original mineralogy. Hence, the geochemical signal in the shell can be used as a fingerprint of food chains, as well as to track early evidence of microbial life and to reconstruct the chemosynthetic behavior among different types of ecosystems in the geologic past (Feng et al. 2018; Wang et al. 2020).

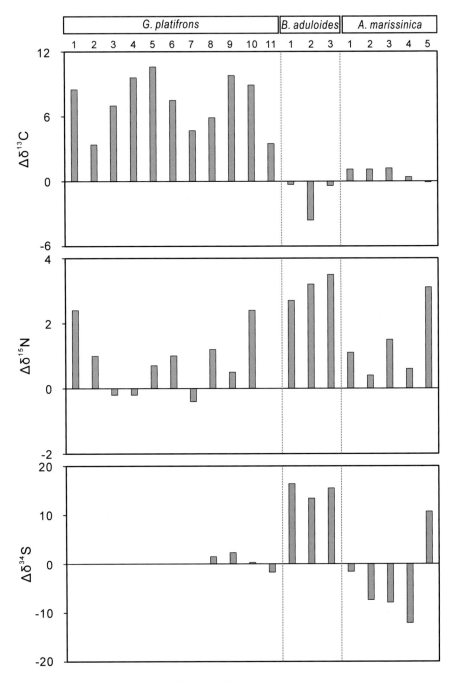

Fig. 7.6 Differences (Δ) in δ13C, δ15N, and δ34S values between the organic shell matrix and mantle of seep bivalves (data from Feng et al. 2018). Abbreviations: *G. platifrons* = *Gigantidas platifrons*; *B. aduloides* = *Bathymodiolus aduloides*; *A. marissinica* = *Archivesica marissinica*

7.5   Summary and Perspective

Research on seep bivalves in the South China Sea (SCS) has acquired a series of profound understandings that have spurred the wave of the development of cold seeps in China in the last 20 years. Studies on the elemental geochemistry, especially rare earth elements of seep bivalves, have further verified the catalysis of light rare earth elements on the aerobic oxidation of methane and provided an excellent example for exploring the coupling (enzymatic) relationship between life activities and trace elements. At the same time, by investigating the carbon, nitrogen, and sulfur isotopes of seep bivalves, temporally and spatially integrated trophic estimates used to understand and define trophic linkages among species and communities are provided.

However, it must be admitted that our knowledge of the geochemical behavior of seep bivalves is restricted since the study of the seep bivalves of the SCS is still in the initial stage of discovery and research. The existing issues include but are not limited to the following: (1) there is a large gap in the research on different species of seep bivalves. Intensive studies on the geochemical characteristics of seep bivalves focus on *G. platifrons*, *G. haimaensis* and *A. marissinica*, while few studies have been conducted on other species. The trace element data of soft tissues are particularly scarce. (2) The mechanism that controls the behavior of carbon, nitrogen, and sulfur isotopes in chemosymbiotic bivalves remains understudied. In addition to the interaction between host and symbiotic bacteria, facultative filter feeding among seep bivalves remains a hard-to-define factor that may obstruct predictions of the behavior of carbon, nitrogen, and sulfur isotopes in these bivalves. (3) Due to the technical difficulties of sampling the deep sea, species of nondominant bivalves are typically found only as bycatch rather than as the result of focused and systematic sampling. Promising in this respect is the Haima cold seep as it covers a vast area, and Yam is a recently discovered seep area, allowing the discovery and research of new species of bivalves.

The exploration of seep bivalves is of great significance to both biology and geology because the combination of soft tissue and calcareous shells makes them excellent model organisms to bridge contemporary earth surface processes (present) and the sedimentary record (fossil). Future research on seep bivalves should strengthen the combination of biology and geology (especially geochemistry) and further apply this research approach to all seep macrofauna, which is expected to decode the 'vital effect' of symbiotic organisms and their overprint on geological records. We believe that future research on seep bivalves (and all seep macrofauna) in the South China Sea cold seeps will continue to expand our knowledge of the biological activities in this unique ecosystem, and we look forward to the advancement of knowledge in the fields of life evolution, symbiotic relationships and beyond.

Acknowledgements Funding was provided by the National Natural Science Foundation of China (Grant: 42106059), Shanghai Sailing Program (Grant: 21YF1416800) and Chenguang Program of Shanghai Education Development Foundation and Shanghai Municipal Education Commission (Grant: 22CGA58).

References

Alperin MJ, Reeburgh WS, Whiticar MJ (1988) Carbon and hydrogen isotope fractionation resulting from anaerobic methane oxidation. Glob Biogeochem Cycle 2(3):279–288

Åström EKL, Carroll ML, Sen A et al (2019) Chemosynthesis influences food web and community structure in high-Arctic benthos. Mar Ecol-Prog Ser 629:19–42

Baker MC, Ramirez-Llodra EZ, Tyler PA et al (2010) Biogeography, ecology, and vulnerability of chemosynthetic ecosystems in the deep sea. In: McIntyre AD (ed) Life in the World's Oceans. Wiley-Blackwell Publishing, Oxford, pp 161–182

Boetius A, Ravenschlag K, Schubert CJ et al (2000) A marine microbial consortium apparently mediating anaerobic oxidation of methane. Nature 407(6804):623–626

Brooks JM, Kennicutt MC II, Fisher CR et al (1987) Deep-sea hydrocarbon seep communities: evidence for energy and nutritional carbon sources. Science 238(4830):1138–1142

Carlier A, Ritt B, Rodrigues CF et al (2010) Heterogeneous energetic pathways and carbon sources on deep eastern Mediterranean cold seep communities. Mar Biol 157(11):2545–2565

Chen C, Okutani T, Liang Q et al (2018) A noteworthy new species of the family Vesicomyidae from the South China Sea (Bivalvia: glossoidea). Venus 76(1–4):29–37

Childress JJ, Fisher CR, Brooks JM et al (1986) A methanotrophic marine molluscan (Bivalvia, Mytilidae) symbiosis: mussels fueled by gas. Science 233(4770):1306–1308

Cordes EE, Bergquist DC, Fisher CR (2009) Macro-ecology of Gulf of Mexico cold seeps. Annu Rev Mar Sci 1:143–168

Corliss JB, Dymond J, Gordon LI et al (1979) Submarine thermal springs on the Galápagos Rift. Science 203(4385):1073–1083

Decker C, Olu K (2012) Habitat heterogeneity influences cold-seep macrofaunal communities within and among seeps along the Norwegian margin—part 2: contribution of chemosynthesis and nutritional patterns. Mar Ecol 33(2):231–245

Decker C, Zorn N, Potier N et al (2014) Globin's structure and function in vesicomyid bivalves from the Gulf of Guinea cold seeps as an adaptation to life in reduced sediments. Physiol Biochem Zool 87(6):855–869

Demopoulos AWJ, McClain-Counts JP, Bourque JR et al (2019) Examination of *Bathymodiolus childressi* nutritional sources, isotopic niches, and food-web linkages at two seeps in the US Atlantic margin using stable isotope analysis and mixing models. Deep-Sea Res I-Oceanogr Res Pap 148:53–66

Dubilier N, Bergin C, Lott C (2008) Symbiotic diversity in marine animals: the art of harnessing chemosynthesis. Nat Rev Microbiol 6(10):725–740

Duperron S, Gaudron SM, Rodrigues CF et al (2013) An overview of chemosynthetic symbioses in bivalves from the North Atlantic and Mediterranean Sea. Biogeosciences 10(5):3241–3267

Feng D, Cheng M, Kiel S et al (2015) Using *Bathymodiolus* tissue stable carbon, nitrogen and sulfur isotopes to infer biogeochemical process at a cold seep in the South China Sea. Deep-Sea Res I-Oceanogr Res Pap 104:52–59

Feng D, Peckmann P, Li N et al (2018) The stable isotope fingerprint of chemosymbiosis in the shell organic matrix of seep-dwelling bivalves. Chem Geol 479:241–250

Fiala-Médioni A, Boulègue J, Ohta S et al (1993) Source of energy sustaining the *Calyptogena* populations from deep trenches in subduction zones off Japan. Deep-Sea Res I-Oceanogr Res Pap 40:1241–1258

Fisher CR (1995) Toward an appreciation of hydrothermal-vent animals: their environment, physiological ecology, and tissue stable isotope values. In: Humphris SE, Zierenberg RA, Mullineaux LS, Thomson RE (eds) Seafloor hydrothermal systems: physical, chemical, biological, and geochemical interactions. Geophysical Monographs Series 91. Blackwell, Washington, DC, pp 297–316 (Geophysical Monograph Series)

Gebruk AV, Krylova EM, Lein AY et al (2003) Methane seep community of the Håkon Mosby mud volcano (the Norwegian Sea): composition and trophic aspects. Sarsia 88(6):394–403

Han XQ, Suess E, Huang YY et al (2008) Jiulong methane reef: Microbial mediation of seep carbonates in the South China Sea. Mar Geol 249(3–4):243–256

Hashimoto J, Okutani T (1994) Four new mytilid mussels associated with deepsea chemosynthetic communities around Japan. Venus 53:61–83

Jiang J, Huang Y, Liang Q et al (2019) Description of two new species (Bivalvia: Vesicomyidae, Verticordiidae) from a cold seep in the South China Sea. Nautilus 133(3–4):94–101

Ke Z, Li R, Chen Y et al (2022) A preliminary study of macrofaunal communities and their carbon and nitrogen stable isotopes in the Haima cold seeps, South China Sea. Deep-Sea Res Part I-Oceanogr Res Pap 184:103774

Kuo M-Y, Kang D-R, Chang C-H et al (2019) New records of three deep-sea bathymodiolus mussels (Bivalvia: Mytilida: Mytilidae) from hydrothermal vent and cold seeps in Taiwan. J Mar Sci Technol 27(4):352–358

Lin Y-T, Kiel S, Xu T et al (2022) Phylogenetic placement, morphology and gill-associated bacteria of a new genus and species of deep-sea mussel (Mytilidae: Bathymodiolinae) from the South China Sea. Deep-Sea Res Part I-Oceanogr Res Pap 190:103894

Levin LA, Mendoza GF, Grupe BM et al (2015) Biodiversity on the rocks: macrofauna inhabiting authigenic carbonate at Costa Rica methane seeps. PLoS ONE 10(7):e0131080

Levin LA, Mendoza GF, Grupe BM (2016) Methane seepage effects on biodiversity and biological traits of macrofauna inhabiting authigenic carbonates. Deep-Sea Res II-Top Stud Oceanogr 137:26–41

Lorion J, Kiel S, Faure BM et al (2013) Adaptive radiation of chemosymbiotic deep-sea mussels. Proc R Soc B-Biol Sci 280(1776):20131243

Lösekann T, Robador A, Niemann H et al (2008) Endosymbioses between bacteria and deep-sea siboglinid tubeworms from an Arctic cold seep (Haakon Mosby Mud Volcano, Barents Sea). Environ Microbiol 10(12):3237–3254

MacAvoy SE, Fisher CR, Carney RS et al (2005) Nutritional associations among fauna at hydrocarbon seep communities in the Gulf of Mexico. Mar Ecol-Prog Ser 292:51–60

McCutchan JH Jr, Lewis WM Jr, Kendall C et al (2003) Variation in trophic shift for stable isotope ratios of carbon, nitrogen, and sulfur. Oikos 102(2):378–390

Okutani T, Fujikura K, Sasaki T (2003) Two New Species of Bathymodiolus (Bivalvia: Mytilidae) from Methane Seeps on the Kuroshima Knoll off the Yaeyama Islands, Southwestern Japan. Venus 62:97–110

Olu-Le Roy K, Caprais J-C, Fifis A et al (2007) Cold-seep assemblages on a giant pockmark off West Africa: spatial patterns and environmental control. Mar Ecol 28(1):115–130

Paull CK, Hecker B, Commeau R et al (1984) Biological communities at the Florida Escarpment resemble hydrothermal vent taxa. Science 226(4677):965–967

Paull CK, Jull AJT, Toolin LJ et al (1985) Stable isotope evidence for chemosynthesis in an abyssal seep community. Nature 317(6039):709–711

Peketi A, Mazumdar A, Sawant B et al (2022) Biogeochemistry and trophic structure of a cold seep ecosystem, offshore Krishna-Godavari basin (east coast of India). Mar Pet Geol 138:105542

Peterson BJ, Fry B (1987) Stable isotopes in ecosystem studies. Ann Rev Ecol Syst 18:293–320

Petersen JM, Dubilier N (2009) Methanotrophic symbioses in marine invertebrates. Environ Microbiol Rep 1(5):319–335

Petersen JM, Kemper A, Gruber- Vodicka H et al (2016) Chemosynthetic symbionts of marine invertebrate animals are capable of nitrogen fixation. Nat Microbiol 2(1):16195

Portail M, Olu K, Escobar-Briones E et al (2015) Comparative study of vent and seep macrofaunal communities in the Guaymas Basin. Biogeosciences 12(18):5455–5479

Pourmand A, Dauphas N, Ireland TJ (2012) A novel extraction chromatography and MC-ICP-MS technique for rapid analysis of REE, Sc and Y: revising CI-chondrite and Post-Archean Australian Shale (PAAS) abundances. Chem Geol 291:38–54

Ritt B, Duperron S, Lorion J et al (2012) Integrative study of a new cold-seep mussel (Mollusca:Bivalvia) associated with chemosynthetic symbionts in the Marmara. Deep-Sea Res I Oceanogr Res Pap 67:121–132

Rodrigues CF, Hilario A, Cunha MR (2013) Chemosymbiotic species from the Gulf of Cadiz (NE Atlantic): distribution, life styles and nutritional patterns. Biogeosciences 10(4):2569–2581

Sellanes J, Quiroga E, Neira C (2008) Megafauna community structure and trophic relationships at the recently discovered concepción methane seep area, Chile, ~36°S. ICES J Mar Sci 65(7):1102–1111

Sun J, Zhang Y, Xu T et al (2017) Adaptation to deep-Sea chemosynthetic environments as revealed by mussel genomes. Nat Ecol Evol 1(5):0121

Thurber AR, Kröger K, Neira C et al (2010) Stable isotope signatures and methane use by New Zealand cold seep benthos. Mar Geol 272(1–4):260–269

Vander Zanden MJ, Rasmussen JB (2001) Variation in $\delta^{15}N$ and $\delta^{13}C$ trophic fractionation: implications for aquatic food web studies. Limnol Oceanogr 46(8):2061–2066

Wang F, Wu Y, Feng D (2022a) Different nitrogen sources fuel symbiotic mussels at cold seeps. Front Mar Sci 9:869226

Wang X, Li C, Zhou L (2017) Metal concentrations in the mussel *Bathymodiolus platifrons* from a cold seep in the South China Sea. Deep-Sea Res I Oceanogr Res Pap 129:80–88

Wang X, Barrat J-A, Bayon G et al (2020) Lanthanum anomalies as fingerprints of methanotrophy. Geochem Perspect Lett 14:26–30

Wang X, Fan D, Kiel S et al (2022b) Archives of short-term fluid flow dynamics and possible influence of human activities at methane seeps: evidence from high-resolution element geochemistry of chemosynthetic bivalve shells. Front Mar Sci 9:960338

Wang X, Guan H, Qiu J-W et al (2022c) Macro-ecology of cold seeps in the south China Sea. Geosyst Geoenviron 1(3):100081

Whiticar MJ (1999) Carbon and hydrogen isotope systematics of bacterial formation and oxidation of methane. Chem Geol 161(1–3):291–314

Xu T, Feng D, Tao J et al (2019) A new species of deep-sea mussel (Bivalvia: Mytilidae: *Gigantidas*) from the South China Sea: morphology, phylogenetic position, and gill-associated microbes. Deep-Sea Res I Oceanogr Res Pap 146:79–90

Zhao Y, Xu T, Law YS et al (2020) Ecological characterization of cold-seep epifauna in the South China Sea. Deep-Sea Res I Oceanogr Res Pap 163:103361

Zhou L, Cao L, Wang X et al (2020) Metal adaptation strategies of deep-sea *Bathymodiolus* mussels from a cold seep and three hydrothermal vents in the West Pacific. Sci Total Environ 707:136046

Chapter 8
Pore Water Geochemistry and Quantification of Methane Cycling

Yu Hu

Abstract Owing to numerous scientific cruises in the past two decades, pore water data from more than 250 sites within gas hydrate and cold seep areas of the South China Sea have been reported. These investigated sites are mainly distributed in the Dongsha–Taixinan, Shenhu, and Qiongdongnan areas of the northern South China Sea, together with a few sites from the Beikang Basin of the southern South China Sea. Pore water geochemical profiles at these sites have been used to indicate fluid sources that are linked to gas hydrates and methane seepage, to distinguish the anaerobic oxidation of methane (AOM) from organoclastic sulfate reduction, to reveal fluid flow patterns, and to quantify the rates of AOM. As the pore water data accumulate over a broad area of the SCS, recent attempts have been made to quantify regional sulfate and methane cycling in the subseafloor of the northern South China Sea. This quantitative assessment on a regional scale highlights the importance of deep-sourced methane in governing subseafloor carbon and sulfur cycling along continental margins.

8.1 Introduction

Early diagenesis occurs once sediment particles are buried below the seafloor. Early diagenesis is mainly driven by the degradation of organic matter and oxidation of hydrocarbons. Discrimination between the primary signals from the water column and the products of early diagenetic reactions is critical when using marine sediments for paleoenvironmental reconstructions. The solid-phase and pore water analysis of marine sediments are two fundamental approaches for investigating early diagenesis (Schulz 2006). Solid-phase analysis has advantages in sample collection and storage and analytical procedures are generally easier to perform, thus producing more reliable results. However, this approach in most cases cannot distinguish whether diagenetic reactions are on-going or occurred in the past and provides no information about the rates and kinetics of the diagenetic reactions (Schulz 2006). This large gap

Y. Hu (✉)
College of Marine Sciences, Shanghai Ocean University, Shanghai 201306, China
e-mail: huyu@shou.edu.cn

© The Author(s) 2023
D. Chen and D. Feng (eds.), *South China Sea Seeps*,
https://doi.org/10.1007/978-981-99-1494-4_8

129

can be filled by pore water analysis, which provides a more immediate biogeochemical archive during early diagenesis. For instance, pore water analysis can be used to quantify the rates of methane-consuming and -producing reactions during early diagenesis. The methane-consuming reaction in marine sediments mainly refers to the anaerobic oxidation of methane (AOM) coupled with sulfate reduction, which is mediated by anaerobic methanotrophic archaea and associated sulfate-reducing bacteria (e.g., Boetius et al. 2000). AOM consumes more than 90% of ascending methane in marine sediments, thereby serving as an efficient filter that prevents the release of methane, a powerful greenhouse gas, into the ocean and atmosphere (Reeburgh 2007).

During the past two decades, extensive efforts on pore water analysis have been made to detect gas hydrates and methane seeps in the South China Sea (SCS). Consequently, a vast number of sediment cores with pore water data were collected from the SCS during numerous research cruises. Due to the significance of pore water sulfate data for tracing methane seepages and AOM, more than 250 sites with available pore water sulfate data on the continental slope and rise of the SCS have been investigated during the past two decades (Fig. 8.1; see references in Table 8.1). These investigated sites mainly focused on the areas where indications for cold seeps and gas hydrates were found, including the Dongsha–Taixinan, Shenhu, and Qiongdongnan areas in the northern SCS and the Beikang Basin in the southern SCS (Fig. 8.1; Table 8.1).

Early pore water geochemical work was mainly dedicated to tracing the sources of pore fluid linked to gas hydrates and methane seepages in the northern SCS (e.g., Jiang et al. 2005; Suess, 2005; Chuang et al. 2006, 2010; Deng et al. 2006; Lin et al. 2006; Yang et al. 2006, 2008a; Wu et al. 2007, 2010, 2011, 2013a, b; Huang et al. 2008). Later work with more analyzed pore water parameters allowed the identification of AOM from organoclastic sulfate reduction in the SCS (e.g., Luo et al. 2013; Hu et al. 2015, 2018, 2020; Liu et al. 2020; Gong et al. 2021). With the increase in the kinds of measured pore water parameters, numerical simulations, mainly reaction–transport models, have been used to quantify the rates of AOM in the SCS (e.g., Chuang et al. 2013; Ye et al. 2016; Feng et al. 2018b, 2019, 2020; Hu et al. 2018, 2019; Zhang et al. 2019; Liu et al. 2020; Zha et al. 2022). Reaction–transport models have also been applied to reveal fluid flow patterns by reproducing the complex shapes of pore water sulfate profiles (Chuang et al. 2013; Hu et al. 2019; Feng et al. 2021). Compared with the numerous investigated sites from the northern SCS, recent pore water studies have extended to the Beikang Basin of the southern SCS (Fig. 8.1; Feng et al. 2018b, 2021; Huang et al. 2022). With the accumulation of pore water data from a vast number of sites (Fig. 8.1), recent work has attempted to quantify regional sulfate and methane cycling in the subseafloor of the northern SCS (Zhang et al. 2019; Hu et al. 2022). This chapter provides an overview of the progress of pore water geochemistry from the continental slope and rise of the SCS that is used for the indication of fluid sources, for the identification of AOM and fluid flow patterns, and for quantifying AOM rates and regional methane cycling.

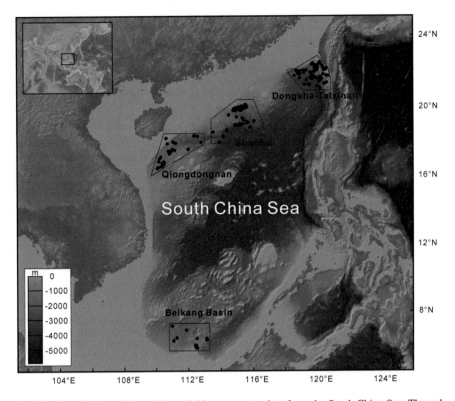

Fig. 8.1 Compilation of sites with available pore water data from the South China Sea. The red dot marks the position of the Haima cold seeps. Data sources were taken from the references listed in Table 8.1. The map was created using GeoMapApp software

Table 8.1 Published pore water data with positional information of gravity, piston, and push cores from the South China Sea. The sites of these cores are presented in Fig. 8.1

Areas	Site number	Core types	References
Dongsha–Taixinan	111	Gravity/piston	[1–14]
Shenhu	85	Gravity/piston/drilling	[15]
Qiongdongnan	46	Gravity/piston/drilling/push	[16–31]
Beikang Basin	11	Piston	[32–34]

Notes [1] Huang et al. 2008; [2] Wu et al. 2013a; [3] Suess, 2005; [4]Ye et al. 2016; [5] Hu et al. 2015; [6] Zhang et al. 2019; [7] Deng et al. 2006; [8] Zhang et al. 2014; [9] Hu et al. 2018; [10] Cao and Lei 2012; [11] Chen et al. 2017; [12] Chuang et al. 2013; [13] Hu et al. 2017; [14] Chuang et al. 2010; [15] Hu et al. 2022; [16] Feng et al. 2019; [17] Feng et al. 2020; [18] Wei et al. 2019; [19] Liu et al. 2020; [20] Yang et al. 2013; [21] Wu et al. 2010; [22] Feng et al. 2018a; [23] Wu et al. 2007; [24] Yang et al. 2006; [25] Jiang et al. 2005; [26] Luo et al. 2013; [27] Hu et al. 2019; [28] Hu et al. 2020; [29] Wang et al. 2018; [30] Hu et al. 2021; [31] Jin et al. 2022; [32] Feng et al. 2018b; [33] Feng et al. 2021; [34] Huang et al. 2022

8.2 Indication of Fluid Sources

Pore water geochemistry is a useful tool to indicate pore fluid sources in submarine cold seep environments. Because methane is the key component of the pore fluid, tracing methane sources and their linkage to gas hydrates thus becomes common issues of concern in areas influenced by methane seepages. Recent years have witnessed the discovery of gas hydrates by sampling and pore water geochemistry in the Dongsha, Shenhu, and Qiongdongnan areas (Yang et al. 2008b, 2015; Luo et al. 2014; Sha et al. 2015; Zhang et al. 2015; Liang et al. 2017; Hu et al. 2019; Wei et al. 2019; Ye et al. 2019). For instance, macroscopic gas hydrates have been found in shallow and deep marine sediments via piston and drill cores from the Qiongdongnan area (Fig. 8.2). Negative anomalies of pore water chloride concentrations relative to seawater value result from the dissociation of gas hydrates during sampling (Hu et al. 2019). In contrast, positive anomalies of pore water chloride concentrations reflect ongoing gas hydrate formation (Fig. 8.2; Wei et al. 2019). Once gas hydrates form in the sediment pore space from ambient water and dissolved methane, dissolved ions, such as pore water chloride, would be excluded as water molecules are incorporated into gas hydrates. This process increases the concentrations of pore water chloride, which in turn indicates the in situ formation of gas hydrates in the sediment pore space.

In shallow marine sediments, methane and other short-chain hydrocarbons mainly exist in the dissolved form in pore waters. These dissolved hydrocarbons can be collected using headspace vials for measuring their compositions. The carbon stable isotope ($\delta^{13}C$) values of methane are quite low, varying from $-93‰$ to $-35‰$ in marine sediments from the SCS (Fig. 8.3). The molecular ratios between methane and other short-chain hydrocarbons ($C_1/(C_1 + C_2)$) span four orders of magnitude (Fig. 8.3). The diagram of $C_1/(C_1 + C_2)$ ratios versus $\delta^{13}C$ values of methane is commonly used to distinguish biogenic methane that is bacterially generated in the methanogenic zone from thermogenic methane ascending from greater depths (Fig. 8.3). Methane in marine sediments of the southern SCS mainly originates from biogenic gas, while methane sources of the northern SCS are complex, with contributions from both biogenic and thermogenic methane (Fig. 8.3; Zhang et al. 2021; Huang et al. 2022). The ascending methane in marine sediments is largely consumed at the sulfate–methane transition zone (SMTZ), producing peak concentrations and $\delta^{13}C$ values of dissolved inorganic carbon (DIC) by AOM (Fig. 8.4; Hu et al. 2019). Based on the linear regression of $\delta^{13}C \times DIC$ versus DIC in cores QDN-14A and R1 of the Haima cold seeps, the estimated $\delta^{13}C$ values of DIC added to the pore water are $-65‰$ and $-62‰$, respectively (Fig. 8.5). These $\delta^{13}C$ values are consistent with the $\delta^{13}C$ values of methane at the Haima cold seeps (Figs. 8.3 and 8.5). This observation suggests that DIC at the SMTZ is almost completely derived from highly ^{13}C-depleted methane by AOM at the Haima cold seeps.

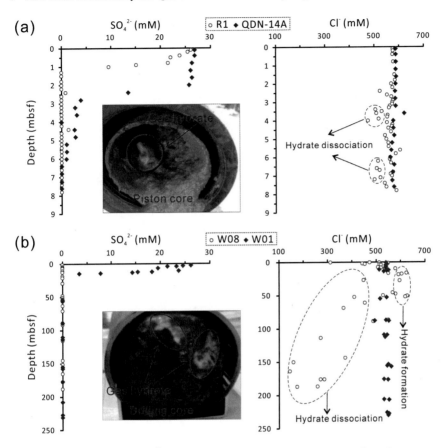

Fig. 8.2 Pore-water sulfate (SO_4^{2-}) and chloride (Cl^-) concentration profiles of (**a**) piston cores and (**b**) drill cores from the South China Sea. Piston cores R1 and QDN-14A were collected from the Haima cold seep, and drill cores W08 and W01 were recovered from the northeastern part of the Qiongdongnan Basin. The pore water data of the piston cores and drill cores were taken from Hu et al. (2019) and Wei et al. (2019), respectively. Photographs showing the presence of gas hydrates in the piston cores and drill cores were taken from Wei et al. (2019) and Liang et al. (2017), respectively. mbsf: meters below the seafloor

8.3   Identification of AOM

The extremely negative $\delta^{13}C$ values of DIC at the SMTZ could be used to identify the on-going AOM process (Fig. 8.4). However, this signature may be obscured by other DIC sources from organoclastic sulfate reduction and/or methanogenesis. Numerical models constrained by pore water sulfate, calcium, DIC, and nutrition concentrations and other parameters can be used to identify AOM and quantify the rates of AOM (e.g., Hu et al. 2018). Sulfur and oxygen isotopic patterns ($\delta^{34}S_{SO4}$ versus $\delta^{18}O_{SO4}$) of pore water sulfate have also been used to distinguish AOM from organoclastic sulfate reduction (e.g., Antler et al. 2015; Hu et al. 2020; Gong et al. 2021). However, the

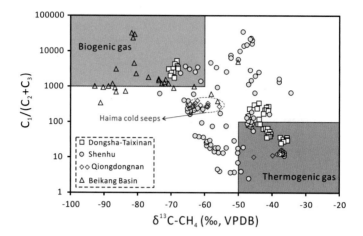

Fig. 8.3 Hydrocarbon compositions of sediment core samples compiled from numerous sites in the South China Sea. Molecular ratios between methane (CH_4) and other short-chain hydrocarbons ($C_1/(C_1 + C_2)$) and $\delta^{13}C$ data of CH_4 were taken from Zhang et al. (2021) and Huang et al. (2022)

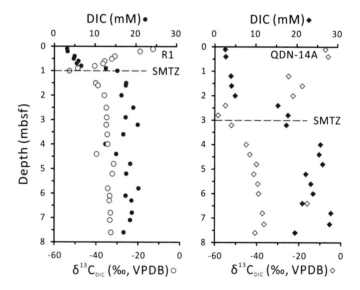

Fig. 8.4 Dissolved inorganic carbon (DIC) concentrations and carbon isotopic compositions of DIC ($\delta^{13}C_{DIC}$) in depth profiles of piston cores collected from the Haima cold seeps from the Qiongdongnan area (data from Hu et al. 2019). Sulfate–methane transition zone (SMTZ). mbsf: meters below the seafloor

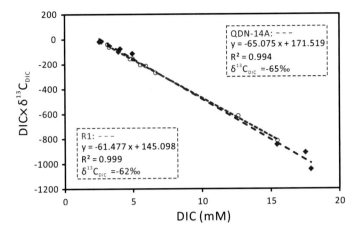

Fig. 8.5 Linear regression between dissolved inorganic carbon (DIC) and the $\delta^{13}C \times DIC$ of cores R1 and QDN-14A from the Haima cold seeps

numerical model approach requires numerous kinds of parameters, and the $\delta^{34}S_{SO4}$ versus $\delta^{18}O_{SO4}$ approach requires advanced analytical technology. The diagram of the produced DIC versus consumed sulfate ratios ($R_{C:S}$) after correcting for carbonate precipitation is widely used to identify AOM (Fig. 8.6; e.g., Luo et al. 2013; Hu et al. 2015, 2018, 2020 and references therein). This $R_{C:S}$ approach requires concentrations of only pore water sulfate, calcium, magnesium, and DIC.

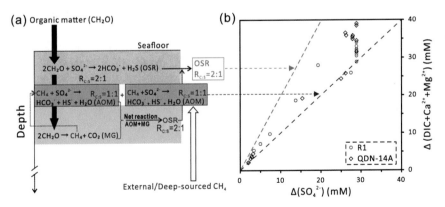

Fig. 8.6 Main biogeochemical processes in marine sediment and their control on the ratios ($R_{C:S}$) between produced dissolved inorganic carbon (DIC) corrected for carbonate precipitation and consumed sulfate (SO_4^{2-}). **a** Main biogeochemical processes responsible for $R_{C:S}$, including organoclastic sulfate reduction (OSR), anaerobic oxidation of methane coupled with sulfate reduction (AOM), and methanogenesis (MG). **b** Produced DIC corrected for carbonate precipitation ($\Delta(DIC + Ca^{2+} + Mg^{2+})$) versus consumed SO_4^{2-} (ΔSO_4^{2-}) relative to typical seawater values (2.1 mM for DIC, 10.3 mM for Ca^{2+}, 53.2 mM for Mg^{2+}, and 28.9 mM for SO_4^{2-}). Pore water data of R1 and QDN-14A were taken from Hu et al. (2019)

The $R_{C:S}$ approach is based on the reaction equations of organoclastic sulfate reduction and AOM that produce $R_{C:S}$ ratios equal to 2:1 and 1:1, respectively (Fig. 8.6a). However, this approach might be compromised by upward diffusive DIC from the methanogenic zone because methanogenesis below the SMTZ during preceding fermentation increases the DIC content of pore waters (Fig. 8.6a; e.g., Chatterjee et al. 2011; Kim et al. 2011; Hong et al. 2013; Hu et al. 2018). In addition to the addition of DIC from local methanogenesis to the SMTZ, methane produced by local methanogenesis is consumed by AOM when this methane diffuses into the SMTZ (Fig. 8.6a). The combined effect of these methane production and consumption processes around the SMTZ generates an $R_{C:S}$ of 2:1, which equals the $R_{C:S}$ of organoclastic sulfate reduction (Fig. 8.6a; Komada et al. 2016; Hu et al. 2018). Decoupled from local methanogenesis, methane without DIC at great depth (namely, external/deep-sourced methane) can migrate toward the SMTZ (c.f. Burdige and Komada 2013; Komada et al. 2016; Hu et al. 2018, 2022). This external/deep-sourced methane would be consumed by AOM when migrating into the SMTZ, thus generating an $R_{C:S}$ of 1:1 (Fig. 8.6a). Despite the potential interference from local methanogenesis, the $R_{C:S}$ approach is still useful to judiciously discriminate AOM from organoclastic sulfate reduction and/or methanogenesis (Fig. 8.6). If local methanogenesis occurs, the $R_{C:S}$ approach would result in an underestimation of AOM, thus representing a conservative approach to identify the presence of AOM and the deep-sourced methane (Fig. 8.6; Hu et al. 2018). Discrimination using the $R_{C:S}$ approach has been conducted in the areas of Dongsha–Taixinan (Hu et al. 2015, 2018), Shenhu (Hu et al. 2018, 2020), and Qiongdongnan (Luo et al. 2013; Liu et al. 2020; this study is shown in Fig. 8.6b). The results from these areas show much lower $R_{C:S}$ ratios of less than 2:1 or $R_{C:S}$ ratios approaching 1:1, revealing the occurrence of AOM (Luo et al. 2013; Hu et al. 2015, 2018, 2020). For instance, piston cores from the Haima cold seeps of the Qiongdongnan area exhibit $R_{C:S}$ ratios approaching 1:1, suggesting the dominance of AOM at the SMTZ (Fig. 8.6b). This inference is in accordance with the observation of extremely ^{13}C-depleted DIC at the SMTZ that almost completely results from AOM.

8.4 Implication for Fluid Flow Patterns

The changes in methane flux, physical mixing due to bioirrigation or bubble irrigation, and mass-wasting events create non-steady state depositional and/or environmental conditions (e.g., Fossing et al. 2000; Zabel and Schulz 2001; Hensen et al. 2003; Kasten et al. 2003; Hu et al. 2019 and references therein). This methane seepage variability and fluid flow patterns can be imprinted in pore water compositions in depth profiles (Hu et al. 2019). In particular, pore water sulfate profiles can exhibit changeable gradients ranging from kink to concave-up to s-types (e.g., Zabel and Schulz 2001; Hensen et al. 2003; Haeckel et al. 2007). Therefore, insights into the geochemical trends of pore water profiles can in turn reveal fluid flow patterns and possible sedimentary events (e.g., Hensen et al. 2003; Hu et al. 2019). Sulfate and iodide concentrations remain nearly unchanged relative to seawater values in the upper 1–3 m of sediment cores at many sites from the Dongsha–Taixinan area of the SCS (Chuang et al. 2013). This irrigation-like feature could be attributed to meter-scale bioirrigation of macrofauna or seawater intrusion during methane ebullition (Fossing et al. 2000). However, this possibility is eliminated since bioirrigation usually occurs only on a decimeter scale in surface sediments (c.f. Haeckel et al. 2007; Hu et al. 2019). The irrigation-like feature is simulated by numerical modeling, assuming methane gas bubble irrigation with seawater, as proposed by Haeckel et al. (2007). The irrigation-like feature of pore water from the Dongsha–Taixinan areas is thus interpreted to reflect the irrigation of seawater by methane gas bubbles that rise through soft surface sediments (Chuang et al. 2013).

Similar irrigation-like features have also been observed in the upper 2 m of pore water profiles at the Haima cold seeps from the Qiongdongnan area (Fig. 8.7; Hu et al. 2019). The numerical model of methane bubble irrigation reproduced measured parameters in pore water profiles (Hu et al. 2019). The core near the seepage center (R1) with a higher methane flux is more susceptible to gas ebullition than cores that are relatively far away from the seepage center (QDN-14A and QDN-14B). It is paradoxical that the core near the seepage center has less indication of gas bubble irrigation (Fig. 8.7), making bubble irrigation unlikely to be the cause of the irrigation-like feature. Instead, the irrigation-like feature is attributed to enhanced methane fluxes from a combination of the lateral and upward migration of methane-rich fluids (Fig. 8.7). This situation is also reproduced by a non-steady-state numerical model (Fig. 8.7a). The presence of gas hydrates in core R1 can serve as a barrier for the upward movement of methane-rich fluids toward the seafloor and can laterally redirect the fluid flow (Fig. 8.7b). This lateral fluid flow is also promoted by the occurrence of coarser sediments in intervals directly below the SMTZ in cores QDN-14B and QDN-14A (Fig. 8.7b; Hu et al. 2019). In other words, the sealing of gas hydrate layers in core R1 redirected the pathway of upward-migrated methane-rich fluids from below the hydrate layers. The lateral migration of methane along with upward-migrated methane suddenly increased methane fluxes in nearby cores, resulting in the observed irrigation-like feature (Fig. 8.7b; Hu et al. 2019).

8.5 Quantifying AOM Rates and Regional Methane Cycling

The anaerobic oxidation of methane (AOM) consumes sulfate, with a stoichiometry of 1:1, thereby competing with organoclastic sulfate reduction above the SMTZ (Fig. 8.6a). Therefore, the values of AOM rates are usually lower than those of total sulfate reduction or sulfate diffusive flux. Methane concentration gradients below the SMTZ can be used to estimate AOM rates. Studies of only some sites from limited areas of the SCS have reported methane concentration data (e.g., Wu et al. 2011, 2013b; Chuang et al. 2013; Chen et al. 2017; Hu et al. 2020; Huang et al. 2022). On the other hand, the risk of underestimating AOM rates exists by using methane concentration gradients due to the rapid decrease in pressure and thereby possible

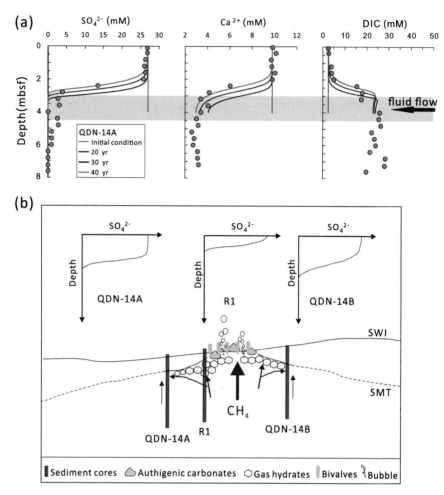

Fig. 8.7 Impact of flow fluid patterns on the shape of pore water sulfate (SO_4^{2-}) profiles (Modified from Hu et al. 2019). **a** Evolution of SO_4^{2-}, calcium (Ca^{2+}), and dissolved inorganic carbon (DIC) profiles over time from non-steady-state simulations of core QDN-14A that assume enhanced methane (CH_4) fluxes resulting from vertical and lateral fluid migration. **b** Schematic diagram showing flow fluid patterns as revealed by the shape of pore water SO_4^{2-} profiles. mbsf: meters below the seafloor. Reprinted from Marine and Petroleum Geology, 103, Hu et al. (2019) Pore fluid compositions and inferred fluid flow patterns at the Haima cold seeps of the South China Sea, 29–40, Copyright (2019), with permission from Elsevier

loss of methane during core recovery. Consequently, the attempt to use methane concentration gradients to constrain AOM rates is not common in the SCS (e.g., Chen et al. 2017; Hu et al. 2020). The rates of AOM can be more directly determined by a radiotracer approach using ^{14}C-labeled methane; however, this approach is rarely used in the SCS (Zhuang et al. 2019). In contrast, the reaction–transport model has been commonly used to quantify the AOM rates at sites in the Dongsha–Taixinan,

Shenhu, Qiongdongnan, and Beikang Basins by fitting numerous kinds of pore water parameters (e.g., Chuang et al. 2013; Ye et al. 2016; Feng et al. 2018b, 2019, 2020; Hu et al. 2018, 2019; Zhang et al. 2019; Liu et al. 2020; Zha et al. 2022). Of all the reported AOM rates, depth-integrated AOM rates at the Haima cold seeps are the highest (Hu et al. 2019; Liu et al. 2020). Due to shallow SMTZs at the Haima cold seeps from ~0.2 to 1 m below the seafloor (mbsf), the AOM rates exceed one thousand mmol m^{-2} a^{-1} (Hu et al. 2019; Liu et al. 2020).

Despite more precision for quantifying AOM rates using reaction–transport models, very few sites from the SCS have been investigated (Zhang et al. 2019). 253 sites with available pore water data have been compiled from the SCS (Fig. 8.1; Table 8.1); however, most of these sites have only sulfate data, which limits the application of the reaction–transport model. These 253 sites all show a clear depth trend in sulfate concentration profiles, thereby allowing for the calculation of the SMTZ depth and reliable diffusive sulfate fluxes. Because sulfate fluxes generally represent the upper limit of AOM rates, the distribution of sulfate fluxes is helpful to constrain the regional AOM rates and methane cycling in the SCS. Sulfate profiles at 85 sites from the Shenhu area reveal that the depths of the SMTZ range from 6.6 to 88.5 mbsf (Hu et al. 2022). By using Fick's first law, estimated sulfate fluxes vary from 2.4 to 30.7 mmol m^{-2} a^{-1}, with an average of 14.3 mmol m^{-2} a^{-1} (Fig. 8.8a; Hu et al. 2022). Based on sulfate profiles of 111 sites from the Dongsha–Taixinan area (Fig. 8.1), the depths of the SMTZ range from ~0.4 to 43.0 mbsf, and estimated sulfate fluxes average 34.8 mmol m^{-2} a^{-1}, varying from 2.9 to 170.7 mmol m^{-2} a^{-1} (Hu et al. 2023). From sulfate profiles at 46 sites in the Qiongdongnan area (Fig. 8.1), the depths of the SMTZ range from 0.2 to 111.2 mbsf, and estimated sulfate fluxes vary from 1.6 to 1203.4 mmol m^{-2} a^{-1}, with an average of 136.4 mmol m^{-2} a^{-1} (Hu et al. 2023). The depths of the SMTZ at 11 sites from the Beikang Basin range from 3.3 to 8.8 mbsf (Fig. 8.1; Feng et al. 2018b, 2021; Huang et al. 2022). By spatial interpolation, the regional sulfate flux estimated from sulfate profiles in the Shenhu area is 0.54×10^{-3} Tmol a^{-1} (Hu et al. 2022). In contrast, regional sulfate fluxes estimated from sulfate profiles at 111 sites from the Dongsha–Taixinan area and 46 sites from Qiongdongnan area are much higher, reaching 1.07×10^{-3} and 2.65×10^{-3} Tmol a^{-1}, respectively (Hu et al. 2023; Figs. 8.9, 8.10 and 8.11).

Deep-sourced methane can be identified by the $R_{C:S}$ approach, as shown in the diagram in Fig. 8.6. The amount of such deep-sourced methane on a regional scale has been recently quantified by the difference between the regional sulfate flux estimated from sedimentation rates and that from sulfate profiles (Hu et al. 2022). Sedimentation rates can be used to estimate sulfate fluxes based on the common good global correlation between diffusive sulfate fluxes and sedimentation rates, generating a global map for the distribution of sulfate diffusive flux (Egger et al. 2018). Sedimentation rates reflect the control of organic matter and sediment ages on the diffusive sulfate flux to the SMTZ, which is determined by the continuous decrease in the reactivity of organic matter with sediment depth (Egger et al. 2018 and references therein). Sulfate fluxes estimated from sedimentation rates thus represent sulfate consumption during organic matter degradation (Hu et al. 2022). This sulfate consumption includes (1) sulfate consumed by organoclastic sulfate reduction above the SMTZ

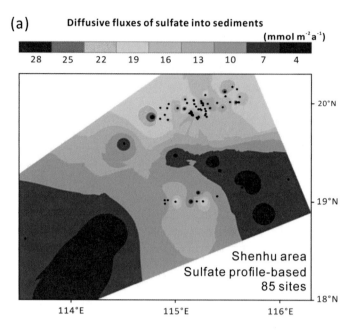

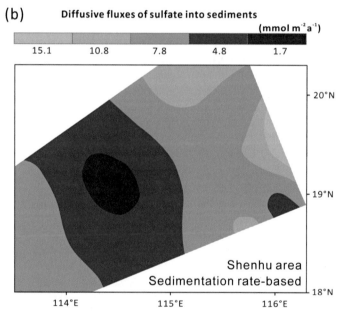

Fig. 8.8 Maps showing the distribution and magnitude of the regional sulfate flux of the Shenhu area based on measured sulfate profiles at 85 sites (**a**) and sedimentation rates (**b**), respectively (Modified from Hu et al. 2022). Reprinted from Science Bulletin, 67, Hu et al. (2022) Enhanced sulfate consumption fueled by deep-sourced methane in a hydrate-bearing area, 122–124, Copyright (2022), with permission from Elsevier

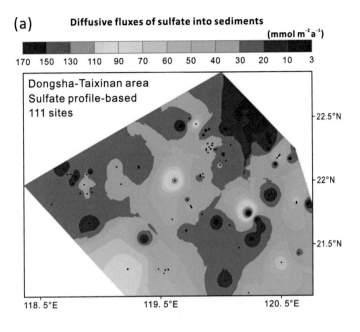

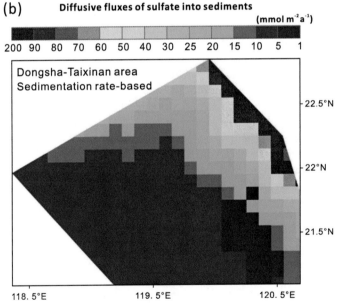

Fig. 8.9 Maps showing the distribution and magnitude of the regional sulfate flux of the Dongsha–Taixinan area based on measured sulfate profiles at 111 sites (**a**) and sedimentation rates (**b**), respectively (Modified from Hu et al. 2023)

and (2) sulfate consumed by AOM at the SMTZ that oxidizes methane fueled by local archaeal methanogenesis through organic matter fermentation below the SMTZ (Hu et al. 2022; Fig. 8.6a). Accordingly, the difference between the regional sulfate fluxes derived from sulfate profiles and those derived from sedimentation rates represents the amount of deep-sourced methane that mainly originated from deep gas hydrate reservoirs and/or deep strata kilometers below the seafloor (Hu et al. 2022).

Regional maps of sulfate flux distribution based on sedimentation rates in the northern SCS were extracted from an established, global map of sulfate flux distribution or generated from locally measured sedimentation rates (Egger et al. 2018; Hu et al. 2022, 2023; Figs. 8.8b, 9.9b and 8.10b). Accordingly, regional sulfate fluxes derived from sedimentation rate in the Dongsha–Taixinan, Shenhu, and Qiongdongnan areas are 0.54×10^{-3}, 0.30×10^{-3}, and 0.39×10^{-3} Tmol a^{-1}, respectively (Hu et al. 2022, 2023; Figs. 8.8–8.11). Based on the discrepancy between regional sulfate flux estimated from sulfate profiles and that estimated from sedimentation rates, estimated fluxes of deep-sourced methane in the Dongsha–Taixinan, Shenhu, and Qiongdongnan areas reach 0.53×10^{-3}, 0.24×10^{-3}, and 2.26×10^{-3} Tmol a^{-1} (Hu et al. 2022, 2023; Fig. 8.11). Summing up these areas, 4.26×10^{-3} Tmol of sulfate and 3.03×10^{-3} Tmol of deep-sourced methane are annually consumed in the subseafloor sediments of the northern SCS for an area of 1.23×10^5 km^2 (Hu et al. 2023; Fig. 8.11).

Ratios between regional flux of deep-sourced methane and regional flux of sulfate based on sulfate profiles represent the contribution of sulfate consumption by ascending deep-sourced methane to total sulfate consumption (Fig. 8.11). These ratios can be used to evaluate the overall impact of ascending deep-sourced methane on the sulfate budget in subseafloor sediments on a regional scale. The ratios reach 50%, 44%, and 85% in the Dongsha–Taixinan, Shenhu, and Qiongdongnan areas, respectively, with the overall ratio for the three study areas accounting for 71% (Hu et al. 2023; Fig. 8.11). These ratios indicate that the contribution of ascending deep-sourced methane to sulfate consumption is similar to or even higher than the contribution of organoclastic sulfate reduction on regional scales in the subseafloor sediments of the northern SCS (Hu et al. 2023). By extrapolating the results of the northern SCS to global slope and rise sediments, 2.18 to 2.65 Tmol of deep-sourced methane is oxidized with sulfate by AOM annually in marine sediments (Hu et al. 2023). Despite a rough extrapolation, it becomes obvious that deep-sourced methane plays a crucial role in subseafloor carbon and sulfur cycling, supporting chemosynthesis-based ecosystems along the continental slope and rise sediments worldwide (Hu et al. 2023).

8.6 Summary and Perspectives

Pore water profiles from more than 250 sites within gas hydrate and cold seep areas of the SCS in the past two decades were investigated to indicate fluid sources, to identify the AOM and fluid flow patterns, and to quantify AOM rates and regional methane

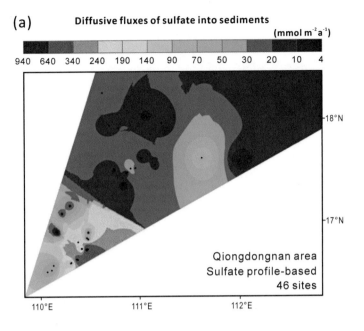

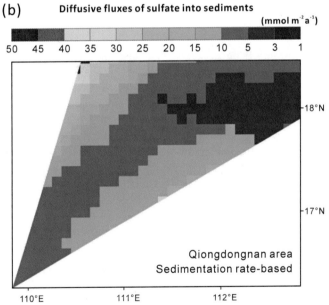

Fig. 8.10 Maps showing the distribution and magnitude of the regional sulfate flux of the Qiong-dongnan area based on measured sulfate profiles at 46 sites (**a**) and sedimentation rates (**b**), respectively (Modified from Hu et al. 2023)

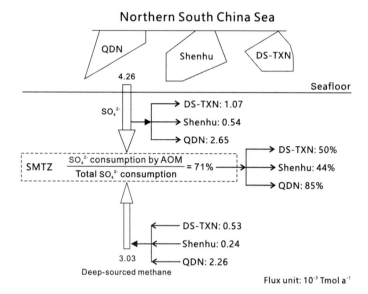

Fig. 8.11 A summary of regional sulfate (SO_4^{2-}) and methane budgets and the role of deep-sourced methane in the subseafloor sediments of the northern South China Sea (Modified from Hu et al. 2023). SMTZ: sulfate-methane transition zone. AOM: anaerobic oxidation of methane. DS-TXN: Dongsha-Taixinan. QDN: Qiongdongnan

cycling. Regional sulfate budgets suggest that deep-sourced methane plays a crucial role in governing subseafloor carbon and sulfur cycling along continental margins. A unique feature of the SCS is indicated by the development of a transition from an active to a passive continental margin. Does this tectonic feature in the Dongsha–Taixinan area influence the distribution and fluxes of sulfate and methane across the tectonic transition? Addressing this issue would provide a great opportunity to understand upward diffusion and/or seepages of methane on regional scales in response to tectonic activity. Although the amount of deep-sourced methane in the northern SCS has been quantified, the impact of such an amount of deep-sourced methane on subseafloor carbon cycling has not been quantitatively assessed. More work is required to quantitatively evaluate the amounts of authigenic carbonate that is produced and the dissolved inorganic carbon that enters the ocean after the deep-sourced methane is oxidized with sulfate by AOM.

Acknowledgements The author gratefully acknowledges the scientists, crews, and organizers of the research vessel *Haiyang-04* and *Haiyang-06* for sampling during several research cruises in the past decade. Min Luo is acknowledged for constructive comments on this chapter. The work is partially supported by the Shanghai Pujiang Program (Grant No 21PJ1404700).

References

Antler G, Turchyn AV, Herut B et al (2015) A unique isotopic fingerprint of sulfate-driven anaerobic oxidation of methane. Geology 43(7):619–622

Boetius A, Ravenschlag K, Schubert CJ et al (2000) A marine microbial consortium apparently mediating anaerobic oxidation of methane. Nature 407(6804):623–626

Burdige DJ, Komada T (2013) Using ammonium pore water profiles to assess stoichiometry of deep remineralization processes in methanogenic continental margin sediments. Geochem Geophys Geosyst 14(5):1626–1643

Cao C, Lei H (2012) Geochemical characteristics of pore water in shallow sediments from north continental slope of South China Sea and their significance for natural gas hydrate occurrence. Procedia Environmental Sciences 12, Part B:1017–1023

Chatterjee S, Dickens GR, Bhatnagar G et al (2011) Pore water sulfate, alkalinity, and carbon isotope profiles in shallow sediment above marine gas hydrate systems: A numerical modeling perspective. J Geophys Res-Solid Earth 116:B09103

Chen N, Yang T, Hong W et al (2017) Production, consumption, and migration of methane in accretionary prism of southwestern Taiwan. Geochem Geophys Geosyst 18(8):2970–2989

Chuang PC, Dale AW, Wallmann K et al (2013) Relating sulfate and methane dynamics to geology: Accretionary prism offshore SW Taiwan. Geochem Geophy Geosyst 14(7):2523–2545

Chuang P, Yang T, Hong W et al (2010) Estimation of methane flux offshore SW Taiwan and the influence of tectonics on gas hydrate accumulation. Geofluids 10(4):497–510

Chuang P, Yang T, Lin S et al (2006) Extremely high methane concentration in bottom water and cored sediments from offshore southwestern Taiwan. Terr Atmos Ocean Sci 17(4):903–920

Deng X, Fu S, Huang Y et al (2006) Geochemical characteristics of sediments at Site HD196 in Dongsha Islands, the North of the South China Sea, and their implication for gas hydrates. Geoscience 20(1):92–102 (in Chinese with English abstract)

Egger M, Riedinger N, Mogollón JM et al (2018) Global diffusive fluxes of methane in marine sediments. Nat Geosci 11(6):421–425

Feng J, Yang S, Sun X et al (2018a) Geochemical tracers for methane microleakage activity in the Qiongdongnan Basin. Journal of Southwest Petroleum University (science & Technology Edition) 40:63–75 (in Chinese with English abstract)

Feng J, Yang S, Liang J et al (2018b) Methane seepage inferred from the porewater geochemistry of shallow sediments in the Beikang Basin of the southern South China Sea. J Asian Earth Sci 168:77–86

Feng J, Yang S, Wang H et al (2019) Methane source and turnover in the shallow sediments to the west of Haima cold seeps on the northwestern slope of the South China Sea. Geofluids 2019:1010824

Feng J, Li N, Luo M et al (2020) A quantitative assessment of methane-derived carbon cycling at the cold seeps in the northwestern South China Sea. Minerals 10(3):256

Feng J, Li N, Liang J (2021) Using multi-proxy approach to constrain temporal variations of methane flux in methane-rich sediments of the southern South China Sea. Mar Pet Geol 132:105152

Fossing H, Ferdelman TG, Berg P (2000) Sulfate reduction and methane oxidation in continental margin sediments influenced by irrigation (South-East Atlantic off Namibia). Geochim Cosmochim Acta 64(5):897–910

Gong S, Feng D, Peng Y et al (2021) Deciphering the sulfur and oxygen isotope patterns of sulfate-driven anaerobic oxidation of methane. Chem Geol 581:120394

Haeckel M, Boudreau BP, Wallmann K et al (2007) Bubble-induced porewater mixing: A 3-D model for deep porewater irrigation. Geochim Cosmochim Acta 71(21):5135–5154

Hensen C, Zabel M, Pfeifer K et al (2003) Control of sulfate pore-water profiles by sedimentary events and the significance of anaerobic oxidation of methane for the burial of sulfur in marine sediments. Geochim Cosmochim Acta 67(14):2631–2647

Hu CY, Yang TF, Burr GS et al (2017) Biogeochemical cycles at the sulfate-methane transition zone (SMTZ) and geochemical characteristics of the pore fluids offshore southwestern Taiwan. J Asian Earth Sci 149:172–183

Hu T, Luo M, Xu Y et al (2021) Production of labile protein-like dissolved organic carbon associated with anaerobic methane oxidization in the Haima cold seeps, South China Sea. Front Mar Sci 8:797084

Hu Y, Luo M, Chen L et al (2018) Methane source linked to gas hydrate system at hydrate drilling areas of the South China Sea: Porewater geochemistry and numerical model constraints. J Asian Earth Sci 168:87–95

Hu Y, Feng D, Liang Q et al (2015) Impact of anaerobic oxidation of methane on the geochemical cycle of redox-sensitive elements at cold-seep sites of the northern South China Sea. Deep-Sea Res Part II-Top Stud Oceanogr 122:84–94

Hu Y, Luo M, Liang Q et al (2019) Pore fluid compositions and inferred fluid flow patterns at the Haima cold seeps of the South China Sea. Mar Pet Geol 103:29–40

Hu Y, Feng D, Peckmann J et al (2020) The impact of diffusive transport of methane on pore-water and sediment geochemistry constrained by authigenic enrichments of carbon, sulfur, and trace elements: A case study from the Shenhu area of the South China Sea. Chem Geol 553:119805

Hu Y, Feng D, Peckmann J et al (2023) The crucial role of deep-sourced methane in maintaining the subseafloor sulfate budget. Geosci Front 14:101530

Hu Y, Zhang X, Feng D et al (2022) Enhanced sulfate consumption fueled by deep-sourced methane in a hydrate-bearing area. Sci Bull 67(2):122–124

Huang Y, Suess E, Wu N et al (2008) Methane and gas hydrate geology of the Northern South China Sea: Sino-German Cooperative SO-177 Cruise Report (in Chinese). Geological Publishing House, Beijing

Hong WL, Torres ME, Kim JH et al (2013) Carbon cycling within the sulfate-methane-transition-zone in marine sediments from the Ulleung Basin. Biogeochemistry 115(1–3):129–148

Huang W, Meng M, Zhang W et al (2022) Geological, geophysical, and geochemical characteristics of deep-routed fluid seepage and its indication of gas hydrate occurrence in the Beikang Basin, Southern South China Sea. Mar Pet Geol 139:105610

Jiang S, Yang T, Xue Z et al (2005) Chlorine and sulfate concentrations in pore-waters from marine sediments in the north margin of the South China Sea and their implications for gas hydrate exploration. Geoscience 19(1):45–54 (in Chinese with English abstract)

Jin M, Feng D, Huang K et al (2022) Magnesium isotopes in pore water of active methane seeps of the South China Sea. Front Mar Sci 9:858860

Kasten S, Zabel M, Heuer V et al (2003) Processes and signals of nonsteady-state diagenesis in deep-sea sediments and their pore waters. In: Wefer G, Mulitza S, Ratmeyer V (eds) The South Atlantic in the Late Quaternary: Reconstruction of Material Budget and Current Systems. Springer, Berlin, pp 431–459

Kim JH, Park MH, Chun JH et al (2011) Molecular and isotopic signatures in sediments and gas hydrate of the central/southwestern Ulleung Basin: high alkalinity escape fuelled by biogenically sourced methane. Geo-Mar Lett 31(1):37–49

Komada T, Burdige DJ, Magen C et al (2016) Recycling of organic matter in the sediments of Santa Monica Basin California Borderland. Aquat Geochem 22(5–6):593–618

Lin S, Hsieh W, Lim YC et al (2006) Methane migration and its influence on sulfate reduction in the Good Weather Ridge region, South China Sea continental margin sediments. Terr Atmos Ocean Sci 17(17):883–902

Liu W, Wu Z, Xu S et al (2020) Pore-water dissolved inorganic carbon sources and cycling in the shallow sediments of the Haima cold seeps, South China Sea. J Asian Earth Sci 201:104495

Luo M, Chen L, Wang S et al (2013) Pockmark activity inferred from pore water geochemistry in shallow sediments of the pockmark field in southwestern Xisha Uplift, northwestern South China Sea. Mar Pet Geol 48:247–259

Luo M, Chen L, Tong H et al (2014) Gas hydrate occurrence inferred from dissolved Cl⁻ concentrations and $\delta^{18}O$ values of pore water and dissolved sulfate in the shallow sediments of the pockmark field in southwestern Xisha uplift, northern South China Sea. Energies 7(6):3886–3899

Liang Q, Hu Y, Feng D et al (2017) Authigenic carbonates from newly discovered active cold seeps on the northwestern slope of the South China Sea: Constraints on fluid sources, formation environments, and seepage dynamics. Deep-Sea Res Part I-Oceanogr Res Pap 124:31–41

Reeburgh WS (2007) Oceanic methane biogeochemistry. Chem Rev 107(2):486–513

Sha Z, Liang J, Zhang G et al (2015) A seepage gas hydrate system in northern South China Sea: Seismic and well log interpretations. Mar Geol 366:69–78

Schulz HD (2006) Schulz HD, Zabel M (Eds) Quantification of early diagenesis: dissolved constituents in pore water and signals in the solid phase. Marine Geochemistry. Berlin, Germany, pp. 73–124

Suess E (2005) RV SONNE Cruise Report SO 177, Sino-German Cooperative Project, South China Sea Continental Margin: Geological Methane Budget and Environmental Effects of Methane Emissions and Gas Hydrates. IFM-GEOMAR Reports. https://doi.org/10.3289/ifm-geomar_rep_4_2005

Wang X, Li N, Feng D et al (2018) Using chemical compositions of sediments to constrain methane seepage dynamics: A case study from Haima cold seeps of the South China Sea. J Asian Earth Sci 168:137–144

Wei J, Liang J, Lu J et al (2019) Characteristics and dynamics of gas hydrate systems in the northwestern South China Sea – Results of the fifth gas hydrate drilling expedition. Mar Pet Geol 110:287–298

Wu D, Wu N, Zhang M et al (2013a) Relationship of sulfate-methane interface (SMI), methane flux and the underlying gas hydrate in Dongsha area, Northern South China Sea. Earth Science-J China Univ Geosci 38(6):1309–1320 (in Chinese with English abstract)

Wu LS, Yang SX, Liang JQ et al (2013b) Variations of pore water sulfate gradients in sediments as indicator for underlying gas hydrate in Shenhu Area, the South China Sea. Sci China-Earth Sci 56(4):530–540

Wu LS, Yang SX, Liang JQ et al (2010) Geochemical characteristics of sediments at site HQ-48PC in Qiongdongnan Area, the north of the South China Sea, and their implication for gas hydrates. Geoscience 24(3):534–544 (in Chinese with English abstract)

Wu N, Ye Y, Wu D et al (2007) Geochemical characteristic of sediments from Southeast Hainan Basin, South China Sea and micro-methane-seep activity. Geology Res South China Sea 00:40–47 (in Chinese with English abstract)

Wu N, Zhang H, Yang S et al (2011) Gas Hydrate System of Shenhu Area, Northern South China Sea: Geochemical Results. J Geol Res 2011:370298

Yang T, Jiang S, Ge L et al (2006) Geochemical characteristics of sediment pore water from Site XS-01 in the Xisha trough of South China Sea and their significance for gas hydrate occurrence. Quaternary Sciences 3:442–448 (in Chinese with English abstract)

Yang T, Jiang SY, Yang JH et al (2008a) Dissolved inorganic carbon (DIC) and its carbon isotopic composition in sediment pore waters from the Shenhu area, northern South China Sea. J Oceanogr 64(2):303–310

Yang SX, Zhang HQ, Wu NY et al (2008b) High concentration hydrate in disseminated forms obtained in Shenhu area, North Slope of South China Sea. Vancouver BC. (Canada) In: 6th International Conference on Gas Hydrates (ICGH 2008), 6–10 July 2008, pp 1–6

Yang T, Jiang SY, Ge L et al (2013) Geochemistry of pore waters from HQ-1PC of the Qiongdongnan Basin, northern South China Sea, and its implications for gas hydrate exploration. Sci China-Earth Sci 56(4):521–529

Yang S, Zhang M, Liang J et al (2015) Preliminary results of China's third Gas hydrate drilling expedition: a critical step from discovery to development in the South China Sea. Fire Ice 15:1–5

Ye H, Yang T, Zhu G et al (2016) Pore water geochemistry in shallow sediments from the northeastern continental slope of the South China Sea. Mar Pet Geol 75:68–82

Ye J, Wei J, Liang J et al (2019) Complex gas hydrate system in a gas chimney, South China Sea. Mar Pet Geol 104:29–39

Zabel M, Schulz HD (2001) Importance of submarine landslides for non-steady state conditions in pore water systems – lower Zaire (Congo) deep-sea fan. Mar Geol 176(1–4):87–99

Zha R, Yang T, Shi X et al (2022) Quantitative assessment of dissolved inorganic carbon cycling in marine sediments from gas hydrate-bearing areas in the South China Sea. Mar Pet Geol 145:105881

Zhang J, Lei H, Ou W et al (2014) Research of the sulfate-methane transition zone (SMTZ) in sediments of 973–4 column in continental slope of Northern South China Sea. Nat Gas Geosci 25(11):1811–1820 (in Chinese with English abstract)

Zhang G, Liang J, Lu J et al (2015) Geological features, controlling factors and potential prospects of the gas hydrate occurrence in the east part of the Pearl River Mouth Basin, South China Sea. Mar Pet Geol 67:356–367

Zhang W, Liang J, Liang Q et al (2021) Gas hydrate accumulation and occurrence associated with cold seep systems in the northern South China Sea: An overview. Geofluids 2021:5571150

Zhang Y, Luo M, Hu Y et al (2019) An areal assessment of subseafloor carbon cycling in cold seeps and hydrate-bearing areas in the northern South China Sea. Geofluids 2019:1–14

Zhuang GC, Xu L, Liang Q et al (2019) Biogeochemistry, microbial activity, and diversity in surface and subsurface deep-sea sediments of South China Sea. Limnol Oceanogr 64(5):2252–2270

Chapter 9
Stable Isotope Signatures of Authigenic Minerals from Methane Seeps

Shanggui Gong, Jörn Peckmann, and Dong Feng

Abstract Authigenic minerals forming at marine seeps constitute an excellent archive of past methane seepage and biogeochemical processes. Over the past two decades, authigenic carbonate and sulfur-bearing minerals from methane seeps of the South China Sea (SCS) have been widely investigated, providing insight into fluid sources and seepage dynamics and facilitating the establishment of geochemical proxies to trace sulfate-driven anaerobic oxidation of methane (SD-AOM). Authigenic carbonates from all seep sites in the SCS commonly exhibit low δ^{13}C and high δ^{18}O values, confirming the incorporation of methane-derived carbon and oxygen from a pore water pool probably affected by gas hydrate dissociation. Pyrite is a common authigenic mineral at methane seeps, also forming at low methane flux where authigenic carbonate tends to be absent. The identification of methane seepage and SD-AOM activity consequently benefited from the advancement of sulfur isotope geochemistry, particularly from in situ measurements of $\delta^{34}S_{pyrite}$ values using nanoSIMS and multiple sulfur isotopes. Quantification of carbon and sulfur fluxes in the course of SD-AOM in modern and ancient marine sedimentary environments remains challenging, highlighting the need for more field-based research and modeling work. Furthermore, other elemental cycles and biogeochemical processes at methane seeps archived in authigenic minerals, such as nitrogen-based metabolisms, remain largely unknown. We highlight that SCS seeps are fascinating natural laboratories to better understand methane-driven biogeochemical processes and their signatures in authigenic minerals, representing a rewarding but also challenging object of research in the field of geomicrobiology.

S. Gong (✉) · D. Feng
College of Marine Sciences, Shanghai Ocean University, Shanghai 201306, China
e-mail: sggong@shou.edu.cn

D. Feng
e-mail: dfeng@shou.edu.cn

J. Peckmann
Institute for Geology, Center for Earth System Research and Sustainability, Universität Hamburg, 20146 Hamburg, Germany
e-mail: joern.peckmann@uni-hamburg.de

D. Chen and D. Feng (eds.), *South China Sea Seeps*,
https://doi.org/10.1007/978-981-99-1494-4_9

9.1 Introduction

Authigenic minerals resulting from sulfate-driven anaerobic methane oxidation (SD-AOM) constitute a unique archive of past methane seepage and biogeochemical processes. As the key process at marine seeps, SD-AOM represents the main methane sink in marine sedimentary environments (Boetius et al. 2000; Reeburgh 2007; Egger et al. 2018), providing the local chemotrophic community with energy (Campbell 2006; Suess et al. 2018; Yang et al. 2020), regulating greenhouse gas emissions at the seafloor (Olson et al. 2016), and representing a hotspot of the marine carbon and sulfur cycles (Peckmann and Thiel 2004; Hu et al. 2022). SD-AOM releases dissolved bicarbonate and hydrogen sulfide, thereby favoring the precipitation of authigenic carbonate and sulfide minerals in the shallow sedimentary subsurface at marine seeps (Boetius et al. 2000). In turn, authigenic carbonate and sulfide and other sulfur-bearing minerals archive locally prominent biogeochemical process in the form of diagnostic stable isotope, trace element, and lipid biomarker patterns (Peckmann and Thiel 2004; Feng et al. 2016; Smrzka et al. 2019, 2020; Lin et al. 2022). These authigenic minerals record modes of methane transport and early diagenetic environments and allow the exploration of the role of methane in Earth's surface environments by tracing SD-AOM activity (Peckmann and Thiel 2004; Feng et al. 2016, 2018; Gong et al. 2022).

Authigenic carbonates with different mineralogies and carbon and oxygen isotope compositions have been widely reported from the South China Sea (SCS; Chen et al. 2005; Han et al. 2008, 2014; Tong et al. 2013; Wang et al. 2014; Feng and Chen 2015; Liang et al. 2017; Huang et al. 2022b). The morphologies of these seep carbonates vary, including crusts, mounds, pipes, tubes, and highly irregular bodies, reflecting different seepage intensities and the interaction of burrowing megafauna with fluid migration (Fig. 9.1; Han et al. 2013; Feng and Chen 2015; Sun et al. 2020b; Lu et al. 2021). Multiple carbonate mineral phases have been identified in the SCS, including aragonite, low-Mg calcite, high-Mg calcite, and dolomite (Fig. 9.2). These carbonates mainly consist of microcrystalline minerals, particularly calcite and dolomite (Fig. 9.3), with larger cement crystals typically represented by aragonite (Feng and Chen 2015). The formation of carbonate is governed by supersaturation, dissolved species concentration (Ca^{2+} and Mg^{2+}, SO_4^{2-}, and PO_4^{3-}), and microbial activity, all of which are highly variable in space and time due to changing methane flux (Peckmann et al. 2001; Luff and Wallmann 2003; Feng and Chen 2015; Gong et al. 2018a; Tong et al. 2019; Lu et al. 2021). In general, the occurrence of aragonite reflects a relatively high methane flux, where high levels of carbonate supersaturation and sulfate concentration as well as relatively low levels of sulfide favor aragonite over calcite precipitation (Burton 1993; Luff and Wallmann 2003). Combined with mineralogical analysis, trace element and lipid biomarker inventories allow to constrain the dynamics of the seep activity by reconstructing redox conditions (Feng and Chen 2015; Guan et al. 2016; Liang et al. 2017, 2022; Gong et al. 2018a; Smrzka et al. 2020). The carbon and oxygen isotope compositions of authigenic carbonates are established proxies to reconstruct the composition and temperature of fluids from

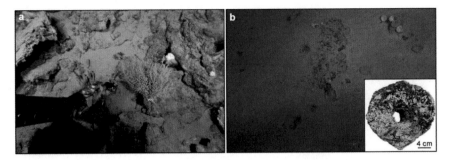

Fig. 9.1 Seafloor images of typical seep manifestations. **a** Massive carbonate crusts; **b** tubular carbonate exposed on the seafloor. Image collected at the Jiulong methane reef (water depth: 684 m) during *ROPOS* dives 2070 and 2073 in 2018

Fig. 9.2 Composition of the carbonate fraction in seep carbonates at South China Sea. Data are from Tong et al. (2013), Han et al. (2014), Feng and Chen (2015), Lu et al. (2015) and Liang et al. (2017)

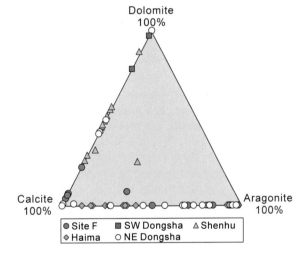

which carbonates precipitated, promoting the understanding of the sources and potential forces of methane seepage in the South China Sea (Chen et al. 2005; Han et al. 2008, 2013; Feng et al. 2015; Liang et al. 2017; Feng et al. 2018).

The extremely negative $\delta^{13}C$ values of authigenic seep carbonates are recognized as the most distinctive geological feature of SD-AOM inherited from the ^{13}C depletion of biogenic methane (−110‰ to −50‰) and thermogenic methane (−50‰ to −30‰; Sackett 1978; Whiticar 1999; Peckmann et al. 2001; Chen et al. 2005). The $\delta^{13}C_{carbonate}$ signature of SD-AOM can be masked by admixture of dissolved inorganic carbon (DIC) from sources other than methane oxidation: DIC sourced from organoclastic sulfate reduction (OSR), seawater DIC with a $\delta^{13}C$ value of 0‰, and a residual ^{13}C-enriched pool after methanogenesis (Feng et al. 2018). Admixture of DIC from other sources than methane oxidation was probably more common in paleo-oceans before the early Paleozoic Era, which were characterized by high seawater

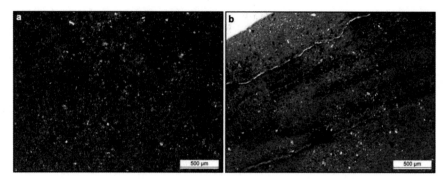

Fig. 9.3 Thin section photomicrographs of seep carbonate obtained from Site F, plane-polarized light (details in Feng and Chen 2015). **a** Typical microcrystalline carbonate matrix with enclosed terrigenous sediment. **b** Microcrystalline carbonate with abundant pyrite framboids (dark) enclosed

DIC and/or low seawater sulfate levels resulting in a lower ratio of SD-AOM-sourced DIC and seawater DIC (Bristow and Grotzinger 2013). Furthermore, the $\delta^{13}C$ proxy typically cannot be employed for tracing SD-AOM in methane diffusion-limited settings, as authigenic carbonate tends to form only in settings with relatively high methane flux (Luff and Wallmann 2003; Hu et al. 2020).

Fortunately, sulfur-bearing minerals, benefiting from recent advances in sulfur isotope biogeochemistry, can be used to identify SD-AOM even in low flux settings and to constrain the sulfur cycle in methane-bearing environments (Jørgensen et al. 2004; Chen et al. 2006; Feng and Robert 2011; Lin et al. 2015, 2016a, b, 2017; Li et al. 2016; Gong et al. 2018a, b; Liu et al. 2022a). Innovative approaches have been applied and new understanding has been obtained thanks to research on seepage in the SCS: (1) environmental controls on the morphology and $\delta^{34}S$ of SD-AOM-derived pyrite (Chen et al. 2005; Lin et al. 2016a, b, 2017; Li et al. 2016; Gong et al. 2018a); (2) the extremely high variability of $\delta^{34}S$ in SD-AOM-derived pyrite via nanoSIMS analysis (Lin et al. 2016a, b); (3) a carbonate-based proxy for SD-AOM (Feng et al. 2016); and (4) diagnostic multiple sulfur isotope systematics of SD-AOM (Lin et al. 2017; Gong et al. 2018b, 2022; Liu et al. 2020, 2022a). These achievements provide a robust approach for SD-AOM tracing in the subrecent marine sedimentary record and the older rock record and promise deeper future insight into the mechanisms of pyritization during early diagenesis, which are key requirements for reconstructing the global sulfur cycle (Wang et al. 2021; Peng et al. 2022).

Overall, methane-derived authigenic carbonate and sulfide minerals provide a useful geological archive of fluid composition, past SD-AOM activity, and early diagenetic environments. In this chapter, we review the current knowledge of the key biogeochemical processes archived in methane-derived authigenic carbonate and sulfide minerals from a stable carbon, oxygen, and sulfur isotope geochemistry perspective, focusing on (1) the recognition of biogeochemical processes and fluid sources archived in authigenic carbonates and (2) the sulfur isotope systematics of SD-AOM in modern marine sediments and its implication for tracing past SD-AOM.

9.2   Fluid Sources and Biogeochemical Processes Archived in Authigenic Carbonate Minerals

9.2.1   C–O Isotope Signatures of Methane-Derived Authigenic Carbonate

[13]C-depleted carbonates found at all seep sites of the SCS revealed that precipitation was predominantly driven by microbial oxidation of methane (Fig. 9.4; Chen et al. 2005; Han et al. 2008, 2013, 2014; Tong et al. 2013; Wang et al. 2014; Feng and Chen 2015; Liang et al. 2017; Lu et al. 2017; Yang et al. 2018). Biological methane sources have been identified across seep sites in the SCS based on the $\delta^{13}C$ values of methane and DIC (Chuang et al. 2013; Zhuang et al. 2016; Hu et al. 2018, 2019; Jin et al. 2022). However, the $\delta^{13}C$ values of carbonates in the SCS are typically higher than $-50‰$, indicating significant admixture of DIC from other sources than methane oxidation, including seawater, decomposition of organic matter, and residual DIC after methanogenesis (Peckmann and Thiel 2004; Feng et al. 2018; Huang et al. 2022a). The great variability of $\delta^{13}C$ values indicates that the degree of mixing among the different DIC pools varies in both space and time. Therefore, the $\delta^{13}C$ values of lipid biomarkers are better proxies to identify different methane sources (Himmler et al. 2015; Guan et al. 2016, 2018). Furthermore, high resolution measurement $\delta^{13}C$ values of carbonate with nanoSIMS analysis is another promising tool to determine methane sources (Feng et al. 2018).

The oxygen isotope composition of authigenic carbonates provides the means to calculate the temperature during precipitation and the $\delta^{18}O$ value of the parent fluid (Naehr et al. 2007; Han et al. 2014). The $\delta^{18}O$ value of authigenic carbonate is controlled by a combination of factors, including (1) mineralogy and chemistry, (2) ambient temperature, and (3) $\delta^{18}O$ value of the parent fluid (Anderson and Arthur 1983; Grossman and Ku 1986; Kim and O'Neil 1997; Mavromatis et al. 2012). Many seep carbonates collected from the SCS exhibit $\delta^{18}O$ values higher than the calculated equilibrium values based on the mineral type, bottom water temperature, and $\delta^{18}O$ value of seawater (Feng and Chen 2015; Liang et al. 2017; Yang et al. 2018). This [18]O enrichment can be explained by the addition of ^{18}O-rich fluids resulting from gas hydrate dissociation (Bohrmann et al. 1998; Han et al. 2013; Feng and Chen 2015; Liang et al. 2017). However, ^{18}O-rich fluids could also originate from clay mineral dehydration (Hesse 2003) and deep-sourced fluids modified by mineral–water interactions (Holser et al. 1979; Giggenbach 1992). Therefore, the sources of ^{18}O-rich fluids and their diagnostic signatures require further study. Reconstruction of the temperature during carbonate precipitation can provide additional information on the environmental settings at methane seeps. However, due to the variable $\delta^{18}O$ values of parent fluids, the paleo-temperature during carbonate precipitation cannot be calculated using $\delta^{18}O$ values alone. Recently, a carbonate clumped isotope (Δ_{47}) thermometer has been explored for methane-derived authigenic carbonates (Wacker

Fig. 9.4 Compilation of published $\delta^{13}C$ and $\delta^{18}O$ values of seep carbonates retrieved from the South China Sea. Data are from Chen et al. (2005); Han et al. (2008, 2014); Tong et al. (2013); Feng and Chen (2015); Lu et al. (2015, 2018); Liang et al. (2017); Huang et al. (2022a) and Liu et al. (2022b)

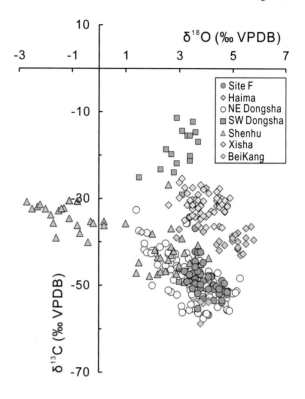

et al. 2014; Loyd et al. 2016; Zhang et al. 2019; Thiagarajan et al. 2020). Both equilibrium and disequilibrium clumped isotope values have been reported for methane-derived authigenic carbonates, highlighting that additional proxies (e.g., Δ_{48}) are needed to further constrain the factors (e.g., the kinetic isotope effect) affecting the Δ_{47} value, hopefully allowing for more accurate paleotemperature reconstructions in the future.

9.2.2 Diagnostic $\delta^{18}O_{SO4}$ Versus $\delta^{34}S_{SO4}$ Patterns of SD-AOM

Since the signature of ^{13}C depletion of SD-AOM can be diluted by admixture of DIC from other sources than methane oxidation, a new carbonate-based proxy for SD-AOM has been established by Feng et al. (2016). The work of these authors emphasized that the isotopic signal of porewater sulfate can be preserved in authigenic carbonate in the form of carbonate-associated sulfate. For a given porewater sulfate profile, the slope of the tangent along the gradient of $\delta^{18}O_{SO4}$ and $\delta^{34}S_{SO4}$ values (referred to as the $\delta^{18}O_{SO4}/\delta^{34}S_{SO4}$ slope) is related to the net sulfate reduction rate (Böttcher et al. 1998, 1999; Aharon and Fu 2000; Antler et al. 2013; Turchyn et al.

2016). This relationship has been interpreted as a decrease in the ratio of reverse and forward fluxes during intracellular enzymatic steps with increasing sulfate reduction rate (Brunner et al. 2005; Antler et al. 2013). The net sulfate reduction rate at methane seeps is several orders of magnitude higher than that in OSR-dominated settings, resulting in the diagnostic small $\delta^{18}O_{SO4}/\delta^{34}S_{SO4}$ slope (<0.5) of SD-AOM, which is distinct from the larger $\delta^{18}O_{SO4}/\delta^{34}S_{SO4}$ slope (>0.7) in OSR-dominated settings (Aharon and Fu 2000; Feng and Robert 2011; Antler et al. 2015). Although a small $\delta^{18}O_{SO4}/\delta^{34}S_{SO4}$ slope (0.36 ± 0.06) was also observed in organic-rich sediments of mangroves (Crémière et al. 2017), the contribution of SD-AOM to the overall removal of sulfate remained uncertain, and the organic-rich environment in mangroves does not support massive carbonate precipitation (Antler and Pellerin 2018). Thus, a diagnostic small $\delta^{18}O_{SO4}/\delta^{34}S_{SO4}$ slope preserved in carbonate-associated sulfate is a robust proxy for tracing SD-AOM in the geological record (Feng et al. 2016; Tong et al. 2019).

Given the utility of this proxy, Gong et al. (2022) quantified the lowest methane flux (i.e., 200 mmol m^{-2} yr^{-1}) required to produce the diagnostic small $\delta^{18}O_{SO4}/\delta^{34}S_{SO4}$ slope of SD-AOM studying the Haima seeps of the SCS, where the contribution of OSR to overall sulfate consumption is negligible. As shown in Fig. 9.5, plotting the $\delta^{18}O_{SO4}/\delta^{34}S_{SO4}$ slope versus net sulfate reduction rates allows to distinguish between OSR– and SD-AOM-dominated settings. These observations indicated that the types of electron donors play a vital role in controlling isotope fractionation during microbial sulfate reduction in marine sediments. The $\delta^{18}O_{SO4}$ versus $\delta^{34}S_{SO4}$ patterns of porewater profiles have been widely used to explore sulfur-based reactions in marine sediments, including OSR, SD-AOM, and sulfide oxidation (Böttcher et al. 1998; Aharon and Fu 2000; Böttcher and Thamdrup 2001; Antler et al. 2014, 2015; Bertran et al. 2020). Generally, methane fluxes and the contribution of SD-AOM to overall sulfate reduction must be considered when using $\delta^{18}O_{SO4}$ versus $\delta^{34}S_{SO4}$ patterns to study the sulfur cycle.

$\delta^{18}O_{SO4}$ versus $\delta^{34}S_{SO4}$ patterns are also controlled by sulfide oxidation and the oxygen isotope composition of sulfate diffusing into the sulfate methane transition zone (SMTZ), highlighting the need for considering the pitfalls and new perspectives of this proxy (Antler and Pellerin 2018; Gong et al. 2021). First, separating SD-AOM from OSR at different depths is challenging in methane-diffusion-limited settings, where OSR in the upper sulfate reduction zone can drive $\delta^{18}O_{SO4}$ values to an apparent equilibrium value before AOM-SR can imprint its signature on the $\delta^{18}O_{SO4}$ versus $\delta^{34}S_{SO4}$ slope (Fig. 9.6). Second, the $\delta^{18}O_{SO4}$ versus $\delta^{34}S_{SO4}$ slope is affected by sulfide reoxidation in two ways: (1) sulfide reoxidation occurring in the whole sulfate reduction zone can increase the $\delta^{18}O_{SO4}$ versus $\delta^{34}S_{SO4}$ slope; (2) quantitative reoxidation of sulfide in the subsurface can alter the initial sulfur and oxygen isotope composition of porewater sulfate. With a higher or lower initial $\delta^{18}O_{SO4}$ value, $\delta^{18}O_{SO4}$ can reach the apparent equilibrium value faster or slower, respectively, consequently resulting in a greater or smaller $\delta^{18}O_{SO4}$ versus $\delta^{34}S_{SO4}$ slope

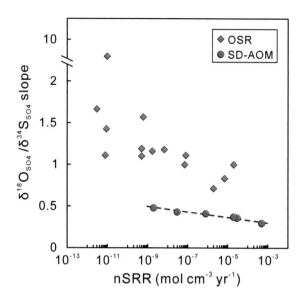

Fig. 9.5 Slope of $\delta^{18}O_{SO4}$ versus $\delta^{34}S_{SO4}$ in the apparent linear phase versus the average net sulfate reduction rate (nSRR; modified after Gong et al. 2021). *Note* OSR and SD-AOM denote organoclastic sulfate reduction and sulfate-driven anaerobic oxidation of methane, respectively. Reprinted from Chemical Geology, 581, Gong et al. (2021) Deciphering the sulfur and oxygen isotope patterns of sulfate-driven anaerobic oxidation of methane, 120394, Copyright (2021), with permission from Elsevier

(Turchyn et al. 2010), respectively. The occurrence and extent of sulfur reoxidation depend on the relative sulfate reduction rate and oxidant replenishment, which vary with the sedimentary environment (Gong et al. 2021). Third, the $\delta^{18}O_{SO4}$ versuss $\delta^{34}S_{SO4}$ slope also depends on the oxygen isotope composition of marine sulfate (Turchyn et al. 2010; Feng et al. 2016; Antler et al. 2017), which has likely changed during Earth history (Claypool et al. 1980). Finally, $\delta^{18}O_{SO4}$ values associated with SD-AOM are diagnostically higher than typical apparent equilibrium $\delta^{18}O_{SO4}$ values in OSR-dominated settings and can serve as a new proxy for the SD-AOM activity (Gong et al. 2021) because the increase in $\delta^{18}O_{SO4}$ is not limited in the course of microbial sulfate reduction with kinetically dominated oxygen isotope fractionation (Turchyn et al. 2010). With the above factors considered, the combined use of $\delta^{18}O_{SO4}$ versus $\delta^{34}S_{SO4}$ is a promising proxy to trace the sulfur cycle in modern and ancient marine sediments.

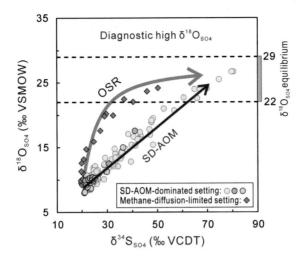

Fig. 9.6 Schematic plots of $\delta^{18}O_{SO4}$ and $\delta^{34}S_{SO4}$ for porewater sulfate at methane seeps. The red arrow denotes a linear correlation between $\delta^{18}O_{SO4}$ and $\delta^{34}S_{SO4}$ with a small slope in an SD-AOM-dominated setting (barite data are marked by gray cycles with data from Feng and Roberts (2011); the colored circles denote porewater data obtained from Gong et al. (2021)). The green diamond is the porewater sulfate profile of core W19-15, representing a methane-diffusion-limited setting (Hu et al. 2020). The gray arrow indicates an increase in $\delta^{18}O_{SO4}$ and $\delta^{34}S_{SO4}$ values at the onset of the curve in OSR-dominated settings, with $\delta^{18}O_{SO4}$ reaching apparent equilibrium values (22–29‰; Wortmann et al. 2007; Turchyn et al. 2016). *Note* OSR and SD-AOM denote organoclastic sulfate reduction and sulfate-driven anaerobic oxidation of methane, respectively

9.3   Biogeochemical Processes Archived in Authigenic Sulfides

9.3.1   High $\delta^{34}S$ Values Indicative of Enhanced Pyrite Formation

[34]S-enriched pyrite preserved in continental-margin sediments has been used to trace the paleo-SMTZ (Jørgensen et al. 2004; Peketi et al. 2012, 2015; Lin et al. 2016a, b, 2017; Wang et al. 2018). During microbial sulfate reduction, [32]S is preferentially distilled into reduced products, resulting in the formation of [34]S-depleted pyrite. Distinct from pyrite derived from OSR, SD-AOM-derived pyrite is generally characterized by a higher $\delta^{34}S_{py}$ values due to (1) the smaller magnitude of sulfur isotope fractionation for SD-AOM (e.g., <40‰) than that for OSR and (2) higher rates of SD-AOM than OSR rates, causing accumulation of dissolved sulfide to high concentrations due to relatively closed system conditions with little sulfate replenishment and high sulfate consumption rates (Aharon and Fu 2000; Deusner et al. 2014; Gong et al. 2018a). However, [34]S-enriched pyrite may not develop in settings with a major contribution of OSR to pyrite formation, low iron availability, and intense sulfide

reoxidation reactions (Borowski et al. 2013; Lin et al. 2016a, b; Formolo and Lyons 2013; Pierre 2017; Feng et al. 2018).

Recent work on the sulfur isotopic signature of SD-AOM-derived pyrite in sediments and carbonates retrieved from seeps of the SCS resulted in a better understanding of pyritization in the SMTZ and the control of dynamic methane fluxes on $\delta^{34}S_{py}$ values (Pu et al. 2007; Li et al. 2016; Lin et al. 2016a, b, 2017; Hu et al. 2017, 2020; Gong et al. 2018a, 2022). The typical $\delta^{34}S$ value of OSR-derived pyrite in the continental slope of the SCS ranges from $-50‰$ to $-20‰$ (Hu et al. 2015, 2018; Lin et al. 2017; Wang et al. 2018), whereas the $\delta^{34}S_{py}$ of SD-AOM-derived pyrite is typically higher than $-20‰$ due to the low sulfur isotope fractionation during SD-AOM ($<40‰$; Aharon and Fu 2000; Deusner et al. 2014; Gong et al. 2021). Combined with the high ratios of total sulfur to total organic carbon, the widely observed high $\delta^{34}S_{py}$ values in the SMTZ indicate extensive methane seepage activity along the continental margin of the SCS (Feng et al. 2018). The extremely high nanoSIMS $\delta^{34}S$ values of SD-AOM-derived pyrite reported for the SCS reach 130.3‰, representing the heaviest stable sulfur isotope composition of pyrite ever reported to our the best of our knowledge in a marine sedimentary environment and reflecting the great variability of $\delta^{34}S$ values of SD-AOM-derived pyrite (Lin et al. 2016b; Guo et al. 2022).

Figure 9.7a provides a schematic diagram of the environmental controls on the $\delta^{34}S$ value of pyrite under high methane flux with the SMTZ close to the seafloor, dissolved sulfide accumulating to high concentrations, and the $\delta^{34}S$ value of porewater sulfide generally increasing with depth from $-20‰$ to approximately 21‰. Under such conditions, pyrite is only moderately ^{34}S-enriched as supported by the typical $\delta^{34}S$ value of pyrite enclosed in methane-derived authigenic carbonates (Feng et al. 2016; Gong et al. 2018b; Crémière et al. 2020; Sun et al. 2020a). Figure 9.7b depicts a methane-diffusion-limited setting with a relatively deep SMTZ, where OSR causes porewater sulfate diffusion into the SMTZ resulting in high $\delta^{34}S$ values. Under such conditions, isotopically super-heavy pyrite can form in the SMTZ. Overall, the $\delta^{34}S$ value of SD-AOM-derived pyrite can be used to trace the relative methane flux and the dynamics of methane seepage (Gong et al. 2018a, 2022). However, potential admixture of early OSR-derived pyrite during the extraction of chromium-reducible sulfides can mask the $\delta^{34}S_{py}$ signatures of SD-AOM. Fortunately, such signatures have been detected for hand-picked pyrite (Lin et al. 2016a, b), via mass-balance calculations (Hu et al. 2020; Gong et al. 2022), and through petrographic study of authigenic pyrite combined with nanoSIMS analysis of stable sulfur isotopes (Lin et al. 2016a, b).

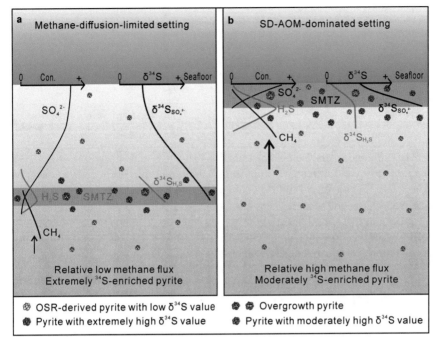

Fig. 9.7 Schematic diagram of the variable sulfur isotope composition of pyrite at methane seeps. **a** Under a relatively low methane flux with a deep SMTZ (i.e., a methane-diffusion-limited setting), the OSR progress at shallow depth causes ^{34}S-enriched sulfate to diffuse into the SMTZ, leading to a high δ^{34}S value of SD-AOM-derived pyrite exceeding the value of seawater sulfate. **b** Under a relatively high methane flux with the whole sulfate reduction zone dominated by SD-AOM (i.e., SD-AOM-dominated setting), SD-AOM-derived pyrite is moderately ^{34}S-enriched, with δ^{34}S values generally ranging from $-20‰$ to ca. 21‰. *Note* OSR, SD-AOM and SMTZ denote organoclastic sulfate reduction, sulfate-driven anaerobic oxidation of methane, and sulfate methane transition zone, respectively

9.3.2 Multiple Sulfur Isotope Fingerprints of SD-AOM

Recently, multiple sulfur isotopes have been applied to identify SD-AOM and to constrain the sulfur cycle in methane-bearing settings; such approach became necessary because of the common overlap of $\delta^{34}S_{py}$ values between OSR- and SD-AOM-derived pyrite (Lin et al. 2017, 2018; Gong et al. 2018b, 2022; Crémière et al. 2020; Liu et al. 2020, 2022a). The multiple sulfur isotope proxy relies on the fact that sulfur-based reactions experience varying dependencies on the expression of mass that can yield small deviations (Eq. 9.1) from thermodynamic equilibrium predictions at 0.515 (Farquhar et al. 2003; Johnston 2011). The small deviation of δ^{33}S from the mass-dependent fractionation law can be expressed with the following capital delta notation (Eq. 9.3):

$$^{33}\theta = \frac{ln^{33}\alpha}{ln^{34}\alpha} \qquad (9.1)$$

$$\delta^{3i} = (^{3i}R / ^{3i}R_{VCDT} - 1) * 1000, i = 3, 4 \qquad (9.2)$$

$$\Delta^{33}S = \delta^{33}S - 1000 * ((\delta^{34}S + 1)^{0.515} - 1) \qquad (9.3)$$

The difference in the isotope fractionation factor ($^{34}\alpha$ and $^{33}\theta$) among sulfur-based reactions can yield different relationships among multiple sulfur isotope compositions (^{32}S, ^{33}S, and ^{34}S) of sulfur-bearing compounds, which are expressed by $\Delta^{33}S$-$\delta^{34}S$ patterns. Combined $\Delta^{33}S$ and $\delta^{34}S$ analysis provides further constraints on the processes that contribute to sulfur cycling and help differentiating among various sulfur-based reactions including OSR, SD-AOM, sulfide oxidation, and sulfur disproportionation (Farquhar et al. 2003; Johnston et al. 2005; Lin et al. 2017; Gong et al. 2018b, 2022; Liu et al. 2022a). Below, we conclude that the diagnostic multiple sulfur isotope signatures of SD-AOM expressed in sulfur-bearing minerals are controlled by the low sulfur isotope fractionation during SD-AOM and the mass-transport effect on the isotope composition of dissolved sulfate and dissolved sulfide.

For a given porewater profile, the $\Delta^{33}S_{SO4} - \delta^{34}S_{SO4}$ pattern is mainly determined by the isotope fractionation of net sulfate reduction, with lower $1000ln^{34}\alpha$ and higher $^{33}\theta$ values leading to a larger slope of $\Delta^{33}S_{SO4} - \delta^{34}S_{SO4}$ (Gong et al. 2018b; Masterson et al. 2018). In OSR-dominated settings (Fig. 9.8a), the positive $\Delta^{33}S_{SO4}$ - $\delta^{34}S_{SO4}$ pattern can be attributed to the low $1000ln^{34}\alpha$ value and high $^{33}\theta$ value close to the equilibrium values at $-70\%_o$ and 0.515, respectively (Gong et al. 2018b, 2022; Masterson et al. 2018, 2022; Liu et al. 2022a). In SD-AOM-dominated settings (Fig. 9.8b), the relatively high $1000ln^{34}\alpha$ ($> -40\%_o$) and $^{33}\theta$ values (<0.5125) yield negative $\Delta^{33}S_{SO4} - \delta^{34}S_{SO4}$ correlations according to the simplified reaction-transport model (Gong et al. 2018b). In methane-diffusion-limited settings (Fig. 9.8c), the pore-water sulfate profile attains a positive $\Delta^{33}S_{SO4}$ - $\delta^{34}S_{SO4}$ correlation in the upper sulfate reduction zone dominated by OSR, which switches to a negative $\Delta^{33}S_{SO4} - \delta^{34}S_{SO4}$ correlation in the SMTZ. Overall, diagnostic negative $\Delta^{33}S_{SO4} - \delta^{34}S_{SO4}$ patterns of porewater sulfate profile in the SMTZ are distinguishable from the positive trajectory in OSR-dominated settings, highlighting that multiple sulfur isotope fractionation during microbial sulfate reduction is affected by the electron donor type, which facilitates the use of this proxy to identify SD-AOM (Gong et al. 2018b, 2022; Crémière et al. 2020; Liu et al. 2022a).

Under high methane fluxes, such as methane seeps with advective transport, the diagnostic negative $\Delta^{33}S_{SO4} - \delta^{34}S_{SO4}$ correlation of SD-AOM can be preserved in barite and carbonate-associated sulfate, thus serving as a useful proxy for SD-AOM in the rock record (Gong et al. 2018b; Crémière et al. 2020). Sulfate, for example recovered from carbonate rock in the form of carbonate-associated sulfate, does not represent a single steady-state pore-water profile but rather different stages of pore water evolution indicating various successive mixtures of porewater sulfate at

Fig. 9.8 Schematic plots of $\Delta^{33}S$ and $\delta^{34}S$ for porewater sulfate in different marine settings. **a** OSR-dominated setting (data from Strauss et al. (2012); Pellerin et al. (2015); Lin et al. (2017); Masterson et al. (2018); Gong et al. (2022); Liu et al. (2022a)); **b** SD-AOM-dominated setting, such as methane seeps (data from Gong et al. (2018b); Crémière et al. (2020)). **c** Methane-diffusion-limited setting (data retrieved from Gong et al. (2022); Liu et al. (2022a)). The black square denotes the seawater sulfate value of Tostevin et al. (2014). *Note* OSR and SD-AOM denote organoclastic sulfate reduction and sulfate-driven anaerobic oxidation of methane, respectively. Reprinted from Chemical Geology, 581, Gong et al. Reprinted from Earth and Planetary Science Letters, 597, Gong et al. (2022) Multiple sulfur isotope systematics of pyrite for tracing sulfate-driven anaerobic oxidation of methane, 117827, Copyright (2022), with permission from Elsevier

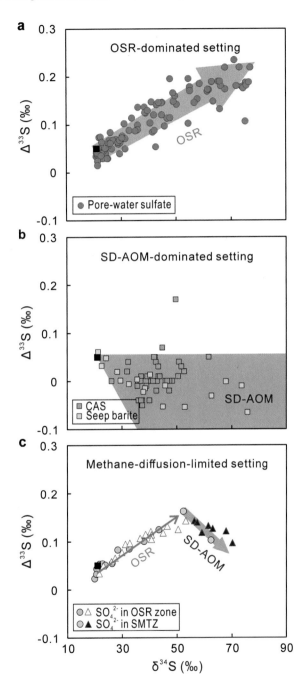

different depths. Consequently, negative $\Delta^{33}S_{SO4} - \delta^{34}S_{SO4}$ correlations cannot be expected to be necessarily archived in barite and carbonate-associated sulfate even if SD-AOM was a prominent process. The mixture of two endmembers with different $\delta^{34}S$ values can lead to a lower $\Delta^{33}S$ value than that of at least one of the endmembers (Ono et al. 2006; Johnston 2011). Thus, $\Delta^{33}S_{SO4}$ values more negative than that of seawater coupled with high $\delta^{34}S_{SO4}$ values preserved in the rock record constitute a diagnostic signature of SD-AOM at methane seeps, sites typified by advective transport of methane (Gong et al. 2022). Globally, methane-diffusion-limited environments are more widely distributed along continental margins than methane seeps (Egger et al. 2018; Hu et al. 2022). Due to the absence of methane-derived authigenic carbonates, however, the diagnostic multiple sulfur isotope signature of porewater sulfate in a methane-diffusion-limited setting can hardly be preserved in carbonate rock or barite.

Case studies on methane-bearing environments of the SCS indicated that multiple sulfur isotope compositions of pyrite can be used for tracing SD-AOM in methane diffusion-limited settings (Lin et al. 2017; Gong et al. 2022). The instantaneously produced sulfide inherits the same $\Delta^{33}S_{SO4} - \delta^{34}S_{SO4}$ trajectory of porewater sulfate, i.e., positive and negative correlations in OSR– and SD–AOM-dominated settings, respectively. However, the porewater sulfide accumulating in the course of sulfate reduction shows similar $\Delta^{33}S - \delta^{34}S$ patterns and approaches the sulfur isotopic composition of seawater sulfate in both organic compound-rich settings (Liu et al. 2022a) and, thus, is expected to occur at methane seeps (Gong et al. 2022). This phenomenon can be explained by the independent diffusion of ^{32}S, ^{33}S, and ^{34}S in sulfide in a setting where dissolved sulfide accumulates to high concentrations (e.g., >1 mmol/l; Jørgensen et al. 2004; Liu et al. 2022a; Masterson et al. 2022). Therefore, multiple sulfur isotopes of pyrite cannot trace SD-AOM activity at methane seeps and in organic-rich settings, as evidenced by the overlap of the $\Delta^{33}S_{py} - \delta^{34}S_{py}$ areas of successively formed, composite pyrite derived from SD-AOM and OSR (Fig. 9.9b).

However, the diffusion effect of porewater sulfide is limited in organic-poor deeper settings such as the continental slope sediment of the SCS, where low sulfate reduction rates led to an excess of buried reactive iron and a low concentration of dissolved sulfide. Under this circumstance, the diffusion effect on the isotope composition of porewater sulfide and the contribution of OSR-derived sulfide to the sulfide pool in the SMTZ is limited. Consequently, the increased $\Delta^{33}S$ and $\delta^{34}S$ signatures of porewater sulfate diffusing into the SMTZ can be archived in pyrite, with the $\Delta^{33}S$ value of the produced sulfide reaching as high as 0.3‰ in the upper SMTZ. Furthermore, the $\Delta^{33}S$ values of instantaneously produced sulfide within the SMTZ decrease with increasing $\delta^{34}S$ value, thus leading to a pronounced negative $\Delta^{33}S_{py}$ value falling out of the $\Delta^{33}S_{py} - \delta^{34}S_{py}$ area of the OSR. Therefore, the diagnostic larger $\Delta^{33}S - \delta^{34}S$ field of pyrite relative to OSR-derived pyrite allows tracing of SD-AOM in continental slope settings (Fig. 9.9c).

Fig. 9.9 Schematic plots of $\Delta^{33}S$ and $\delta^{34}S$ for porewater sulfide and authigenic pyrite in different marine settings: **a** OSR-dominated setting (data from Johnston et al. (2008); Strauss et al. (2012); Lin et al. (2017); Gong et al. (2022); Liu et al. (2022a)); **b** SD-AOM-dominated setting, such as methane seeps (data obtained from Crémière et al. (2020)). **c** Methane-diffusion-limited setting (data from Lin et al. (2017); Gong et al. (2022); Liu et al. (2022a)). The black square indicates the seawater sulfate value of Tostevin et al. (2014). The gray shaded area denotes the largest $\Delta^{33}S$–$\delta^{34}S$ field of pyrite in the OSR-dominated setting, and the purple area denotes the $\Delta^{33}S$–$\delta^{34}S$ field of SD-AOM-derived pyrite (modified from Gong et al. (2022)). *Note* OSR and SD-AOM denote organoclastic sulfate reduction and sulfate-driven anaerobic oxidation of methane, respectively. Reprinted from Earth and Planetary Science Letters, 597, Gong et al. (2022) Multiple sulfur isotope systematics of pyrite for tracing sulfate-driven anaerobic oxidation of methane, 117827, Copyright (2022), with permission from Elsevier

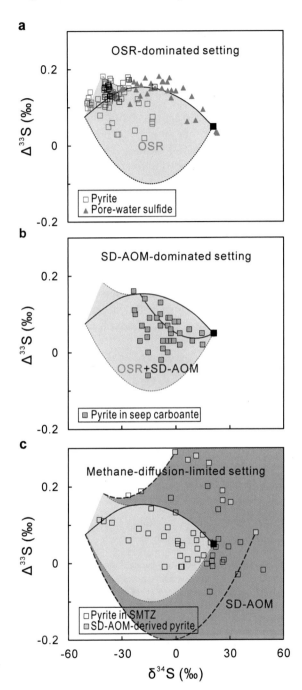

9.4 Summary and Future Studies

Widespread methane seepage along the continental margins of the South China Sea provides the opportunity to study methane-related biogeochemical processes and their fingerprints in the sedimentary record, allowing to trace the occurrence and strength of methane seepage and its effect on local to global marine environments through geologic time. Previous work on authigenic minerals formed at methane seeps in the SCS mainly aimed to reconstruct the origin of seep fluids and seepage dynamics and to establish proxies for tracing past SD-AOM activity. One of the outstanding achievements in this endeavor is our improved understanding of the sulfur isotope systematics of SD-AOM, owed to in situ nanoSIMS analysis and multiple sulfur isotopes. The latter analytical approaches allow confident identification of the origin of early diagenetic pyrite, the most common mineral at marine seeps. With respect to the South China Sea–now representing one of the best studied seepage provinces–further research targeting quantification of sulfur cycling will probably provide fundamental knowledge of the effect of SD-AOM on ocean margin sediments. Other biogeochemical processes taking place in association with SD-AOM, such as processes involving nitrogen, need to be further investigated to fully unravel the interaction between carbon, sulfur, and nitrogen cycling at marine methane seeps.

Acknowledgements Funding was provided by the NSF of China (Grants: 42225603, 41906046, and 42176056). Yu Hu and Min Luo are thanked for their constructive comments, which have greatly improved the quality of the chapter.

References

Aharon P, Fu BS (2000) Microbial sulfate reduction rates and sulfur and oxygen isotope fractionations at oil and gas seeps in deepwater Gulf of Mexico. Geochim Cosmochim Acta 64(2):233–246

Anderson TF, Arthur MA (1983) Stable isotopes of oxygen and carbon and their application to sedimentologic and paleoenvironmental problems. In: Stable Isotopes in Sedimentary Geology. SEPM (Society for Sedimentary Geology): Dallas TX USA pp 1–151

Antler G, Pellerin A (2018) A critical look at the combined use of sulfur and oxygen isotopes to study microbial metabolisms in methane-rich environments. Front Microbiol 9:519

Antler G, Turchyn AV, Rennie V et al (2013) Coupled sulfur and oxygen isotope insight into bacterial sulfate reduction in the natural environment. Geochim Cosmochim Acta 118:98–117

Antler G, Turchyn AV, Herut B et al (2014) Sulfur and oxygen isotope tracing of sulfate driven anaerobic methane oxidation in estuarine sediments. Estuar Coast Shelf Sci 142:4–11

Antler G, Turchyn AV, Herut B et al (2015) A unique isotopic fingerprint of sulfate-driven anaerobic oxidation of methane. Geology 43(7):619–622

Antler G, Turchyn AV, Ono S et al (2017) Combined ^{34}S, ^{33}S and ^{18}O isotope fractionations record different intracellular steps of microbial sulfate reduction. Geochim Cosmochim Acta 203:364–380

Bertran E, Waldeck A, Wing BA et al (2020) Oxygen isotope effects during microbial sulfate reduction: applications to sediment cell abundances. ISME J 14(6):1508–1519

Boetius A, Ravenschlag K, Schubert CJ et al (2000) A marine microbial consortium apparently mediating anaerobic oxidation of methane. Nature 407(6804):623–626

Bohrmann G, Greinert J, Suess E et al (1998) Authigenic carbonates from the Cascadia subduction zone and their relation to gas hydrate stability. Geology 26(7):647–650

Borowski WS, Rodriguez NM, Paull CK et al (2013) Are ^{34}S-enriched authigenic sulfide minerals a proxy for elevated methane flux and gas hydrates in the geologic record? Mar Pet Geol 43:381–395

Bristow TF, Grotzinger JP (2013) Sulfate availability and the geological record of cold-seep deposits. Geology 41(7):811–814

Brunner B, Bernasconi SM, Kleikemper J et al (2005) A model for oxygen and sulfur isotope fractionation in sulfate during bacterial sulfate reduction processes. Geochim Cosmochim Acta 69(20):4773–4785

Burton EA (1993) Controls on marine carbonate cement mineralogy-review and reassessment. Chem Geol 105(1–3):163–179

Böttcher ME, Thamdrup B (2001) Anaerobic sulfide oxidation and stable isotope fractionation associated with bacterial sulfur disproportionation in the presence of MnO_2. Geochim Cosmochim Acta 65(10):1573–1581

Böttcher ME, Brumsack HJ, de Lange GJ (1998) Sulfate reduction and related stable isotope (^{34}S, ^{18}O) variations in interstitial waters from the eastern Mediterranean. In: Robertson AHF, Emeis KC, Richter C et al (Eds) Proc. ODP, Sci. Results, Vol. 16. Ocean Drilling Program, College Station, TX, pp 365–373

Böttcher ME, Bernasconi SM, Brumsack HJ (1999) Carbon, sulfur, and oxygen isotope geochemistry of interstitial waters from the western Mediterranean. In: Zahn R, Comas M C, Klaus A (Eds) Proc. ODP, Sci. Results, Vol. 161. Ocean Drilling Program, College Station, TX, pp 413–421

Campbell KA (2006) Hydrocarbon seep and hydrothermal vent paleoenvironments and paleontology: Past developments and future research directions. Paleogeogr Paleoclimatol Paleoecol 232(2–4):362–407

Chen D, Huang Y, Yuan X et al (2005) Seep carbonates and preserved methane oxidizing archaea and sulfate reducing bacteria fossils suggest recent gas venting on the seafloor in the Northeastern South China Sea. Mar Pet Geol 22(5):613–621

Chen D, Feng D, Su Z et al (2006) Pyrite crystallization in seep carbonates at gas vent and hydrate site. Mater Sci Eng C-Mater Biol Appl 26(4):602–605

Chuang PC, Dale AW, Wallmann K et al (2013) Relating sulfate and methane dynamics to geology: Accretionary prism offshore SW Taiwan. Geochem Geophy Geosy 14(7):2523–2545

Claypool GE, Holser WT, Kaplan IR et al (1980) The age curves of sulfur and oxygen isotopes in marine sulfate and their mutual interpretation. Chem Geol 28(3–4):199–260

Crémière A, Strauss H, Sebilo M et al (2017) Sulfur diagenesis under rapid accumulation of organic-rich sediments in a marine mangrove from Guadeloupe (French West Indies). Chem Geol 454:67–79

Crémière A, Pellerin A, Wing BA et al (2020) Multiple sulfur isotopes in methane seep carbonates track unsteady sulfur cycling during anaerobic methane oxidation. Earth Planet Sci Lett 532:115994

Deusner C, Holler T, Arnold GL et al (2014) Sulfur and oxygen isotope fractionation during sulfate reduction coupled to anaerobic oxidation of methane is dependent on methane concentration. Earth Planet Sci Lett 399:61–73

Egger M, Riedinger N, Mogollon JM et al (2018) Global diffusive fluxes of methane in marine sediments. Nat Geosci 11(6):421–425

Farquhar J, Johnston DT, Wing BA et al (2003) Multiple sulphur isotopic interpretations of biosynthetic pathways: implications for biological signatures in the sulphur isotope record. Geobiology 1(1):27–36

Feng D, Chen D (2015) Authigenic carbonates from an active cold seep of the northern South China Sea: New insights into fluid sources and past seepage activity. Deep-Sea Res Part II-Top Stud Oceanogr 122:74–83

Feng D, Roberts HH (2011) Geochemical characteristics of the barite deposits at cold seeps from the northern Gulf of Mexico continental slope. Earth Planet Sci Lett 309(1–2):89–99

Feng D, Peng Y, Bao H et al (2016) A carbonate-based proxy for sulfate-driven anaerobic oxidation of methane. Geology 44(12):999–1002

Feng D, Qiu J, Hu Y et al (2018) Cold seep systems in the South China Sea: An overview. J Asian Earth Sci 168:3–16

Formolo MJ, Lyons TW (2013) Sulfur biogeochemistry of cold seeps in the Green Canyon region of the Gulf of Mexico. Geochim Cosmochim Acta 119:264–285

Giggenbach WF (1992) Isotopic shifts in waters from geothermal and volcanic systems along convergent plate boundaries and their origin. Earth Planet Sci Lett 113(4):495–510

Gong S, Hu Y, Li N et al (2018a) Environmental controls on sulfur isotopic compositions of sulfide minerals in seep carbonates from the South China Sea. J Asian Earth Sci 168:96–105

Gong S, Peng Y, Bao H et al (2018b) Triple sulfur isotope relationships during sulfate-driven anaerobic oxidation of methane. Earth Planet Sci Lett 504:13–20

Gong S, Feng D, Peng Y et al (2021) Deciphering the sulfur and oxygen isotope patterns of sulfate-driven anaerobic oxidation of methane. Chem Geol 581:120394

Gong S, Izon G, Peng Y et al (2022) Multiple sulfur isotope systematics of pyrite for tracing sulfate-driven anaerobic oxidation of methane. Earth Planet Sci Lett 597:117827

Grossman EL, Ku TL (1986) Oxygen and carbon isotope fractionation in biogenic aragonite: Temperature effects. Chem Geol 59:59–74

Guan H, Feng D, Wu N et al (2016) Methane seepage intensities traced by biomarker patterns in authigenic carbonates from the South China Sea. Org Geochem 91:109–119

Guan H, Birgel D, Peckmann J et al (2018) Lipid biomarker patterns of authigenic carbonates reveal fluid composition and seepage intensity at Haima cold seeps, South China Sea. J Asian Earth Sci 168:163–172

Guo Z, Liu Y, Qin G et al (2022) Extremely variable sulfur isotopic compositions of pyrites in carbonate pipes from the northern south China sea: Implications for a non-steady microenvironment. Mar Pet Geol 146:105927

Han X, Suess E, Huang Y et al (2008) Jiulong methane reef: Microbial mediation of seep carbonates in the South China Sea. Mar Geol 249(3–4):243–256

Han X, Yang K, Huang Y (2013) Origin and nature of cold seep in northeastern Dongsha area, South China Sea: Evidence from chimney-like seep carbonates. Chin Sci Bull 58(30):3689–3697

Han X, Suess E, Liebetrau V et al (2014) Past methane release events and environmental conditions at the upper continental slope of the South China Sea: constraints by seep carbonates. Int J Earth Sci 103(7):1873–1887

Hesse R (2003) Pore water anomalies of submarine gas-hydrate zones as tool to assess hydrate abundance and distribution in the subsurface - What have we learned in the past decade? Earth-Sci Rev 61(1–2):149–179

Himmler T, Birgel D, Bayon G et al (2015) Formation of seep carbonates along the Makran convergent margin, northern Arabian Sea and a molecular and isotopic approach to constrain the carbon isotopic composition of parent methane. Chem Geol 415:102–117

Holser WT, Kaplan IR, Sakai H et al (1979) Isotope geochemistry of oxygen in the sedimentary sulfate cycle. Chem Geol 25(1–2):1–17

Hu Y, Feng D, Liang Q et al (2015) Impact of anaerobic oxidation of methane on the geochemical cycle of redox-sensitive elements at cold-seep sites of the northern South China Sea. Deep-Sea Res Part II-Top Stud Oceanogr 122:84–94

Hu Y, Chen L, Feng D et al (2017) Geochemical record of methane seepage in authigenic carbonates and surrounding host sediments: A case study from the South China Sea. J Asian Earth Sci 138:51–61

Hu Y, Luo M, Chen LY et al (2018) Methane source linked to gas hydrate system at hydrate drilling areas of the South China Sea: Porewater geochemistry and numerical model constraints. J Asian Earth Sci 168:87–95

Hu Y, Luo M, Liang Q et al (2019) Pore fluid compositions and inferred fluid flow patterns at the Haima cold seeps of the South China Sea. Mar Pet Geol 103:29–40

Hu Y, Feng D, Peckmann J et al (2020) The impact of diffusive transport of methane on pore-water and sediment geochemistry constrained by authigenic enrichments of carbon, sulfur, and trace elements: A case study from the Shenhu area of the South China Sea. Chem Geol 553:119805

Hu Y, Zhang X, Feng D et al (2022) Enhanced sulfate consumption fueled by deep-sourced methane in a hydrate-bearing area. Sci Bull 67(2):122–124

Huang H, Feng D, Guo Y et al (2022a) Organoclastic sulfate reduction in deep-buried sediments: Evidence from authigenic carbonates of the Gulf of Mexico. Chem Geol 610:121094

Huang W, Meng M, Zhang W et al (2022b) Geological, geophysical, and geochemical characteristics of deep-routed fluid seepage and its indication of gas hydrate occurrence in the Beikang Basin, Southern South China Sea. Mar Pet Geol 139:105610

Jin M, Feng D, Huang K et al (2022) Magnesium isotopes in pore water of active methane seeps of the South China Sea. Front Mar Sci 9:858860

Johnston DT (2011) Multiple sulfur isotopes and the evolution of Earth's surface sulfur cycle. Earth-Sci Rev 106(1–2):161–183

Johnston DT, Farquhar J, Wing BA et al (2005) Multiple sulfur isotope fractionations in biological systems: A case study with sulfate reducers and sulfur disproportionators. Am J Sci 305(6–8):645–660

Johnston DT, Farquhar J, Habicht KS et al (2008) Sulphur isotopes and the search for life: strategies for identifying sulphur metabolisms in the rock record and beyond. Geobiology 6(5):425–435

Jørgensen BB, Böttcher ME, Lüschen H et al (2004) Anaerobic methane oxidation and a deep H_2S sink generate isotopically heavy sulfides in Black Sea sediments. Geochim Cosmochim Acta 68(9):2095–2118

Kim ST, O'Neil JR (1997) Equilibrium and nonequilibrium oxygen isotope effects in synthetic carbonates. Geochim Cosmochim Acta 61(16):3461–3475

Li N, Feng D, Chen L et al (2016) Using sediment geochemistry to infer temporal variation of methane flux at a cold seep in the South China Sea. Mar Pet Geol 77:835–845

Liang Q, Hu Y, Feng D et al (2017) Authigenic carbonates from newly discovered active cold seeps on the northwestern slope of the South China Sea: Constraints on fluid sources, formation environments, and seepage dynamics. Deep-Sea Res Part I-Oceanogr Res Pap 124:31–41

Liang Q, Huang H, Sun Y et al (2022) New insights into the archives of redox conditions in seep carbonates from the northern South China Sea. Front Mar Sci 9:945908

Lin Q, Wang J, Fu S et al (2015) Elemental sulfur in northern South China Sea sediments and its significance. Sci China-Earth Sci 58(12):2271–2278

Lin Q, Wang J, Taladay K et al (2016a) Coupled pyrite concentration and sulfur isotopic insight into the paleo sulfate–methane transition zone (SMTZ) in the northern South China Sea. J Asian Earth Sci 115:547–556

Lin Z, Sun X, Peckmann J et al (2016b) How sulfate-driven anaerobic oxidation of methane affects the sulfur isotopic composition of pyrite: A SIMS study from the South China Sea. Chem Geol 440:26–41

Lin Z, Sun X, Lu Y et al (2017) The enrichment of heavy iron isotopes in authigenic pyrite as a possible indicator of sulfate-driven anaerobic oxidation of methane: Insights from the South China Sea. Chem Geol 449:15–29

Lin Z, Sun X, Strauss H et al (2018) Multiple sulfur isotopic evidence for the origin of elemental sulfur in an iron-dominated gas hydrate-bearing sedimentary environment. Mar Geol 403:271–284

Lin Z, Sun X, Chen K et al (2022) Effects of sulfate reduction processes on the trace element geochemistry of sedimentary pyrite in modern seep environments. Geochim Cosmochim Acta 333:75–94

Liu J, Pellerin A, Izon G et al (2020) The multiple sulphur isotope fingerprint of a sub-seafloor oxidative sulphur cycle driven by iron. Earth Planet Sci Lett 536:116165

Liu J, Pellerin A, Wang J et al (2022a) Multiple sulfur isotopes discriminate organoclastic and methane-based sulfate reduction by sub-seafloor pyrite formation. Geochim Cosmochim Acta 316:309–330

Liu Y, Wei J, Li Y et al (2022b) Seep dynamics as revealed by authigenic carbonates from the eastern Qiongdongnan Basin. South China Sea. Mar Pet Geol 142:105738

Loyd SJ, Sample J, Tripati RE et al (2016) Methane seep carbonates yield clumped isotope signatures out of equilibrium with formation temperatures. Nat Commun 7:12274

Lu Y, Sun X, Lin Z et al (2015) Cold seep status archived in authigenic carbonates: Mineralogical and isotopic evidence from Northern South China Sea. Deep-Sea Res Part II-Top Stud Oceanogr 122:95–105

Lu Y, Liu Y, Sun X et al (2017) Intensity of methane seepage reflected by relative enrichment of heavy magnesium isotopes in authigenic carbonates: A case study from the South China Sea. Deep-Sea Res Part I-Oceanogr Res Pap 129:10–21

Lu Y, Sun X, Xu H et al (2018) Formation of dolomite catalyzed by sulfate-driven anaerobic oxidation of methane: Mineralogical and geochemical evidence from the northern South China Sea. Am Miner 103(5):720–734

Lu Y, Yang X, Lin Z et al (2021) Reducing microenvironments promote incorporation of magnesium ions into authigenic carbonate forming at methane seeps: Constraints for dolomite formation. Sedimentology 68(7):2945–2964

Luff R, Wallmann K (2003) Fluid flow, methane fluxes, carbonate precipitation and biogeochemical turnover in gas hydrate-bearing sediments at Hydrate Ridge, Cascadia Margin: Numerical modeling and mass balances. Geochim Cosmochim Acta 67(18):3403–3421

Masterson A, Alperin MJ, Berelson WM et al (2018) Interpreting multiple sulfur isotope signals in modern anoxic sediments using a full diagenetic model (California-Mexico Margin: Alfonso Basin). Am J Sci 318(5):459–490

Masterson AL, Alperin MJ, Arnold GL et al (2022) Understanding the isotopic composition of sedimentary sulfide: A multiple sulfur isotope diagenetic model for Aarhus Bay. Am J Sci 322(1):1–27

Mavromatis V, Schmidt M, Botz R et al (2012) Experimental quantification of the effect of Mg on calcite-aqueous fluid oxygen isotope fractionation. Chem Geol 310:97–105

Naehr TH, Eichhubl P, Orphan VJ et al (2007) Authigenic carbonate formation at hydrocarbon seeps in continental margin sediments: A comparative study. Deep-Sea Res Part II-Top Stud Oceanogr 54(11–13):1268–1291

Olson SL, Reinhard CT, Lyons TW (2016) Limited role for methane in the mid-Proterozoic greenhouse. Proc Natl Acad Sci U S A 113(41):11447–11452

Ono S, Wing B, Johnston D et al (2006) Mass-dependent fractionation of quadruple stable sulfur isotope system as a new tracer of sulfur biogeochemical cycles. Geochim Cosmochim Acta 70(9):2238–2252

Peckmann J, Thiel V (2004) Carbon cycling at ancient methane-seeps. Chem Geol 205(3–4):443–467

Peckmann J, Reimer A, Luth U et al (2001) Methane-derived carbonates and authigenic pyrite from the northwestern Black Sea. Mar Geol 177(1–2):129–150

Peketi A, Mazumdar A, Joshi RK et al (2012) Tracing the Paleo sulfate-methane transition zones and H_2S seepage events in marine sediments: An application of C-S-Mo systematics. Geochem Geophys Geosyst 13:Q10007

Peketi A, Mazumdar A, Joao HM et al (2015) Coupled C-S-Fe geochemistry in a rapidly accumulating marine sedimentary system: Diagenetic and depositional implications. Geochem Geophys Geosyst 16(9):2865–2883

Pellerin A, Bui TH, Rough M et al (2015) Mass-dependent sulfur isotope fractionation during reoxidative sulfur cycling: A case study from Mangrove Lake, Bermuda. Geochim Cosmochim Acta 149:152–164

Peng Y, Bao H, Jiang G et al (2022) A transient peak in marine sulfate after the 635-Ma snowball Earth. Proc Natl Acad Sci U S A 119(19):e2117341119

Pierre C (2017) Origin of the authigenic gypsum and pyrite from active methane seeps of the southwest African Margin. Chem Geol 449:158–164

Pu X, Zhong S, Yu W et al (2007) Authigenic sulfide minerals and their sulfur isotopes in sediments of the northern continental slope of the South China Sea and their implications for methane flux and gas hydrate formation. Chin Sci Bull 52(3):401–407

Reeburgh WS (2007) Oceanic methane biogeochemistry. Chem Rev 107(2):486–513

Sackett WM (1978) Carbon and hydrogen isotope effects during the thermocatalytic production of hydrocarbons in laboratory simulation experiments. Geochim Cosmochim Acta 42:571–580

Smrzka D, Zwicker J, Bach W et al (2019) The behavior of trace elements in seawater, sedimentary pore water, and their incorporation into carbonate minerals: a review. Facies 65(4):1–47

Smrzka D, Feng D, Himmler T et al (2020) Trace elements in methane-seep carbonates: Potentials, limitations, and perspectives. Earth-Sci Rev 208:103263

Strauss H, Bast R, Cording A et al (2012) Sulphur diagenesis in the sediments of the Kiel Bight, SW Baltic Sea, as reflected by multiple stable sulphur isotopes. Isot Environ Healt Stud 48(1):166–179

Suess E (2018) Marine cold seeps: Background and recent advances. In: Wilkes H (ed) Hydrocarbons, Oils and Lipids: Diversity, Origin, Chemistry and Fate. Springer, Cham, pp 1–21

Sun Y, Gong S, Li N et al (2020a) A new approach to discern the hydrocarbon sources (oil vs methane) of authigenic carbonates forming at marine seeps. Mar Pet Geol 114:1–9

Sun Y, Peckmann J, Hu Y et al (2020b) Formation of tubular carbonates within the seabed of the northern South China Sea. Minerals 10(9):1–17

Thiagarajan N, Cremiere A, Blattler C et al (2020) Stable and clumped isotope characterization of authigenic carbonates in methane cold seep environments. Geochim Cosmochim Acta 279:204–219

Tong H, Feng D, Cheng H et al (2013) Authigenic carbonates from seeps on the northern continental slope of the South China Sea: New insights into fluid sources and geochronology. Mar Pet Geol 43:260–271

Tong H, Feng D, Peckmann J et al (2019) Environments favoring dolomite formation at cold seeps: A case study from the Gulf of Mexico. Chem Geol 518:9–18

Tostevin R, Turchyn AV, Farquhar J et al (2014) Multiple sulfur isotope constraints on the modern sulfur cycle. Earth Planet Sci Lett 396:14–21

Turchyn AV, Brüchert V, Lyons TW et al (2010) Kinetic oxygen isotope effects during dissimilatory sulfate reduction: A combined theoretical and experimental approach. Geochim Cosmochim Acta 74(7):2011–2024

Turchyn AV, Antler G, Byrne D et al (2016) Microbial sulfur metabolism evidenced from pore fluid isotope geochemistry at Site U1385. Glob Planet Change 141:82–90

Wacker U, Fiebig J, Todter J et al (2014) Empirical calibration of the clumped isotope paleothermometer using calcites of various origins. Geochim Cosmochim Acta 141:127–144

Wang SH, Yan W, Chen Z et al (2014) Rare earth elements in cold seep carbonates from the southwestern Dongsha area, northern South China Sea. Mar Pet Geol 57:482–493

Wang X, Li N, Feng D et al (2018) Using chemical compositions of sediments to constrain methane seepage dynamics: A case study from Haima cold seeps of the South China Sea. J Asian Earth Sci 168:137–144

Wang W, Hu YL, Muscente AD et al (2021) Revisiting Ediacaran sulfur isotope chemostratigraphy with in situ nanoSIMS analysis of sedimentary pyrite. Geology 49(6):611–616

Whiticar MJ (1999) Carbon and hydrogen isotope systematics of bacterial formation and oxidation of methane. Chem Geol 161(1–3):291–314

Wortmann UG, Chernyavsky B, Bernasconi SM et al (2007) Oxygen isotope biogeochemistry of pore water sulfate in the deep biosphere: Dominance of isotope exchange reactions with ambient water during microbial sulfate reduction (ODP Site 1130). Geochim Cosmochim Acta 71(17):4221–4232

Yang K, Chu F, Zhu Z et al (2018) Formation of methane-derived carbonates during the last glacial period on the northern slope of the South China Sea. J Asian Earth Sci 168:173–185

Yang S, Lv Y, Liu X et al (2020) Genomic and enzymatic evidence of acetogenesis by anaerobic methanotrophic archaea. Nat Commun 11(1):3941

Zhang N, Lin M, Snyder GT et al (2019) Clumped isotope signatures of methane-derived authigenic carbonate presenting equilibrium values of their formation temperatures. Earth Planet Sci Lett 512:207–213

Zhuang C, Chen F, Cheng S et al (2016) Light carbon isotope events of foraminifera attributed to methane release from gas hydrates on the continental slope, northeastern South China Sea. Sci China-Earth Sci 59(10):1981–1995

Chapter 10
Non-traditional Stable Isotope Geochemistry of Seep Deposits

Meng Jin and Dong Feng

Abstract Non-traditional stable isotope geochemistry is a useful tool for revealing element migration, transformation and circulation in geological processes. A series of biogeochemical processes result in special and variable sedimentary environments in seep systems. Many elements are impacted and involved in different biogeochemical processes in seep systems, especially the formation of authigenic minerals, making seep deposits archives for studying elemental and isotopic behaviors in natural environments. Iron (Fe) and molybdenum (Mo) are involved in the formation of pyrite, and magnesium (Mg) and calcium (Ca) are closely related to the precipitation of authigenic carbonate. Research on the Fe, Mo, Mg and Ca isotopic compositions of different seep deposits from the South China Sea has been conducted in recent years. Preliminary studies have provided new insights into the mechanisms of isotopic fractionation and element cycling during early diagenesis. In this chapter, we provide an overview of the current understanding of the Fe, Mo, Mg and Ca isotope geochemistry of seep deposits from the South China Sea, targeting authigenic minerals, sediments and pore fluids.

10.1 Introduction

Traditional stable isotope geochemistry usually refers to isotopes of C, H, O, N and S. With the development of multicollector inductively coupled plasma–mass spectrometry (MC–ICP–MS), high-precision measurements of more isotopes have led to the thriving of so-called non-traditional stable isotope geochemistry (Halliday

M. Jin (✉)
MNR Key Laboratory of Marine Mineral Resources, Guangzhou Marine Geological Survey, Guangzhou 511458, China
e-mail: jinmeng17@mails.ucas.ac.cn

National Engineering Research Center for Gas Hydrate Exploration and Development, Guangzhou 511458, China

D. Feng
College of Marine Sciences, Shanghai Ocean University, Shanghai 201306, China
e-mail: dfeng@shou.edu.cn

171
D. Chen and D. Feng (eds.), *South China Sea Seeps*,
https://doi.org/10.1007/978-981-99-1494-4_10

et al. 1995). The distinctive geochemical features of these non-traditional stable isotopes, such as diverse geochemical activity, various concentrations in different geological reservoirs, redox sensitivity, and the links to biological activities, make them unique tracers for different geological processes (Teng et al. 2017).

The formation of authigenic minerals at seeps involves element migration and corresponding isotope fractionation, such as Fe and Mo for pyrite and Mg and Ca for carbonate. In addition, the geochemical processes in the cold seep system leads to changes in the redox state in the environment and involves the isotope fractionation of redox-sensitive elements, such as Mo. The seep system is an excellent "laboratory" for studying the behaviors of these isotopes in the natural environment. These isotopic compositions of different seep deposits have been successively analyzed in recent years. The South China Sea is one of the most investigated seep systems, and various non-traditional stable isotope geochemical techniques have been used to explore various geochemical processes. Iron isotopic compositions of anaerobic oxidation of methane (AOM)-derived pyrite from several sites provide constraints on the relationship with AOM activity and pyrite formation (Lin et al. 2017, 2022). The Mo enrichment and isotopic composition of sediments and carbonates provide new insights into the geochemical cycling of Mo (Lin et al. 2021). Magnesium isotopic compositions of pore water from "Haima" and seep carbonate from the Dongsha area put forward a new understanding of the behavior of Mg isotopes during carbonate precipitation (Jin et al. 2021, 2022). Studies of the Ca isotope geochemistry of dolomite from the Dongsha area have constrained the process of carbonate formation (Wang et al. 2012, 2013).

10.2   Fe Isotopes

Iron is a major widely distributed element in the silicate Earth. It has three oxidation states, namely, metallic iron, ferrous iron and ferric iron, which lead to its complex chemical and isotopic behaviors. Iron has four stable isotopes, with mass numbers of 54, 56, 57 and 58, which represent 5.845%, 91.754%, 2.119%, and 0.282% of the total mass, respectively (Meija et al. 2016). The Fe isotope ratios are usually reported as $\delta^{56}Fe$:

$$\delta^{56}Fe = \left[\frac{\left(^{56}Fe/^{54}Fe\right)_{sample}}{\left(^{56}Fe/^{54}Fe\right)_{standard}} - 1 \right] \times 1000$$

where the standard usually refers to IRMM-14. Another standard, the average of terrestrial igneous rocks (IgRx), was previously used to define the isotopic composition of Fe (Beard and Johnson 1999). The conversion of $\delta^{56}Fe$ values relative to the two standards can be adjusted according to Beard et al. (2003a):

$$\delta^{56}Fe_{IRMM-14} = \delta^{56}Fe_{IgRx} + 0.09\%_{00}$$

in the following text, $\delta^{56}Fe$ values refer to $\delta^{56}Fe_{IRMM-14}$ unless noted otherwise.

Pyrite is a common mineral in anoxic marine sediments, and its sulfur isotopic composition serves as an archive for constraining the biochemical cycle of sulfur and the evolution of ocean chemistry (Canfield and Teske 1996; Farquhar et al. 2000; Bekker et al. 2004; Johnston et al. 2006; Pasquier et al. 2017). The geochemical cycles of sulfur and iron are intimately coupled and recorded in sedimentary pyrite (Rouxel et al. 2005; Johnson et al. 2008; Hofmann et al. 2009; Liu et al. 2020). Pyrite in seep systems can present as framboids, overgrowths and euhedral crystals in sediments or enclosed in seep carbonates (e.g., Peckmann et al. 2001; Lin et al. 2016). The variable sulfur isotopic compositions of pyrite are closely related to AOM activity (e.g., Jørgensen et al. 2004; Borowski et al. 2013). The study of Fe isotope geochemistry may be useful to better understand the formation process of pyrite in seep systems. The Fe isotopic composition of pyrite depends on (1) the Fe isotopic composition of Fe sources and (2) the fractionation of Fe isotopes during the formation of pyrite (Butler et al. 2005; Guilbaud et al. 2011; Dauphas et al. 2017). Terrigenous active iron minerals are the main iron source for pyrite formation in continental margin sediments (Berner 1970; Poulton et al. 2004). Both theoretical calculations and experiments show that ferrous iron is ^{56}Fe-depleted compared to ferric iron that is either dissolved or contained in minerals (Johnson et al. 2002; Welch et al. 2003). Aqueous Fe^{2+} is released into pore water during dissimilatory iron reduction (DIR), with fractionation of $-0.5‰$ to $-2.95‰$ between aqueous Fe^{2+} and Fe oxides (Dauphas et al. 2017 and references therein). In addition, abiotic reductive dissolution by sulfide and iron-mediated anaerobic oxidation of methane (Fe-AOM) can also release isotopically lighter Fe into pore water (Riedinger et al. 2014; Egger et al. 2015). Hydrogen sulfide produced by organoclastic sulfate reduction and AOM reacts with dissolved iron or reactive iron minerals to form metastable intermediates, such as mackinawite, pyrrhotite and greigite, which are finally converted into pyrite (Jørgensen and Kasten 2006). The $\Delta^{56}Fe_{Fe(II)-FeS}$ are $-0.52 \pm 0.16‰$ in equilibrium at 2 °C and $+0.85 \pm 0.3‰$ in kinetic fractionation (Butler et al. 2005; Wu et al. 2012). The fractionation of Fe between FeS and pyrite is $+2.20 \pm 0.70‰$ (Guilbaud et al. 2011). Combining all these processes, potential fractionation between aqueous Fe^{2+} and pyrite can vary from $-3.1‰$ to $+0.5‰$ depending on different extents of pyritization (Guilbaud et al. 2011).

To date, the reported $\delta^{56}Fe$ values of bulk pyrite from seeps in the South China Sea range from $-1.48‰$ to $+0.38‰$ (n = 42) (Lin et al. 2017, 2018), similar to the range of variation in $\delta^{56}Fe$ values of other sedimentary pyrites (Fig. 10.1). The Fe isotopic compositions of bulk sediment at seeps from the South China Sea ($\delta^{56}Fe = +0.038‰$ to $+0.122‰$, n = 17; Fig. 10.1) are also close to those of other modern marine sediments ($\delta^{56}Fe = -0.12‰$ to $+0.23‰$; Beard et al. 2003b; Severmann et al. 2006; Fehr et al. 2010). However, in situ analyses of the Fe isotopic compositions of pyrite aggregates of different types from seeps by LA–MC–ICP–MS revealed large variations in $\delta^{56}Fe$ values ranging from $-1.09‰$ to $+1.90‰$ (n = 160) (Lin et al. 2018).

The iron pools in marine sediments include iron oxyhydroxides (Fe_{ox}), magnetite (Fe_{mag}), carbonate (Fe_{carb}), silicate (Fe_{sil}), and pyrite (Fe_{py}). Highly reactive iron

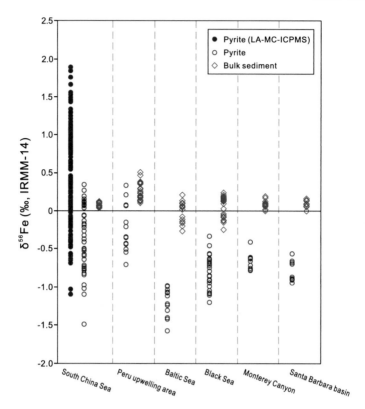

Fig. 10.1 Comparison of iron isotopic compositions of bulk sediment and pyrite from modern marine environments. Data from the South China Sea are from Lin et al. (2017, 2018). Data for the Peru upwelling area are from Scholz et al. (2014). Data from the Baltic Sea are from Fehr et al. (2008, 2010). Data from the Black Sea and Santa Barbara are from Severmann et al. (2006). Note that the pyrites from the South China Sea were hand-picked, while those from other studies were obtained through a chemical leaching procedure; see details in Severmann et al. (2006)

(Fe$_{HR}$) is defined as the sum of Fe$_{ox}$, Fe$_{mag}$, Fe$_{carb}$ and Fe$_{py}$, and Fe$_{py}$/Fe$_{HR}$ is introduced to indicate the extent of pyritization (e.g., Scholz et al. 2014; Lin et al. 2017, 2018). Researchers have suggested that iron oxyhydroxides and magnetite are the major iron sources for the precipitation of pyrite at seeps in the South China Sea (Lin et al. 2017, 2018). The enrichment of ^{56}Fe in pyrite is consistent with the increased degree of pyritization in the South China Sea (Fig. 10.2a). It has been suggested that δ^{56}Fe values of pyrite are controlled by both sulfide availability and iron pools in seep systems (Lin et al. 2018). A positive correlation between δ^{56}Fe and δ^{34}S values of pyrite has been observed, and pyrite affected by AOM and organoclastic sulfate reduction can be distinguished by different Fe and S isotopic compositions (Fig. 10.2b; Lin et al. 2017). In situ analysis of δ^{56}Fe indicates that the later-stage overgrowths and euhedral pyrites are more enriched in ^{56}Fe than the earlier formed

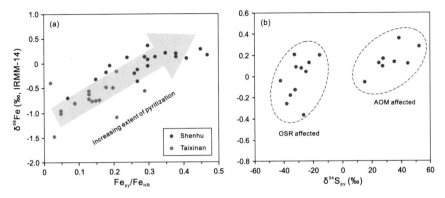

Fig. 10.2 Enrichment of ^{56}Fe in pyrite with increasing extent of pyritization (**a**) and plot of δ^{56}Fe against δ^{34}S of pyrite from the Shenhu area in the South China Sea (**b**). Modified after Lin et al. (2017, 2018)

framboids (Lin et al. 2018). This finding has been explained by Rayleigh-type distillation owing to the relatively Fe-limited environment within sediment during the formation of later-stage pyrite (Lin et al. 2022). The combination of Fe and S isotopic compositions of pyrite from seeps serves as a potential proxy to trace the different biogeochemical processes in early diagenesis.

10.3 Mo Isotopes

Molybdenum is a transition element with a wide range of redox states ($-$IV to $+$VIII), leading to its high degree of chemical reactivity. Molybdenum (IV) and Mo (VI) are most common on Earth's surface. Molybdenum has seven stable isotopes (mass numbers of 92, 94, 95, 96, 97, 98 and 100) with similar abundances of approximately 10% to 25% (Hlohowskyj et al. 2021). The isotopic composition of Mo is expressed as δ^{98}Mo relative to NIST-SRM-3134 $= 0.25‰$ (Nägler et al. 2014):

$$\delta^{98}\text{Mo} = \left[\frac{\left(^{98}\text{Mo}/^{95}\text{Mo}\right)_{\text{sample}}}{\left(^{98}\text{Mo}/^{95}\text{Mo}\right)_{\text{NIST-SRM-3134}}} - 1 \right] \times 1000 + 0.25‰$$

where $+0.25‰$ accounts for the offset from the in-house standards used previously (Nägler et al. 2014).

Molybdenum has a long residence time of ~0.44 to 0.8 Ma and is generally ubiquitous (~105 nM) in oxygenated oceans (Miller et al. 2011; Nakagawa et al. 2012; Hlohowskyj et al. 2021). The Mo isotopic composition (~2.3‰) is also homogenous in modern open seawater (Hlohowskyj et al. 2021). Molybdenum is redox sensitive and has characteristic isotopic fractionation (Kendall et al. 2017). The Mo isotopic signatures in sediments have been used as a proxy to reconstruct marine redox conditions (e.g., Poulson et al. 2006; Scott and Lyons 2012; Kendall et al. 2017 and references therein). In oxic environments, Mo mainly presents as molybdate (MoO_4^{2-}) in seawater (Siebert et al. 2003). The dissolved Mo can easily absorb to metal oxides, leading to $\Delta^{98}Mo_{seawater\text{-}solid}$ of approximately 0.8‰ to 3.2‰ (e.g., Barling et al. 2001; Siebert et al. 2003; Wasylenki et al. 2008; Scholz et al. 2017). Under anoxic/sulfidic conditions, MoO_4^{2-} transforms to thiomolybdates ($MoO_xS_{(4\text{-}x)}^{2-}$) facilitated by hydrogen sulfide, which is easily captured by iron sulfide or sulfur-rich organic matter and then stored in sediments (Helz et al. 1996, 2011; Tribovillard et al. 2006). The removal of Mo from aqueous environments is mainly controlled by the availability of dissolved hydrogen sulfide (Kendall et al. 2017). At $[H_2S]_{aq}$ > 11 μM, Mo is quantitatively transformed into tetrathiomolybdate and scavenges into sediments (Erickson and Helz 2000). The sediments record the Mo isotopic composition of ambient water (Nägler et al. 2011). Under conditions with $[H_2S]_{aq}$ < 11 μM, sediment yields a large variation in Mo isotopic composition due to incomplete sulfurization from molybdate to tetrathiomolybdate (Erickson and Helz 2000; Neubert et al. 2008; Nägler et al. 2011).

The AOM in seep systems create strongly sulfidic environmental conditions in pore water, with $[H_2S]_{aq}$ easily exceeding 11 μM in the case of high methane flux (e.g., Gieskes et al. 2005, 2011; Joye et al. 2010). This facilitates the accumulation of authigenic Mo in sediments or seep carbonates. The enrichment of Mo in seep sediment and the covariation of Mo and U in seep carbonate are strongly associated with redox states and the intensity of AOM (Hu et al. 2014; Chen et al. 2016). Seeps may serve as an important sink in the marine geochemical cycle of Mo (Hu et al. 2015). Therefore, Mo isotope geochemistry can provide new insights into tracing the sedimentary environment in seep systems. To date, only sparse data on the Mo isotopic composition of sediments and carbonates at seeps have been reported (Lin et al. 2021). The $\delta^{98}Mo$ values of bulk sediment and carbonate nodules from the South China Sea range from +0.2‰ to +2.0‰ (n = 36) and +1.4‰ to +2.1‰ (n = 5), respectively (Fig. 10.3; Lin et al. 2021). The large variation in Mo isotopic compositions of sediment was ascribed to the mixing of authigenic Mo sequestered in the course of AOM and detrital Mo in sediment unaffected by seepage. In addition, the later-formed pyrite may also cause additional Mo accumulation and influence the $\delta^{98}Mo$ of bulk sediment (Lin et al. 2021). For seep carbonate with a narrow range of $\delta^{98}Mo$ values, the offset between $\delta^{98}Mo_{carbonate}$ and $\delta^{98}Mo_{seawater}$ is close to that of strongly euxinic conditions (Lin et al. 2021). Considering the much higher H_2S concentration in active seepage than the critical Mo speciation threshold of ~11 μM, $\delta^{98}Mo_{carbonate}$ was thought to have potential for tracing the Mo isotopic composition of seawater (Lin et al. 2021).

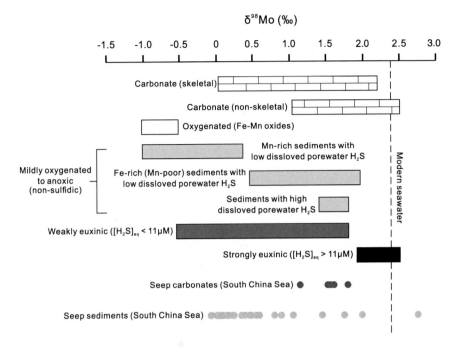

Fig. 10.3 Molybdenum isotopic compositions of seep carbonates and sediments from the South China Sea (Lin et al. 2021) and other sinks of Mo in the modern oceans (modified after Kendall et al. 2017)

10.4 Mg Isotopes

Magnesium is widely distributed in the silicate Earth, hydrosphere and biosphere. It has three stable isotopes, i.e., ^{24}Mg, ^{25}Mg and ^{26}Mg, with typical abundances of 78.97%, 10.01% and 11.03%, respectively (Meija et al. 2016). The isotopic composition of Mg is usually reported as δ^{26}Mg relative to the DSM3 standard (Young and Galy 2004):

$$\delta^{26}Mg = \left[\frac{\left(^{26}Mg/^{24}Mg \right)_{sample}}{\left(^{26}Mg/^{24}Mg \right)_{DSM3}} - 1 \right] \times 1000$$

Magnesium is the second most abundant cation in seawater, with a long residence time of ~13 Ma (Li 1982). The Mg isotopic composition of modern seawater is −0.83 ± 0.09‰ (Ling et al. 2011). Carbonate precipitation represents a sink of Mg in the ocean and is an important part of the oceanic Mg cycle (Higgins and Schrag 2015). The Mg isotopic composition of marine carbonates serves as a tool for unraveling the geochemical cycling of Mg in the ocean on geological timescales (e.g., Tipper et al. 2006; Higgins and Schrag 2012, 2015; Fantle and Higgins 2014). Carbonate is generally enriched in ^{24}Mg relative to seawater and displays a large variation in δ^{26}Mg

(Teng 2017). Fractionation of Mg isotopes during carbonate precipitation represents a temperature dependence of only approximately $0.01‰ \ °C^{-1}$, while it exhibits a strong mineralogical control (Saenger and Wang 2014). The order of enrichment from light to heavy isotopes is calcite < magnesite < dolomite < aragonite (Saenger and Wang 2014). In addition, the Mg isotopic composition of carbonate could also be influenced by other factors, such as the precipitation rate, inorganic and organic ligands in solution, and biological effects (Saenger and Wang 2014 and references therein; Schott et al. 2016; Mavromatis et al. 2017).

Most seep carbonates from the South China Sea are composed of microcrystalline high-Mg calcite (HMC) and aragonite cement. In addition, a significant amount of microcrystalline dolomite occurs at some seep sites (Feng et al. 2018 and references therein). The content of $MgCO_3$ in some HMCs can even reach 38 mol%, occurring in association with minor amounts of dolomite (Han et al. 2008). Previous studies have confirmed that the presence of hydrogen sulfide provided by AOM can prominently affect the incorporation of Mg into carbonate minerals (e.g., Naehr et al. 2007; Zhang et al. 2012; Lu et al. 2018, 2021; Tong et al. 2019). Therefore, seep deposits are good subjects for understanding the behavior of Mg isotopes during carbonate precipitation in natural environments. The Mg^{2+} in seawater diffusing into pore water is the main source of Mg in seep carbonates. The $\delta^{26}Mg$ of pore water from an active seep in the South China Sea ranges from $-0.88‰$ to $-0.71‰$ (from 0 to 8 meter below the seafloor), similar to that of seawater (Jin et al. 2022). A slight increase in the pore water $\delta^{26}Mg$ values with depth indicates the precipitation of authigenic carbonate. This occurs because the sufficient replenishment of Mg from seawater is far greater than the consumption of Mg in pore water during precipitation of authigenic carbonate (Jin et al. 2022). Lu et al. (2017) and Jin et al. (2021) analyzed the Mg isotopic composition of seep carbonates from the South China Sea. A comparison of the $\delta^{26}Mg$ values of seep carbonates and other marine carbonates is presented in Fig. 10.4. The $\delta^{26}Mg$ values of seep carbonates display a smaller range of variation relative to those of other marine carbonates. Additionally, the biologically mediated vital effects on the Mg isotopic composition are not obvious compared to biodetrital carbonates based on current data. Carbonates in Lu et al. (2017) consist mainly of dolomite, with relatively consistent Mg isotopic compositions ($\delta^{26}Mg = -2.45‰$ to $-2.15‰$, n = 11). However, the $\delta^{26}Mg$ values of HMC in pipe-like seep carbonates vary from $-3.42‰$ to $-2.63‰$ (n = 10), and a decreasing trend of $\delta^{26}Mg$ values from the periphery to the inner portion of the "pipe" was observed (Jin et al. 2021).

The correlation of inorganic $\delta^{13}C$ and $\delta^{26}Mg$ values of seep carbonates is considered to be related to the dissolved hydrogen sulfide in pore water produced by AOM (Lu et al. 2017; Jin et al. 2021). The existence of dissolved H_2S has been confirmed to promote the incorporation of Mg into carbonate in precipitation experiments (Zhang et al. 2012). Hydrogen sulfide derived from AOM in seep systems has also been deemed to facilitate the formation of carbonate minerals with high Mg content, especially dolomite (Lu et al. 2015; Tong et al. 2019). The observed different trends of $\delta^{13}C$ and $\delta^{26}Mg$ values refer to opposite trends in the degree of Mg isotope fractionation with enhanced intensity of AOM (Fig. 10.5). The dissolved H_2S may affect the Mg isotopic composition of carbonates, although the mechanism is still not well

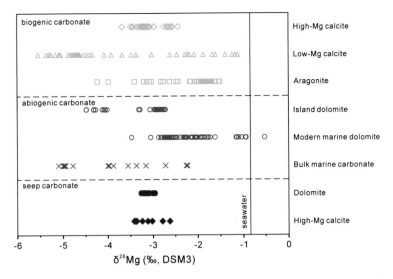

Fig. 10.4 Comparison of magnesium isotopic compositions of carbonates. Data on abiogenic carbonate are from Saenger and Wang (2014) and references therein. Data on island dolomites are from Hu et al. (2022). Data on modern marine dolomites are from Higgins and Schrag (2010), Fantle and Higgins (2014), Mavromatis et al. (2014) and Blättler et al. (2015). Data on dolomite and HMC in seep carbonate are from Lu et al. (2017) and Jin et al. (2021), respectively

understood. This may provide new insights into the application of carbonate Mg isotopic compositions during geological time of widespread oceanic anoxia.

Fig. 10.5 Correlation of inorganic $\delta^{13}C$ and $\delta^{26}Mg$ values of seep carbonates from the South China Sea (modified after Jin et al. 2021). Data from the Shenhu and SW Taiwan areas are from Lu et al. (2017). Data from the Dongsha area are from Jin et al. (2021)

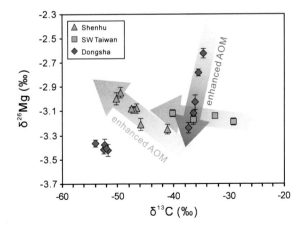

10.5 Ca Isotopes

Calcium is abundant in both terrestrial and marine systems. It has five stable isotopes (^{40}Ca, ^{42}Ca, ^{43}Ca, ^{44}Ca and ^{46}Ca) and a radioactive isotope (^{48}Ca) with an extremely long half-life of 1.9×10^{19} years (Griffith and Fantle 2021). The abundance of isotopes from ^{40} to ^{48}Ca is 96.941%, 0.647%, 0.135%, 2.086%, 0.004% and 0.187%, respectively (Meija et al. 2016). The Ca isotope ratios are expressed as δ^{44}Ca:

$$\delta^{44}Ca = \left[\frac{\left(^{44}Ca/^{40}Ca\right)_{sample}}{\left(^{44}Ca/^{40}Ca\right)_{standard}} - 1 \right] \times 1000$$

Multiple standards are in use to define δ^{44}Ca, such as SRM 915a, SRM 915b, seawater, and bulk silicate Earth (Griffith and Fantle 2021). Data are reported relative to seawater in the following text. The conversion of δ^{44}Ca relative to SRM 915a and seawater is according to Hippler et al. (2003):

$$\delta^{44}Ca_{SRM915a} = \delta^{44}Ca_{seawater} + 1.88\%_{00}$$

The formation of carbonate plays an important role in the cycling of Ca. The calcium isotopic composition of carbonate has been used to quantify rates of carbonate diagenesis and trace global Ca cycling (Griffith and Fantle 2021). The δ^{44}Ca values of modern marine carbonates generally display isotope fractionations between $- 1.8\permil$ and $- 0.8\permil$ from seawater (Blättler et al. 2012; Fantle and Tipper 2014). Similar to Mg isotopes, carbonate mineralogy and precipitation rate are two dominant factors controlling the calcium isotopic compositions of carbonate (Gussone et al. 2005; Tang et al. 2008; Blättler et al. 2012). The high alkalinity of pore water in seep systems facilitates the formation of authigenic carbonate with a fast precipitation rate (Aloisi et al. 2000; Naehr et al. 2007). The calcium isotopic compositions of seep carbonate may have the potential for understanding the formation process of carbonate during early diagenesis. The diffusion of seawater-dissolved Ca^{2+} into pore water is the main source of authigenic carbonate at seeps. Carbonate depleted in ^{44}Ca indicates that the precipitation of seep carbonate prefers light Ca isotopes (Fig. 10.6). Mg-calcite in seep carbonate yields higher δ^{44}Ca values than aragonite (Fig. 10.6). The reported δ^{44}Ca values of dolomite-dominated seep carbonate from the South China Sea ($-0.67\permil$ to $- 0.41\permil$, n = 11) are similar to those of Mg-calcite (Wang et al. 2012, 2013; Thiagarajan et al. 2020; Blättler et al. 2021). Wang et al. (2012) suggested that the variation in δ^{44}Ca values can be attribute to supersaturation, precipitation rate and the degree of Ca consumption in semi-closed systems at seeps. The Rayleigh effect on Ca isotope fractionation during the precipitation of seep carbonate can also be indicated by the downward-increasing δ^{44}Ca values of pore water from the Storfjordrenna Trough in Blättler et al. (2021).

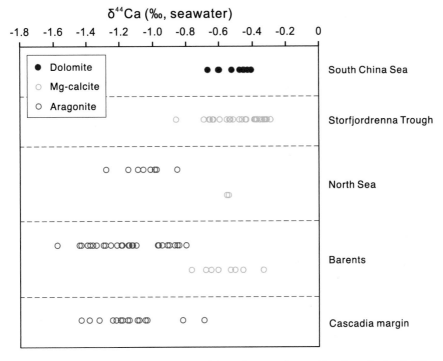

Fig. 10.6 Comparison of Ca isotopic compositions of seep carbonate. Data from the South China Sea are from Wang et al. (2012, 2013). Data from the Storfjordrenna Trough and the North Sea are from Thiagarajan et al. (2020). Data from the Cascadia margin are from Teichert et al. (2005)

10.6 Summary and Future Studies

In summary, the special and complex sedimentary environments in seep systems could impact the behaviors of isotopes, such as Fe, Mo, Mg and Ca. The iron isotopic compositions of pyrite are strongly related to its formation process and ambient environmental conditions. Isotopic signatures of Mo in seep carbonates and sediments are relevant to Mo migration and transformation with a changing redox state. Magnesium and Ca isotopic compositions of seep carbonate reflect the precipitation of authigenic carbonate, and the hydrogen sulfide derived from AOM may significantly influence the fractionation of Mg isotopes. Non-traditional isotope geochemistry in seep systems is still in its infancy. Whether the special environment at seeps affects the fractionation mechanism of these isotopes is unclear. The application potential of these non-traditional isotopic compositions of seep deposits also remains to be developed. Below are some problems in the field of non-traditional isotope geochemistry in seep systems that could be further studied in the future:

1. Pyrite in seep systems is the result of continuous accumulation along with the dynamic activity of methane seepage. The Fe isotopic compositions of pyrite

may be an assemblage of complex biogeochemical processes, especially when multiple seepage events occur. More research is needed to further reveal the information hidden in the Fe isotopic compositions of pyrite.

2. Molybdenum in seep deposits exists in carbonate minerals, sulfide phases, detritus and organic materials. The Mo isotopic signature of bulk seep carbonate is a mixture of the δ^{98}Mo of all these Mo-bearing phases. This causes uncertainty in the application of seep carbonates to trace the Mo isotopic compositions of seawater. Analyzing the Mo isotopic compositions of different phases in seep carbonates by sequential chemical extraction may help to better understand the behavior of Mo isotopes in seep systems.

3. Dissolved hydrogen sulfide plays an important role in the behavior of Mg. Therefore, the AOM in seep systems has been thought to have a possible effect on the fractionation of Mg isotopes during the formation of seep carbonates. The behavior of Mg isotopes during carbonate precipitation under sulfidic conditions could be better constrained by precipitation experiments.

4. Seep carbonates form under a fast precipitation rate. The precipitation rate is an important factor controlling the fractionation of Ca isotopes during carbonate precipitation. However, the Rayleigh effect also obviously influences the δ^{44}Ca of seep carbonates. Therefore, more research is needed to better constrain the precipitation rate of seep carbonate to quantify the effect of the precipitation rate on its Ca isotopic composition.

Acknowledgements The authors are grateful to Yu Hu (Shanghai Ocean University) and Zhiyong Lin (Universität Hamburg) for the constructive comments.

References

Aloisi G, Pierre C, Rouchy JM et al (2000) Methane-related authigenic carbonates of eastern Mediterranean Sea mud volcanoes and their possible relation to gas hydrate destabilisation. Earth Planet Sci Lett 184(1):321–338

Barling J, Arnold GL, Anbar AD (2001) Natural mass-dependent variations in the isotopic composition of molybdenum. Earth Planet Sci Lett 193(3–4):447–457

Beard BL, Johnson CM (1999) High precision iron isotope measurements of terrestrial and lunar materials. Geochim Cosmochim Acta 63(11–12):1653–1660

Beard BL, Johnson CM, Skulan JL et al (2003a) Application of Fe isotopes to tracing the geochemical and biological cycling of Fe. Chem Geol 195(1–4):87–117

Beard BL, Johnson CM, Von Damm KL et al (2003b) Iron isotope constraints on Fe cycling and mass balance in oxygenated Earth oceans. Geology 31(7):629–632

Bekker A, Holland HD, Wang PL et al (2004) Dating the rise of atmospheric oxygen. Nature 427(6970):117–120

Berner RA (1970) Sedimentary pyrite formation. Am J Sci 268:1–23

Blättler CL, Henderson GM, Jenkyns HC (2012) Explaining the Phanerozoic Ca isotope history of seawater. Geology 40(9):843–846

Blättler CL, Hong W-L, Kirsimäe K et al (2021) Small calcium isotope fractionation at slow precipitation rates in methane seep authigenic carbonates. Geochim Cosmochim Acta 298:227–239

Blättler CL, Miller NR, Higgins JA (2015) Mg and Ca isotope signatures of authigenic dolomite in siliceous deep-sea sediments. Earth Planet Sci Lett 419:32–42

Borowski WS, Rodriguez NM, Paull CK et al (2013) Are [34]S-enriched authigenic sulfide minerals a proxy for elevated methane flux and gas hydrates in the geologic record? Mar Pet Geol 43:381–395

Butler IB, Archer C, Vance D et al (2005) Fe isotope fractionation on FeS formation in ambient aqueous solution. Earth Planet Sci Lett 236(1–2):430–442

Canfield DE, Teske A (1996) Late Proterozoic rise in atmospheric oxygen concentration inferred from phylogenetic and sulphur-isotope studies. Nature 382(6587):127–132

Chen F, Hu Y, Feng D et al (2016) Evidence of intense methane seepages from molybdenum enrichments in gas hydrate-bearing sediments of the northern South China Sea. Chem Geol 443:173–181

Dauphas N, John SG, Rouxel O (2017) Iron Isotope Systematics. Rev Mineral Geochem 82:415–510

Egger M, Rasigraf O, Sapart CJ et al (2015) Iron-Mediated Anaerobic Oxidation of Methane in Brackish Coastal Sediments. Environ Sci Technol 49(1):277–283

Erickson BE, Helz GR (2000) Molybdenum (VI) speciation in sulfidic waters: Stability and lability of thiomolybdates. Geochim Cosmochim Acta 64(7):1149–1158

Fantle MS, Higgins J (2014) The effects of diagenesis and dolomitization on Ca and Mg isotopes in marine platform carbonates: Implications for the geochemical cycles of Ca and Mg. Geochim Cosmochim Acta 142:458–481

Fantle MS, Tipper ET (2014) Calcium isotopes in the global biogeochemical Ca cycle: Implications for development of a Ca isotope proxy. Earth-Sci Rev 129:148–177

Farquhar J, Bao H, Thiemens M (2000) Atmospheric Influence of Earth's Earliest Sulfur Cycle. Science 289(5480):756–758

Fehr MA, Andersson PS, Hålenius U et al (2010) Iron enrichments and Fe isotopic compositions of surface sediments from the Gotland Deep, Baltic Sea. Chem Geol 277(3–4):310–322

Fehr MA, Andersson PS, Hålenius U et al (2008) Iron isotope variations in Holocene sediments of the Gotland Deep, Baltic Sea. Geochim Cosmochim Acta 72(3): 807–826

Feng D, Qiu J-W, Hu Y et al (2018) Cold seep systems in the South China Sea: An overview. J Asian Earth Sci 168:3–16

Gieskes J, Mahn C, Day S et al (2005) A study of the chemistry of pore fluids and authigenic carbonates in methane seep environments: Kodiak Trench, Hydrate Ridge, Monterey Bay, and Eel River Basin. Chem Geol 220(3–4):329–345

Gieskes J, Rathburn AE, Martin JB et al (2011) Cold seeps in Monterey Bay, California: Geochemistry of pore waters and relationship to benthic foraminiferal calcite. Appl Geochem 26(5):738–746

Griffith EM, Fantle MS (2021) Calcium Isotopes. Cambridge University Press, Cambridge. https://doi.org/10.1017/9781108853972

Guilbaud R, Butler IB, Ellam RM (2011) Abiotic Pyrite Formation Produces a Large Fe Isotope Fractionation. Science 332(6037):1548–1551

Gussone N, Böhm F, Eisenhauer A et al (2005) Calcium isotope fractionation in calcite and aragonite. Geochim Cosmochim Acta 69(18):4485–4494

Halliday AN, Lee D-C, Christensen JN et al (1995) Recent developments in inductively coupled plasma magnetic sector multiple collector mass spectrometry. Int J Mass Spectrom Ion Process 146:21–33

Han X, Suess E, Huang Y et al (2008) Jiulong methane reef: Microbial mediation of seep carbonates in the South China Sea. Mar Geol 249(3–4):243–256

Helz GR, Bura-Nakić E, Mikac N et al (2011) New model for molybdenum behavior in euxinic waters. Chem Geol 284(3–4):323–332

Helz GR, Miller CV, Charnock JM et al (1996) Mechanism of molybdenum removal from the sea and its concentration in black shales: EXAFS evidence. Geochim Cosmochim Acta 60(19):3631–3642

Higgins JA, Schrag DP (2010) Constraining magnesium cycling in marine sediments using magnesium isotopes. Geochim Cosmochim Acta 74(17):5039–5053

Higgins JA, Schrag DP (2012) Records of Neogene seawater chemistry and diagenesis in deep-sea carbonate sediments and pore fluids. Earth Planet Sci Lett 357:386–396

Higgins JA, Schrag DP (2015) The Mg isotopic composition of Cenozoic seawater – evidence for a link between Mg-clays, seawater Mg/Ca, and climate. Earth Planet Sci Lett 416:73–81

Hippler D, Schmitt A-D, Gussone N et al (2003) Calcium Isotopic Composition of Various Reference Materials and Seawater. Geostand Newsl 27(1):13–19

Hlohowskyj SR, Chappaz A, Dickson AJ (2021) Molybdenum as a Paleoredox Proxy: Past, Present, and Future. Cambridge University Press, Cambridge. https://doi.org/10.1017/9781108993777

Hofmann A, Bekker A, Rouxel O et al (2009) Multiple sulphur and iron isotope composition of detrital pyrite in Archaean sedimentary rocks: A new tool for provenance analysis. Earth Planet Sci Lett 286(3–4):436–445

Hu Y, Feng D, Liang Q et al (2015) Impact of anaerobic oxidation of methane on the geochemical cycle of redox-sensitive elements at cold-seep sites of the northern South China Sea. Deep-Sea Res Part II-Top Stud Oceanogr 122:84–94

Hu Y, Feng D, Peckmann J et al (2014) New insights into cerium anomalies and mechanisms of trace metal enrichment in authigenic carbonate from hydrocarbon seeps. Chem Geol 381:55–66

Hu Z, Shi Z, Li G et al (2022) The Cenozoic Seawater Conundrum: New constraints from Mg isotopes in island dolostones. Earth Planet Sci Lett 595:117755

Jin M, Feng D, Huang K et al (2022) Magnesium Isotopes in Pore Water of Active Methane Seeps of the South China Sea. Front Mar Sci 9:858860

Jin M, Feng D, Huang K et al (2021) Behavior of Mg isotopes during precipitation of methane-derived carbonate: Evidence from tubular seep carbonates from the South China Sea. Chem Geol 567:120101

Johnson CM, Beard BL, Roden EE (2008) The Iron Isotope Fingerprints of Redox and Biogeochemical Cycling in Modern and Ancient Earth. Annu Rev Earth Planet Sci 36:457–493

Johnson CM, Skulan JL, Beard BL et al (2002) Isotopic fractionation between Fe(III) and Fe(II) in aqueous solutions. Earth Planet Sci Lett 195(1–2):141–153

Johnston DT, Poulton SW, Fralick PW et al (2006) Evolution of the oceanic sulfur cycle at the end of the Paleoproterozoic. Geochim Cosmochim Acta 70(23):5723–5739

Jørgensen BB, Böttcher ME, Lüschen H et al (2004) Anaerobic methane oxidation and a deep H_2S sink generate isotopically heavy sulfides in Black Sea sediments. Geochim Cosmochim Acta 68:2095–2118

Jørgensen BB, Kasten S (2006) Sulfur Cycling and Methane Oxidation. In: Schulz HD, Zabel M (eds) Marine Geochemistry. Springer, Berlin Heidelberg, Berlin, Heidelberg, pp 271–309

Joye SB, Bowles MW, Samarkin VA et al (2010) Biogeochemical signatures and microbial activity of different cold-seep habitats along the Gulf of Mexico deep slope. Deep-Sea Res Part II-Top Stud Oceanogr 57(21–23):1990–2001

Kendall B, Dahl TW, Anbar AD (2017) The Stable Isotope Geochemistry of Molybdenum. Rev Mineral Geochem 82:683–732

Li Y-H (1982) A brief discussion on the mean oceanic residence time of elements. Geochim Cosmochim Acta 46(12):2671–2675

Lin Z, Sun X, Chen K et al (2022) Effects of sulfate reduction processes on the trace element geochemistry of sedimentary pyrite in modern seep environments. Geochim Cosmochim Acta 333:75–94

Lin Z, Sun X, Lu Y et al (2018) Iron isotope constraints on diagenetic iron cycling in the Taixinan seepage area, South China Sea. J Asian Earth Sci 168:112–124

Lin Z, Sun X, Peckmann J et al (2016) How sulfate-driven anaerobic oxidation of methane affects the sulfur isotopic composition of pyrite: A SIMS study from the South China Sea. Chem Geol 440:26–41

Lin Z, Sun X, Lu Y et al (2017) The enrichment of heavy iron isotopes in authigenic pyrite as a possible indicator of sulfate-driven anaerobic oxidation of methane: Insights from the South China Sea. Chem Geol 449:15–29

Lin Z, Sun X, Strauss H et al (2021) Molybdenum isotope composition of seep carbonates – Constraints on sediment biogeochemistry in seepage environments. Geochim Cosmochim Acta 307:56–71

Ling M-X, Sedaghatpour F, Teng F-Z et al (2011) Homogeneous magnesium isotopic composition of seawater: an excellent geostandard for Mg isotope analysis. Rapid Commun Mass Spectrom 25(19):2828–2836

Liu J, Pellerin A, Antler G et al (2020) Early diagenesis of iron and sulfur in Bornholm Basin sediments: The role of near-surface pyrite formation. Geochim Cosmochim Acta 284:43–60

Lu Y, Liu Y, Sun X et al (2017) Intensity of methane seepage reflected by relative enrichment of heavy magnesium isotopes in authigenic carbonates: A case study from the South China Sea. Deep-Sea Res Part I-Oceanogr Res Pap 129:10–21

Lu Y, Sun XM, Lin ZY et al (2015) Cold seep status archived in authigenic carbonates: Mineralogical and isotopic evidence from Northern South China Sea. Deep-Sea Res Part II-Top Stud Oceanogr 122:95–105

Lu Y, Sun XM, Xu HF et al (2018) Formation of dolomite catalyzed by sulfate-driven anaerobic oxidation of methane: Mineralogical and geochemical evidence from the northern South China Sea. Am Miner 103(5):720–734

Lu Y, Yang X, Lin Z et al (2021) Reducing microenvironments promote incorporation of magnesium ions into authigenic carbonate forming at methane seeps: Constraints for dolomite formation. Sedimentology 68(7):2945–2964

Mavromatis V, Immenhauser A, Buhl D et al (2017) Effect of organic ligands on Mg partitioning and Mg isotope fractionation during low-temperature precipitation of calcite in the absence of growth rate effects. Geochim Cosmochim Acta 207:139–153

Mavromatis V, Meister P, Oelkers EH (2014) Using stable Mg isotopes to distinguish dolomite formation mechanisms: A case study from the Peru Margin. Chem Geol 385:84–91

Meija J, Coplen TB, Berglund M et al (2016) Isotopic compositions of the elements 2013 (IUPAC Technical Report). Pure Appl Chem 88(3):293–306

Miller CA, Peucker-Ehrenbrink B, Walker BD et al (2011) Re-assessing the surface cycling of molybdenum and rhenium. Geochim Cosmochim Acta 75(22):7146–7179

Naehr TH, Eichhubl P, Orphan VJ et al (2007) Authigenic carbonate formation at hydrocarbon seeps in continental margin sediments: A comparative study. Deep-Sea Res Part II-Top Stud Oceanogr 54(11–13):1268–1291

Nägler TF, Anbar AD, Archer C et al (2014) Proposal for an International Molybdenum Isotope Measurement Standard and Data Representation. Geostand Geoanal Res 38(2):149–151

Nägler TF, Neubert N, Böttcher ME et al (2011) Molybdenum isotope fractionation in pelagic euxinia: Evidence from the modern Black and Baltic Seas. Chem Geol 289(1–2):1–11

Nakagawa Y, Takano S, Firdaus ML et al (2012) The molybdenum isotopic composition of the modern ocean. Geochem J 46(2):131–141

Neubert N, Nägler TF, Böttcher ME (2008) Sulfidity controls molybdenum isotope fractionation into euxinic sediments: Evidence from the modern Black Sea. Geology 36(10):775–778

Pasquier V, Sansjofre P, Rabineau M et al (2017) Pyrite sulfur isotopes reveal glacial–interglacial environmental changes. Proc Natl Acad Sci U S A 114(23):5941–5945

Peckmann J, Reimer A, Luth U et al (2001) Methane-derived carbonates and authigenic pyrite from the northwestern Black Sea. Mar Geol 177(1–2):129–150

Poulson RL, Siebert C, McManus J et al (2006) Authigenic molybdenum isotope signatures in marine sediments. Geology 34(8):617–620

Poulton SW, Krom MD, Raiswell R (2004) A revised scheme for the reactivity of iron (oxyhydr)oxide minerals towards dissolved sulfide. Geochim Cosmochim Acta 68(18):3703–3715

Riedinger N, Formolo MJ, Lyons TW et al (2014) An inorganic geochemical argument for coupled anaerobic oxidation of methane and iron reduction in marine sediments. Geobiology 12(2):172–181

Rouxel OJ, Bekker A, Edwards KJ (2005) Iron Isotope Constraints on the Archean and Paleoproterozoic Ocean Redox State. Science 307:1088–1091

Saenger C, Wang ZR (2014) Magnesium isotope fractionation in biogenic and abiogenic carbonates: implications for paleoenvironmental proxies. Quat Sci Rev 90:1–21

Scholz F, Severmann S, McManus J et al (2014) On the isotope composition of reactive iron in marine sediments: Redox shuttle versus early diagenesis. Chem Geol 389:48–59

Scholz F, Siebert C, Dale AW et al (2017) Intense molybdenum accumulation in sediments underneath a nitrogenous water column and implications for the reconstruction of paleo-redox conditions based on molybdenum isotopes. Geochim Cosmochim Acta 213:400–417

Schott J, Mavromatis V, Fujii T et al (2016) The control of carbonate mineral Mg isotope composition by aqueous speciation: Theoretical and experimental modeling. Chem Geol 445:120–134

Scott C, Lyons TW (2012) Contrasting molybdenum cycling and isotopic properties in euxinic versus non-euxinic sediments and sedimentary rocks: Refining the paleoproxies. Chem Geol 324:19–27

Severmann S, Johnson CM, Beard BL et al (2006) The effect of early diagenesis on the Fe isotope compositions of porewaters and authigenic minerals in continental margin sediments. Geochim Cosmochim Acta 70(8):2006–2022

Siebert C, Nägler TF, von Blanckenburg F et al (2003) Molybdenum isotope records as a potential new proxy for paleoceanography. Earth Planet Sci Lett 211(1–2):159–171

Tang J, Dietzel M, Böhm F et al (2008) Sr^{2+}/Ca^{2+} and $^{44}Ca/^{40}Ca$ fractionation during inorganic calcite formation: II. Ca Isotopes. Geochim Cosmochim Acta 72(15):3733–3745

Teichert BMA, Gussone N, Eisenhauer A et al (2005) Clathrites: Archives of near-seafloor pore-fluid evolution ($\delta^{44}/^{40}Ca$, $\delta^{13}C$, $\delta^{18}O$) in gas hydrate environments. Geology 33(3):213–216

Teng F-Z (2017) Magnesium Isotope Geochemistry. Rev Mineral Geochem 82:219–287

Teng F-Z, Dauphas N, Watkins JM (2017) Non-Traditional Stable Isotopes: Retrospective and Prospective. Rev Mineral Geochem 82:1–26

Thiagarajan N, Crémière A, Blättler C et al (2020) Stable and clumped isotope characterization of authigenic carbonates in methane cold seep environments. Geochim Cosmochim Acta 279:204–219

Tipper E, Galy A, Gaillardet J et al (2006) The magnesium isotope budget of the modern ocean: Constraints from riverine magnesium isotope ratios. Earth Planet Sci Lett 250(1–2):241–253

Tong H, Feng D, Peckmann J et al (2019) Environments favoring dolomite formation at cold seeps: A case study from the Gulf of Mexico. Chem Geol 518:9–18

Tribovillard N, Algeo TJ, Lyons T et al (2006) Trace metals as paleoredox and paleoproductivity proxies: An update. Chem Geol 232(1–2):12–32

Wang S, Yan W, Magalhães HV et al (2012) Calcium isotope fractionation and its controlling factors over authigenic carbonates in the cold seeps of the northern South China Sea. Chin Sci Bull 57(11):1325–1332

Wang S, Yan W, Magalhães VH et al (2013) Factors influencing methane-derived authigenic carbonate formation at cold seep from southwestern Dongsha area in the northern South China Sea. Environ Earth Sci 71(5):2087–2094

Wasylenki LE, Rolfe BA, Weeks CL et al (2008) Experimental investigation of the effects of temperature and ionic strength on Mo isotope fractionation during adsorption to manganese oxides. Geochim Cosmochim Acta 72(24):5997–6005

Welch SA, Beard BL, Johnson CM et al (2003) Kinetic and equilibrium Fe isotope fractionation between aqueous Fe (II) and Fe (III). Geochim Cosmochim Acta 67(22):4231–4250

Wu L, Druschel G, Findlay A et al (2012) Experimental determination of iron isotope fractiona-
tions among Fe_{aq}^{2+}–FeS_{aq}–Mackinawite at low temperatures: Implications for the rock record.
Geochim Cosmochim Acta 89:46–61
Young ED, Galy A (2004) The Isotope Geochemistry and Cosmochemistry of Magnesium. Rev
Mineral Geochem 55:197–230
Zhang FF, Xu HF, Konishi H et al (2012) Dissolved sulfide-catalyzed precipitation of disor-
dered dolomite: Implications for the formation mechanism of sedimentary dolomite. Geochim
Cosmochim Acta 97:148–165

Chapter 11
Biomarker Indicators of Cold Seeps

Hongxiang Guan, Lei Liu, Nengyou Wu, and Sanzhong Li

Abstract Lipid biomarkers of seep carbonates and sediments retrieved from the Dongsha area, Shenhu, Site F and Haima in the South China Sea (SCS) over the last two decades were studied. Biomarker inventories, microbial consortia, seepage dynamics, and biogeochemical processes of anaerobic oxidation of methane (AOM), aerobic oxidation of methane (AeOM), and oxidation of non-methane hydrocarbons, were reconstructed. Authigenic carbonates contained varying contents of ^{13}C-depleted archaeal and bacterial biomarkers, reflecting their formation as a result of AOM under varying conditions. Except for the typical isoprenoids found in various cold seeps worldwide, 3,7,11,15-tetramethyl hexadecan-1,3-diol and two novel *sn2-/sn3*-O-hydroxyphytanyl glycerol monoethers with notable ^{13}C-depletion were observed in authigenic carbonates obtained from Haima, which are most likely hydrolysis products of archaea-specific diethers. Furthermore, molecular fossils, compound-specific δ^{13}C values, and mineralogies, were used to trace dominant microbial consortia, seepage activities, and environmental conditions in the cold seep ecosystems of the SCS. In this chapter, the archaeal and bacterial lipid biomarker

H. Guan (✉) · L. Liu · S. Li
Frontiers Science Center for Deep Ocean Multispheres and Earth System, Key Lab of Submarine Geosciences and Prospecting Techniques, MOE and College of Marine Geosciences, Ocean University of China, Qingdao 266100, China
e-mail: guanhongxiang@ouc.edu.cn

L. Liu
e-mail: liuleifw@163.com

S. Li
e-mail: sanzhong@ouc.edu.cn

H. Guan · L. Liu · N. Wu
Laboratory for Marine Mineral Resources, Qingdao National Laboratory for Marine Science and Technology, Qingdao 266061, China
e-mail: wuny@ms.giec.ac.cn

L. Liu · N. Wu
Key Laboratory of Gas Hydrate, Qingdao Institute of Marine Geology, Ministry of Natural Resources, Qingdao 266071, China

D. Chen and D. Feng (eds.), *South China Sea Seeps*,
https://doi.org/10.1007/978-981-99-1494-4_11

geochemistry of methane seeps is systematically introduced. AOM, AeOM, oxidation of non-methane hydrocarbons, oil degradation, and the diagenetic fate of glycerol ethers, are further summarized.

11.1 Introduction

Marine sediments are the largest methane reservoir on Earth, with greenhouse methane in the form of gas hydrates, free gas, and dissolved methane in pore-water (Wallmann et al. 2012; Joye 2020). At active seep sites, it is common that gas and fluids are transported from deep reservoirs into shallow surface sediments (Wu et al. 2011, 2022; Jin et al. 2022; Wang et al. 2022). However, most of the rising methane is consumed by methanotrophic microbial communities before it can reach the hydrosphere (Reeburgh 2007; Boetius and Wenzhöfer 2013). Methane is primarily consumed via anaerobic oxidation of methane (AOM) and aerobic oxidation of methane (AeOM) (Boetius and Wenzhöfer 2013). Sulfate-driven AOM in cold seep ecosystems is mainly performed by consortia of anaerobic methane-oxidizing archaea (ANME) and sulfate-reducing bacteria (SRB; Boetius et al. 2000), whereas AeOM is performed by aerobic microbes and animal–microbe symbioses (Boetius and Wenzhöfer 2013).

To date, three types of microbial consortia have been demonstrated to mediate sulfate-driven AOM: ANME-1/*Desulfosarcina*/*Desulfococcus* (*DSS*), ANME-2/*DSS*, and ANME-3/*Desulfobulbus* spp. (*DBB*) (Hinrichs et al. 2000; Orphan et al. 2001; Knittel et al. 2003, 2005; Losekann et al. 2007). These archaeal assemblages are closely or loosely associated with their sulfate-reducing partners, varying with the methane and sulfate concentrations, temperature, and oxygen availability (Elvert et al. 2005; Knittel et al. 2005; Nauhaus et al. 2005). Lipid biomarkers and their compound-specific carbon stable isotope compositions are useful tools to monitor the distribution and functioning of microorganisms in seep environments (Birgel et al. 2006; Niemann and Elvert 2008; Stadnitskaia et al. 2008a, b; Blumenberg et al. 2015). This is because different ANME/SRB consortia biosynthesize different lipid biomarkers and produce distinct carbon differences from methane to biomarkers (Niemann and Elvert 2008). For example, ANME-2/*DSS* consortia are commonly characterized by large amounts of 2,6,11,15-tetramethylhexadecane (crocetane), high $sn2$-hydroxyarchaeol/archaeol ratios (>1.1), and low ai-$C_{15:0}$/i-$C_{15:0}$-fatty acid ratios, whereas high glycerol dibiphytanyl glycerol tetraethers (GDGTs) contents, low $sn2$-hydroxyarchaeol/archaeol ratios (<0.8), small amounts or absence of crocetane, and high ai-$C_{15:0}$/i-$C_{15:0}$-fatty acid ratios, are indicators of ANME-1/*DSS* (Niemann and Elvert 2008). Except for specific biomarkers, the maxima in rates of AOM and sulfate reduction are commonly correlated with the highest biomarker abundance and the lowest carbon isotopic signatures (Elvert et al. 2005).

The investigation of cold seeps in the SCS started in 2004 (Chen et al. 2005; Suess 2005). Since then, authigenic carbonates, sediments at active seep sites, and seep-dwelling bivalves have been collected from more than 40 seep sites, covering a wide range of water depths on both the northern and southern continental slopes of the

SCS (Chen et al. 2005; Han et al. 2008; Liang et al. 2017). Lipid biomarkers of seep carbonates retrieved from the SCS were first reported in 2008 (Birgel et al. 2008; Yu et al. 2008). Based on the molecular fossils of seep carbonates, fluid sources, microbial consortia, and seepage dynamics were constrained (Guan et al. 2013, 2016a, b, 2018; Ge et al. 2015). At the Haima cold seeps, ^{13}C-depleted isoprenoidal 3,7,11,15-tetramethylhexadecan-1,3-diol and two novel isoprenoids, namely, $sn2$-/$sn3$-O-hydroxyphytanyl glycerol monoethers, were found as hydrolysis products of isoprenoid diethers in authigenic carbonates (Xu et al. 2022a, b). Furthermore, lipid biomarker patterns combined with the compound-specific carbon stable isotopes of active seep sediments were used to uncover AOM, AeOM and oil degradation processes (Guan et al. 2022). In the seep ecosystems of the SCS, lipid biomarkers and compound-specific carbon stable isotope compositions have been studied in the Site F, northeast Dongsha, southwest Dongsha, Shenhu, and Haima cold seep ecosystems, revealing AOM, sulfate reduction, and oil biodegradation processes (Birgel and Peckmann 2008; Yu et al. 2008; Ge et al. 2010, 2011, 2015; Guan et al. 2013, 2014, 2016a, b, 2018, 2022; Lai et al. 2021; Li et al. 2021). In this chapter, we reviewed current lipid biomarker studies of seep ecosystems in the SCS. Fluid sources, consortia assemblages, and seepage dynamics during carbonate precipitation are summarized in detail.

11.2 Lipid Biomarker Inventories

11.2.1 Archaeal Biomarkers

The biomarker inventories of seep carbonates retrieved from NE Dongsha, Shenhu, Site F, and Haima revealed diagnostic lipids of specific groups of ANME/SRB (Fig. 11.1; Birgel et al. 2008; Yu et al. 2008; Ge et al. 2010, 2015; Guan et al. 2013, 2016a, b, 2018). The biomarker patterns and stable carbon isotope compositions were found to vary among individual study sites and geographic areas. Among these data, ^{13}C-depleted isoprenoid hydrocarbons, crocetane, 2,6,10,15,19-pentamethyleicosane (PMI), and squalane, are common (Fig. 11.2; Birgel et al. 2008; Yu et al. 2008; Ge et al. 2011, 2015; Guan et al. 2013, 2016a, b, 2018). Crocetane is particularly abundant at ANME-2-dominated methane seeps, whereas it is only present in minor amounts or absent in ANME-1-dominated environments (Guan et al. 2013, 2016a, b; Ge et al. 2015). ^{13}C-depleted PMI-acid specific to methanotrophic archaea was found in an authigenic carbonate from NE Dongsha (Guan et al. 2013). Monocyclic diphytanyl glycerol diethers (monocyclic MDGD), possibly derived from methanotrophic archaea, were observed in carbonate pipes in the Shenhu area (Ge et al. 2011). Other isoprenoids, including ^{13}C-depleted phytanyl-glycerolmonoethers $sn2$-/$sn3$-O-phytanyl glycerol ethers, phytanol, and phytanoic acid, are common in authigenic carbonates and sediments (Guan et al. 2013, 2016a,

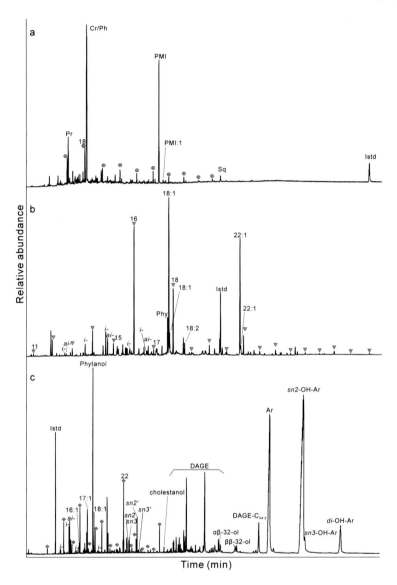

Fig. 11.1 Gas chromatograms of the hydrocarbon fraction (**a**), carboxylic acid fraction (**b**), and alcohol fraction (**c**) from seep carbonates of the South China Sea. Istd: internal standard. **a** Pr: pristane; Cr/Ph: crocetane/phytane; PMI: 2,6,10,15,19-pentamethylicosane; Sq: squalane; gray dots: *n*-alkanes. Phy: phytanoic acid; gray triangles: *n*-fatty acids. DAGE: non-isoprenoidal dialkyl glycerol ether; ββ-32-ol: 17β(H), 21β(H)-C$_{32}$-hopanol; *sn2/sn3*: *sn2-/sn3*-O-phytanyl glycerol monoethers; *sn2'/sn3'*: *sn2-/sn3*-O-hydroxyphytanyl glycerol monoethers; Ar: archaeol; *sn2-/sn3*-OH−Ar: *sn2-/sn3*-hydroxyarchaeol; gray diamonds: *n*-alcohols

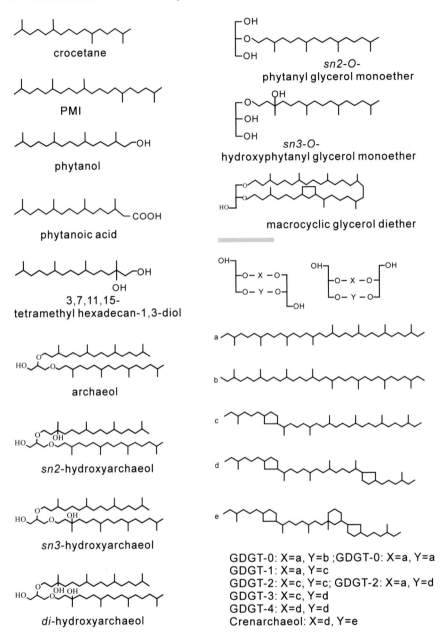

Fig. 11.2 Biomarkers of anaerobic methane-oxidizing archaea

b, 2018, 2022). The GDGT patterns are characterized by the dominance of GDGTs with 0–2 cyclopentane moieties (Guan et al. 2016a).

Biphytanic diacids, including acyclic biphytanic diacid, monocyclic biphytanic diacid, and bicyclic biphytanic diacid, were found in seep carbonates from NE

Dongsha (the Jiulong methane reef; Birgel et al. 2008). Except for the most widespread isoprenoids above, isoprenoidal 3,7,11,15-tetramethylhexadecan-1,3-diol and two novel *sn2-/sn3*-O-hydroxyphytanyl glycerol monoethers hydroxylated at C-3 of the phytanyl moieties were observed in seep carbonates from the Haima seep (Xu et al. 2022a, b). Most archaeal-specific isoprenoids are strongly depleted in ^{13}C with δ^{13}C values ranging from $-140‰$ to $-84‰$ (Fig. 11.3; Birgel et al. 2008; Yu et al. 2008; Ge et al. 2011, 2015; Guan et al. 2013, 2016a, b, 2018, 2022; Li et al. 2021).

11.2.2 Sulfate-Reducing Bacterial Biomarkers

Seep carbonates and seep sediments from the SCS are known to contain various bacterial lipids (Fig. 11.4). The otherwise uncommon fatty acids $C_{16:1\omega5}$ and cyc$C_{17:0\omega5,6}$ were found in high abundances and were assigned to the sulfate-reducing partners of ANME-2 (Elvert et al. 2003; Blumenberg et al. 2004, 2005). The presence of *ai-/i-*$C_{15:0}$-fatty acids with ratios lower than 2 is an indicator of ANME-2/*DSS*-dominated seeps (Niemann and Elvert 2008; Guan et al. 2013). However, the *ai-/i-*$C_{15:0}$-fatty acid ratio does not always discriminate the sulfate-reducing partners of ANME since their contents and δ^{13}C values can be modified by SRBs involved in oil degradation (Guan et al. 2018, 2022; Krake et al. 2022). Non-isoprenoidal monoglycerol ethers (MAGEs), ascribed to *DSS* associated with ANME-2 and *DBB* associated with ANME-3 (Niemann and Elvert 2008), were only abundant in Haima seep carbonates (Xu et al. 2022a, b). SRB-specific fatty acids were less ^{13}C-depleted than those originating from archaea because they can assimilate dissolved inorganic carbon (DIC) as a carbon source (Wegener et al. 2008). The SRB-specific biomarkers yielded δ^{13}C values from $-11133‰$ to $-111‰$ in seep carbonates of the SCS (Fig. 11.3; Guan et al. 2013, 2016a, b, 2018, 2022). In addition, various DAGEs with extremely low δ^{13}C values were found with high contents in authigenic carbonates from Site F, NE Dongsha, and Haima (Fig. 11.5; Guan et al. 2013, 2016a; Xu et al. 2022a, b).

11.2.3 Aerobic Methanotrophic Biomarkers

Another striking feature of the seep carbonates originating from the SCS is the significant amounts of hopanoids and steroids with moderate to significant ^{13}C-depletions (Fig. 11.5; Guan et al. 2016a, b, 2018, 2022). These hopanoids generally include tetrahymanol, diploptene, diplopterol, hopanoic acids (e.g., 17α(H)-21β(H)-bishomo-hopanoic acid and 17β(H)-21β(H)-bishomo-hopanoic acid), and hopanols (e.g., 17α(H)-21β(H)-bishomo-hopanol and 17β(H)-21β(h)-bishomo-hopanol), reflecting a characteristic suitably agreeing with patterns found at other seep sites (Fig. 11.6; Birgel et al. 2008, 2011; Himmler et al. 2015; Natalicchio et al. 2015). However, bacteriohopanepolyols (BHPs) and their related geohopanoids are

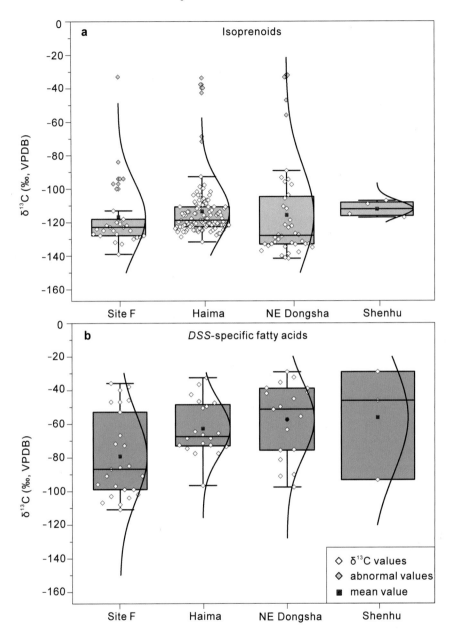

Fig. 11.3 Comparison of the $\delta^{13}C$ values of archaeal biomarkers (**a**) and sulfate-reducing bacterial biomarkers (*DSS* cluster) (**b**) in authigenic carbonates from various seep sites of the SCS. Isoprenoids include crocetane/phytane, PMI, phytanoic acid, phytanol, 3,7,11,15-tetramethylhexadecan-1,3-diol, *sn2-/sn3-O*-phytanyl-glycerolethers, *sn2-/sn3-O*-hydroxyphytanyl glycerol monoethers, archaeol, *sn2-/sn3*-hydroxyarchaeols, and *di*-hydroxyarchaeol. Sulfate-reducing bacterial biomarkers include *i-/ai*-$C_{15:0}$, *n*-$C_{16:1\omega5}$, *n*-$C_{17:1\omega6}$, and *cyc*$C_{17:0\omega5,6}$ fatty acids, the MAGEs were excluded. Data are taken from Birgel et al. (2008), Yu et al. (2008), Guan et al. (2013, 2014, 2016a, b, 2018), Ge et al. (2011, 2015), and Xu et al. (2022a)

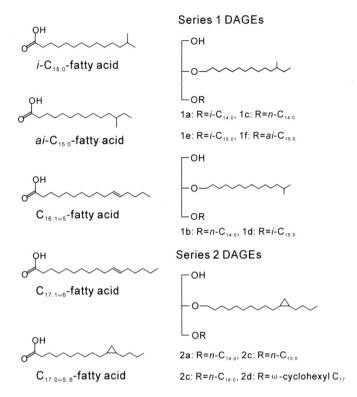

Fig. 11.4 Selected biomarkers of sulfate-reducing bacteria

known to be produced by both aerobic and anaerobic bacteria (Talbot et al. 2001; Sinninghe Damste et al. 2004; Blumenberg et al. 2006; Rush et al. 2016). Therefore, diplopterol, diploptene, hopanols, hopanoic acids, and their possible diagenetic products generally cannot be used alone as reliable indicators of past aerobic methanotrophic bacteria at seeps (Birgel and Peckmann 2008). Generally, ^{13}C-depleted 3β-methylated hopanoids, 4-methylated steroids, and BHPs aminotetrol/aminotriol and further related BHPs are diagnostic biomarkers of aerobic methanotrophic bacteria (Talbot et al. 2001; Birgel and Peckmann 2008; Birgel et al. 2011; Himmler et al. 2015; Natalicchio et al. 2015; Rush et al. 2016). In the SCS, hopanoids were observed in seep carbonates and sediments originating from NE Dongsha, Shenhu, Site F, and Haima, yielding δ^{13}C values ranging from −31‰ to −101‰ (Fig. 11.5; Guan et al. 2013, 2014, 2016a, b, 2018, 2022). Regarding the Haima cold seeps, 4-methylated steroid 4α-methylcholesta-2,8(14),24-triene and two tetracyclic hopanoid ketones, C_{30}-3β-methyl tetracyclic-ketone and C_{34}-3β-methyl tetracyclic-ketone, were found in seep sediments (Guan et al. 2022). In this cold seep ecosystem, both 4-methylated steroid and tetracyclic hopanoid ketones were attributed to Type I and/or X methanotrophs, with precursors most likely including 4α-methylcholesta-8(14),24-dien-3β-ol and 3-methylated hopanoids, respectively (Guan et al. 2022). These hopanoids

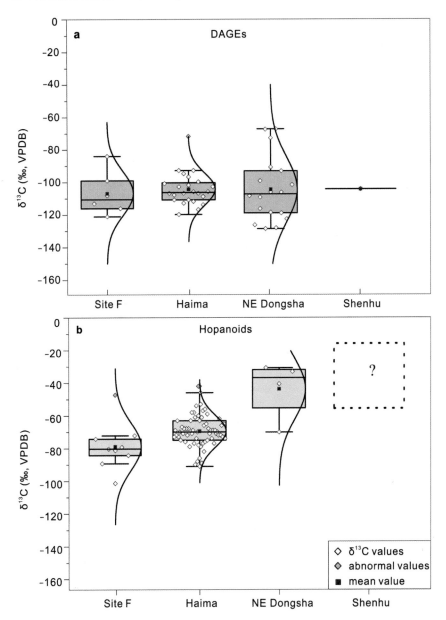

Fig. 11.5 δ^{13}C values of DAGEs (**a**), and hopanoids (**b**) from seep carbonates of the SCS. Hopanoids include diplopterol, 17α(H),21β(H)-32-hopanol, 17β(H),21β(H)-32-hopanol, 17α(H),21β(H)-32-hopanoic acid, 17β(H),21β(H)-31-hopanoic acid, and 17β(H),21β(H)-32-hopanoic acid. Data are taken from Guan et al. (2013, 2014, 2016a, b, 2018); Ge et al. (2011, 2015

along with aerobic methanotrophic bacteria-specific biomarkers are indicators of aerobic methanotrophic bacteria at seeps.

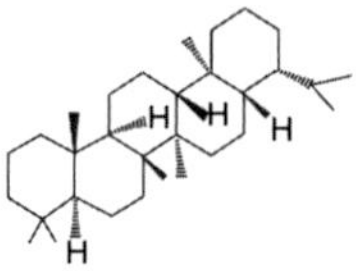

Diploptane

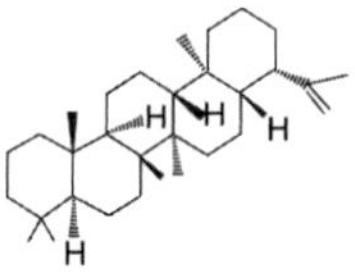

Diploptene

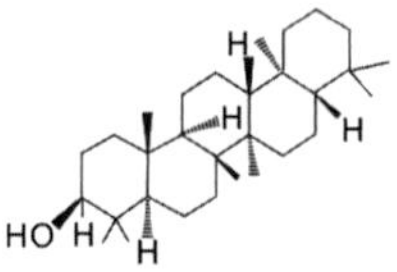

Tetrahymanol

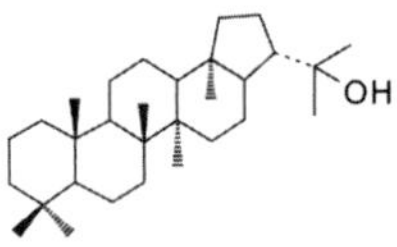

Diplopterol

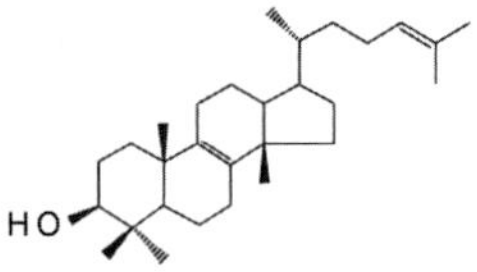

Lanosterol

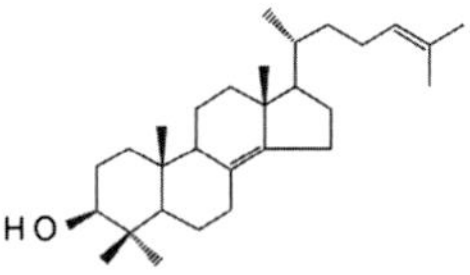

4,4-dimethylcholesta-
8(14),24-dien-3β-ol

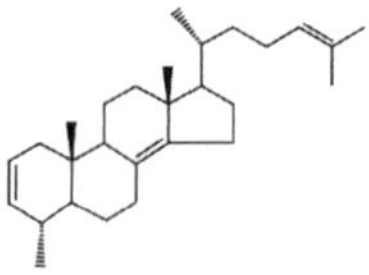

4α-methylcholesta-
2,8(14),24-triene

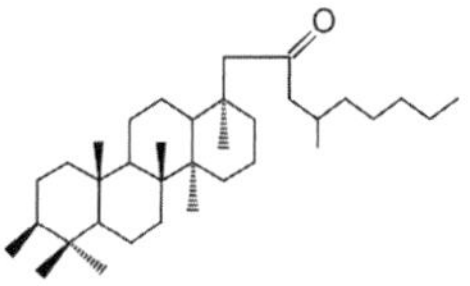

C_{34}-3β-methyl tetracyclic-
ketone

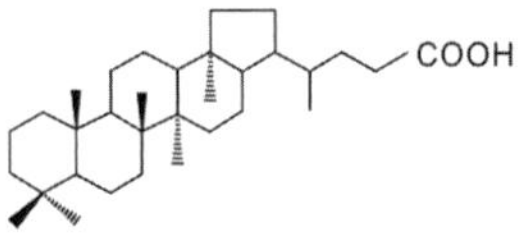

17β(H),21β(H)-
32-hopanoic acid

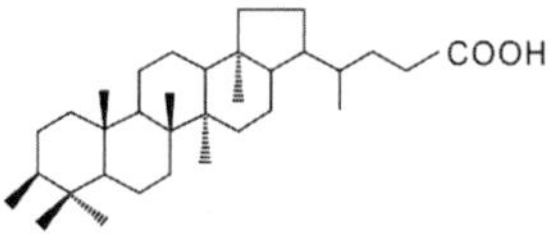

17β(H),21β(H)-
3β-Me-32-hopanoic acid

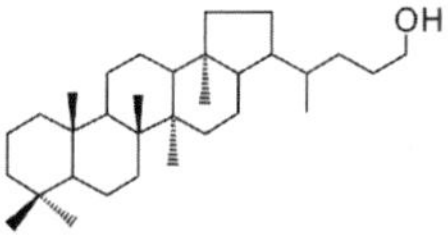

17β(H),21β(H)-
32-hopanol

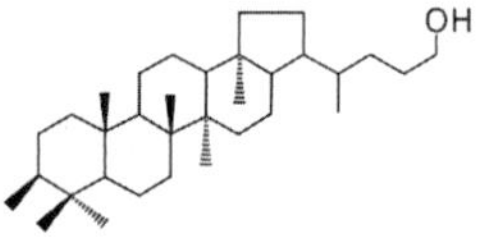

17β(H),21β(H)-
3β-Me-32-hopanol

Fig. 11.6 Biomarkers of aerobic methanotrophic bacteria

11.3 Authigenic Carbonate Formations Constrained by Lipid Biomarkers

11.3.1 Environmental Conditions Inferred from Biomarker Patterns

Microbial activity, diversity and related geochemical processes in cold seep ecosystems are of great concern because the upward ascending methane is mainly consumed by anaerobic methanotrophs in consortia with sulfate-reducing bacteria (Boetius et al. 2000). In the cold seep ecosystems of the SCS, microbial diversity has been reported at the NE Dongsha, Site F, Shenhu, and Haima cold seeps. Methanotrophic archaea (ANME-1, ANME-2, and ANME-3) and sulfate-reducing partners (*Desulfosarcina/Desulfococcus*) were observed, and their abundance varied with the depth and site (Niu et al. 2017; Cui et al. 2019; Zhuang et al. 2019). Processes of methanogenesis, AOM, and sulfate reduction could be archived by microbial activities at active seep sites (Boetius et al. 2000; Blumenberg et al. 2004). However, these advantages of microbial tools disappear during rock periods. Instead, lipid biomarkers are preferred because of their stable carbon skeleton and the carried information on their precursors (Aloisi et al. 2000; Peckmann et al. 2004, 2009; Birgel et al. 2008; Stadnitskaia et al. 2008a, b).

The lipid biomarker inventories of seep carbonates from NE Dongsha, Shenhu, Site F, and Haima revealed diagnostic lipids of specific groups of ANMEs and their sulfate-reducing partners (Fig. 11.1; Birgel et al. 2008; Yu et al. 2008; Guan et al. 2013, 2016a, b, 2018; Ge et al. 2015; Xu et al. 2022b). The biomarker patterns and carbon stable isotope compositions were found to vary among individual study sites and among different geographic areas (Figs. 11.3 and 11.5). Lipid biomarkers of ANMEs and SRBs originating from NE Dongsha and Site F carbonates commonly yielded lower $\delta^{13}C$ values than those originating from Haima carbonates, reflecting differences in fluid compositions and methane types (Fig. 11.2; Guan et al. 2013, 2016a, 2018). NE Dongsha and Site F are located in the convergence zone of the active and passive margins, and seeps appear to be largely controlled by strike-slip faults (Liu et al. 2008, 2015; Chen et al. 2017; Feng et al. 2018). Methane at both sites indicates biogenic sources and is released from gas hydrate dissociation with $\delta^{13}C$ values ranging from $-72.3‰$ to $-69.4‰$ (Zhuang et al. 2016). Regarding the Haima cold seeps, hydrate-bound methane is a mixture of biogenic and thermogenic gas, with most $\delta^{13}C$ values ranging from $-72.3‰$ to $-48.4‰$ (Fang et al. 2019; Ye et al. 2019; Wei et al. 2019, 2021; Lai et al. 2022a, b). Potential methane types during carbonate precipitation could be inferred according to specific microbial consortia and $\delta^{13}C$ values of archaea-specific isoprenoids because the $\delta^{13}C$ differences from methane to archaeal-specific lipids reach approximately $-30‰$ and $-50‰$ for ANME-1 and ANME-2, respectively (Niemann and Elvert 2008; Birgel et al. 2011; Himmler et al. 2015). The $\delta^{13}C$ variability of lipid biomarkers originating from NE Dongsha, Site

F, and Haima carbonates reflected the patterns of their parent methane and carbon fractionations from methane to lipids.

The application of molecular fossils in combination with their compound-specific isotope signatures is an efficient tool to reconstruct seepage intensities. At seep sites, the sulfate methane transition zone (SMTZ) is largely controlled by the seepage intensity (Borowski et al. 1996), which tends to occur shallower within sediments at sites with a high methane flux than that at sites with a relatively low methane flux (Luff and Wallmann 2003). The relatively high sulfate concentration and porosity on or near the seafloor surface favor the precipitation of aragonite over high-Mg-calcite (Fig. 11.7; Greinert et al. 2001; Haas et al. 2010). In regard to microbial consortia, ANME-2 is better adapted to higher methane concentrations and lower temperatures and is less sensitive to oxygen than ANME-1, whereas ANME-1 can tolerate moderate and lower methane concentrations (Elvert et al. 2005; Knittel et al. 2005; Nauhaus et al. 2005). In the SCS, including the Dongsha area, Site F, Shenhu, and Haima cold seeps, aragonite- and calcite-dominated carbonates are commonly formed at sites predominated by ANME-2 and ANME-1, respectively, indicating varying environmental conditions and seepage intensities (Guan et al. 2013, 2016a, b, 2018; Ge et al. 2015). Therefore, lipid biomarkers, combined with mineralogy, petrography, and carbon stable and oxygen isotope compositions, are effective indicators to reconstruct the formation conditions of seep-related carbonates.

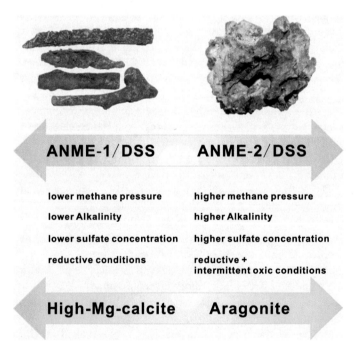

Fig. 11.7 Schematic model of relationship of microbial communities, mineralogies, and environmental conditions

11.3.2 Sulfate Reductions Coupled to AOM and Organoclastic Oxidation

At marine seep sites, both AOM and oxidation of non-methane hydrocarbons can fuel sulfate reduction (Joye et al. 2004; Kallmeyer and Boetius 2004; Bowles et al. 2011; Kleindienst and Knittel 2020). Joye et al. (2004) found that sulfate-reducing rates generally exceeded AOM rates and concluded that the majority of sulfate reduction at Gulf of Mexico hydrocarbon seep sites was likely fueled by the oxidation of other organic matter. At oil-related cold seeps, hydrocarbons other than methane are the most likely reductants. Among sulfate reducers, *DSS* of Deltaproteobacteria are known as the primary non-methane hydrocarbon degraders at methane seeps (Kniemeyer et al. 2007; Kleindienst et al. 2012, 2014; Jaekel et al. 2013). *DSS* involved in sulfate-driven AOM lives in consortia with ANME, using CO_2 as a carbon source and yielding extreme ^{13}C-depletion of *DSS*-derived fatty acids (Wegener et al. 2008, 2015). In contrast, all known *DSS* capable of non-methane hydrocarbon degradation are free-living organisms, coupling substrate oxidization and sulfate reduction in a single-cell process (Kleindienst et al. 2012, 2014; Jaekel et al. 2013; Petro et al. 2019). These *DSS* were demonstrated to assimilate carbon from short-chain hydrocarbons in enrichment cultures (Jaekel et al. 2013; Kleindienst et al. 2014).

Although the $\delta^{13}C$ values of methane and oil-related hydrocarbons overlap to a certain extent ($-110‰$ to $-15‰$ and higher than $-35‰$, respectively; Milkov and Etiope (2018)), methane in the SCS yielded notably lower carbon stable isotopes (lower than $-48‰$) than oil-related hydrocarbons (Zhuang et al. 2016; Yang et al. 2018; Fang et al. 2019; Wei et al. 2019, 2021; Ye et al. 2019; Lai et al. 2022a, b; Liang et al. 2022). Consequently, the lipid biomarkers of *DSS* involved in AOM and oxidation of non-methane hydrocarbons are expected to be different. Guan et al. (2018) analyzed seep carbonates originating from the Haima cold seeps and found that *i-/ai-*$C_{15:0}$-fatty acids were accompanied by less negative $\delta^{13}C$ values, suggesting that the input of *DSS* was involved in the oxidation of non-methane hydrocarbons. Similar observations were also found for sediments obtained from the Haima cold seeps (Guan et al. 2022). These conclusions were supported by lipid biomarker comparisons between methane-seep carbonates and oil-seep carbonates (Krake et al. 2022). The authors found that oil-seep-related carbonates not only yielded a high abundance of DAGEs with more diverse alkyl chains but also heavier $\delta^{13}C$ values on average than those of methane-seep carbonates (Krake et al. 2022).

Unresolved complex mixtures (UCMs) are an additional indicator of the oxidation of non-methane hydrocarbons. UCMs are commonly reported in seep-related carbonates or sediments, especially oil seepages (Sassen et al. 1993; Zhang et al. 2003; Naehr et al. 2009; Birgel et al. 2011; Feng et al. 2014, 2018; Li et al. 2021). UCMs contain naphthenes and other related compounds that cannot be completely resolved by conventional GC analysis (Zhang et al. 2003; Naehr et al. 2009; Birgel et al. 2011). At the Haima cold seeps, UCMs occurred in hydrocarbon fractions of both seep carbonates and sediments, with partially altered hydrocarbons (Guan et al. 2018, 2022). Since *n*-alkanes were not completely altered, these oils in the sediments were attributed to light to middle biodegradation (Peters and Moldowan 1993). Furthermore, the heavier $\delta^{13}C_{TIC}$ values of sediments retrieved from the Haima seeps were found to be correlated with a high total organic carbon content, which again agreed with non-methane hydrocarbons supplied to sulfate-reducing bacteria (*DSS*) and resulted in relatively heavier $\delta^{13}C$ values of *DSS*-derived fatty acids (Guan et al. 2022).

11.3.3 Constraints on the Hydrolysis Process

In lipid biomarker analysis of authigenic carbonates originating from the Haima cold seep, 3,7,11,15-tetramethylhexadecan-1,3-diol and two novel *sn2-/sn3*-O-hydroxyphytanyl glycerol monoethers hydroxylated at C-3 of the phytanyl moieties were identified. Both methanogens and methanotrophic archaea were found to produce hydroxydiethers (Sprott et al. 1990, 1992, 1993; Rossel et al. 2008, 2011). Among methanotrophic archaea, ANME-1 contains abundant GDGTs and small amounts of *sn2-/sn3*-hydroxyarchaeols, whereas ANME-2 and ANME-3 are dominated by phosphate-based polar derivatives of *di*-hydroxyarchaeol, *sn2-/sn3*-hydroxyarchaeols and archaeol (Rossel et al. 2008, 2011). Regarding these Haima seep carbonates, the abundant crocetane and high ratios of *sn2*-hydroxyarchaeol relative to archaeol (>1.1) indicated the dominance of ANME-2 or ANME-3 (Xu et al. 2022b).

Generally, understanding the composition of molecular fossils, biological precursors and past diagenetic evolution process constitutes a key step in applying a fossil lipid indicator (Briggs and Summons 2014). The ^{13}C-depleted short-chain isoprenoids in cold seep ecosystems are putative degradation products of archaeol and *sn2-/sn3*-hydroxyarchaeols, and can be attributed to ANMEs (Thiel et al. 2001a, 2018; Oba et al. 2006). According to Koga et al. (1993) and Oba et al. (2006), all hydroxyarchaeols were converted into monophytanyl glycerol ethers in the hydrolysis process with chloroform/methanol/concentrated HCl. Based on the sequential degradation pathway of hydroxyarchaeols, *sn2-/sn3*-O-hydroxyphytanyl glycerol monoethers are likely hydrolysis products of *di*-hydroxyarchaeol rather than *sn2-/sn3*-hydroxyarchaeols, whereas isoprenoidal 3,7,11,15-tetramethylhexadecan-1,3-diol is most likely a hydrolysis product of *sn2-/sn3*-O-hydroxyarchaeols, *di*-hydroxyarchaeol, and *sn2-/sn3*-O-hydroxyphytanyl glycerol monoethers (Fig. 11.8; Xu et al. 2022b). Except for these isoprenoid ethers, a similar relationship was found for MAGEs, DAGEs and corresponding non-isoprenoid short-chain alcohols in the Haima seep carbonates. The MAGEs, DAGEs, and non-isoprenoid short-chain alcohols yielded similar $\delta^{13}C$ values, chain (side-chain) lengths and unsaturation patterns, indicating that non-isoprenoid alcohols might be either biosynthetic intermediates or degradation products of MAGEs and DAGEs (Xu et al. 2022b).

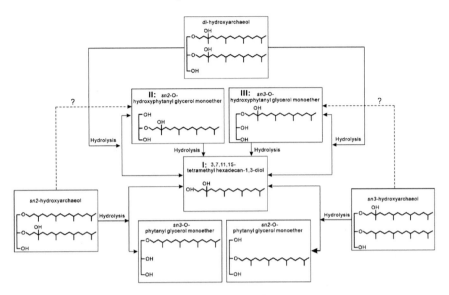

Fig. 11.8 Potential diagenetic pathways of isoprenoid glycerol ethers. Reprinted from Organic Geochemistry, 163, Xu et al. (2022b). Diagenetic fate of glycerol ethers revealed by two novel isoprenoid hydroxyphytanyl glycerol monoethers and non-isoprenoid alkyl glycerolethers, 104, 344, Copyright (2022), with permission from Elsevier

11.4 Summary and Outlook

Authigenic carbonates sampled from the SCS contained varying contents of ^{13}C-depleted archaeal and bacterial biomarkers, indicating their formation as a result of anaerobic oxidation of methane (AOM) via anaerobic methane-oxidizing archaea (ANMEs) and sulfate-reducing bacteria (SRBs). Based on various biomarker patterns, methane was predominantly oxidized by ANME-1/*DSS* and ANME-2/*DSS*, varying from site to site. Generally, ANME-2/*DSS*-specific crocetane, n-$C_{16:1\omega5}$, cyc-$C_{17:0\omega5,6}$-fatty acids, high $sn2$-hydroxyarchaeol/archaea ratios and low *anteiso*-$C_{15:0}$/*iso*-$C_{15:0}$-fatty acid (ai-$/i$-$C_{15:0}$) ratios were found in aragonite-dominated carbonates, suggesting carbonates formed in seep ecosystems with high seepage intensities. Carbonates characterized by GDGTs, small amounts of crocetane, low $sn2$-hydroxyarchaeol/archaea ratios and high ai-$C_{15:0}$/i-$C_{15:0}$-fatty acid ratios were dominated by high-magnesium calcite, reflecting their formation at a greater depth with a relatively low seepage intensity. The oxidation of non-methane hydrocarbons fueled by sulfate reduction was demonstrated by abundant i-$/ai$-$C_{15:0}$ fatty acids with relatively higher δ^{13}C values. Distinct biomarker patterns and respective geochemical processes at cold seeps may be correlated with the microbial community succession sequence during seepage evolution because the different growth rates of aerobic methanotrophs, SRBs, and ANMEs could be recorded in an emerging cold seep ecosystem. Moreover, the ether hydrolysis products isoprenoidal 3,7,11,15-tetramethylhexadecan-1,3-diol and $sn2$-$/sn3$-O-hydroxyphytanyl glycerol monoethers were observed, reflecting a diagenetic process from isoprenoid dieters to isoprenoid monoethers and isoprenoid alcohols.

Although diagnostic biomarkers could be used to discriminate ANME-1/*DSS* and ANME-2/*DSS* consortia, it is unclear what factors control their ability to respond to changing seepage intensities and environments. Since cold seeps comprise a substantial carbon pool worldwide, it is critical to increase the efficiency of hydrocarbon biofilters. In addition to the well-advanced identification of sulfate-AOM, AeOM, and oxidation of non-methane hydrocarbons with biomarkers as outlined above, future studies combining culturing, molecular microbiology, and lipid biomarkers may provide further insight into (1) the modes and variability of hydrocarbon consumption possibly involving different electron acceptors and (2) the allochthonous and autochthonous contributions of seep carbonates to biomarker inventories.

Acknowledgements We sincerely thank the crew and scientists of the ROV *Haima*, *ROPOS* and RV *SONNE*, *Haiyang-06*, and *"Tan Kah Kee"* for their professional work to meet the scientific goal during expeditions. S. Gao (GIG, CAS) and S. Liu (GIEC, CAS) are thanked for their help with technical assistance. This research was supported by Laoshan Laboratory (No. LSKJ202203502), the National Natural Science Foundation of China (Grants: 42276053 and 91958105), and the Young Taishan Scholars Program (Grant No. tsqn202211069).

References

Aloisi G, Pierre C, Rouchy JM et al (2000) Methane–related authigenic carbonates of eastern Mediterranean Sea mud volcanoes and their possible relation to gas hydrate destabilisation. Earth Planet Sci Lett 184(1):321–338

Birgel D, Elvert M, Han XQ et al (2008) 13C–depleted biphytanic diacids as tracers of past anaerobic oxidation of methane. Org Geochem 39(1):152–156

Birgel D, Peckmann J (2008) Aerobic methanotrophy at ancient marine methane seeps: a synthesis. Org Geochem 39(12):1659–1667

Birgel D, Feng D, Roberts HH et al (2011) Changing redox conditions at cold seeps as revealed by authigenic carbonates from Alaminos Canyon, Northern Gulf of Mexico. Chem Geol 285(1–4):82–96

Birgel D, Thiel V, Hinrichs K-U et al (2006) Lipid biomarker patterns of methane–seep microbialites from the Mesozoic convergent margin of California. Org Geochem 37(10):1289–1302

Blumenberg M, Krüger M, Nauhaus K et al (2006) Biosynthesis of hopanoids by sulfate–reducing bacteria (genus Desulfovibrio). Environ Microbiol 8(7):1220–1227

Blumenberg M, Seifert R, Nauhaus K et al (2005) In vitro study of lipid biosynthesis in an anaerobically methane-oxidizing microbial mat. Appl Environ Microbiol 71(8):4345–4351

Blumenberg M, Seifert R, Reitner J et al (2004) Membrane lipid patterns typify distinct anaerobic methanotrophic consortia. Proc Natl Acad Sci USA 101(30):11111–11116

Blumenberg M, Walliser E, Taviani M et al (2015) Authigenic carbonate formation and its impact on the biomarker inventory at hydrocarbon seeps e a case study from the Holocene Black Sea and the Plio-Pleistocene Northern Apennines (Italy). Mar Pet Geol 66:532–541

Boetius A, Ravenschlag K, Schubert CJ et al (2000) A marine microbial consortium apparently mediating anaerobic oxidation of methane. Nature 407(6804):623–626

Boetius A, Wenzhöfer F (2013) Seafloor oxygen consumption fuelled by methane from cold seeps. Nat Geosci 6(9):725–734

Borowski WS, Paul CK, Ussler WU (1996) Marine pore–water sulfate profiles indicate in situ methane flux from underlying gas hydrate. Geology 24(7):655–658

Bowles MW, Samarkin VA, Bowles KM et al (2011) Weak coupling between sulfate reduction and the anaerobic oxidation of methane in methane–rich seafloor sediments during ex situ incubation. Geochim Cosmochim Acta 75(2):500–519

Briggs DEG, Summons RE (2014) Ancient biomolecules: their origins, fossilization, and role in revealing the history of life. BioEssays 36(5):482–490

Chen DF, Huang YY, Yuan XL et al (2005) Seep carbonates and preserved methane oxidizing archaea and sulfate reducing bacteria fossils suggest recent gas venting on the seafloor in the Northeastern South China Sea. Mar Pet Geol 22(5):613–621

Chen WH, Huang CY, Yan Y et al (2017) Stratigraphy and provenance of forearc sequences in the Lichi Melange, Coastal Range: geological records of the active Taiwan arc–continent collision. J Geophys Res-Solid Earth 122(9):7408–7436

Chevalier N, Bouloubassi I, Stadnitskaia A et al (2010) Distributions and carbon isotopic compositions of lipid biomarkers in authigenic carbonate crusts from the Nordic Margin (Norwegian Sea). Org Geochem 41(9):885–890

Chevalier N, Bouloubassi I, Birgel D et al (2013) Microbial methane turnover at Marmara Sea cold seeps: a combined 16S rRNA and lipid biomarker investigation. Geobiology 11(1):55–71

Cui HP, Su X, Chen F et al (2019) Microbial diversity of two cold seep systems in gas hydrate–bearing sediments in the South China Sea. Mar Environ Res 144:230–239

Elvert M, Boetius A, Knittel K et al (2003) Characterization of specific membrane fatty acids as chemotaxonomic markers for sulfate–reducing bacteria involved in anaerobic oxidation of methane. Geomicrobiol J 20(4):403–419

Elvert M, Hopmans EC, Treude T et al (2005) Spatial variations of methanotrophic consortia at cold methane seeps: implications from a high–resolution molecular and isotopic approach. Geobiology 3(3):195–209

Fang Y, Wei J, Lu H et al (2019) Chemical and structural characteristics of gas hydrates from the Haima cold seeps in the Qiongdongnan Basin of the South China Sea. J Asian Earth Sci 182:103924

Feng D, Birgel D, Peckmann J et al (2014) Time integrated variation of sources of fluids and seepage dynamics archived in authigenic carbonates from Gulf of Mexico Gas Hydrate Seafloor Observatory. Chem Geol 385:129–139

Feng D, Qiu J-W, Hu Y et al (2018) Cold seep systems in the South China sea: an overview. J Asian Earth Sci 168:3–16

Ge L, Jiang SY, Blumenberg M et al (2015) Lipid biomarkers and their specific carbon isotopic compositions of cold seep carbonates from the South China sea. Mar Pet Geol 66:501–510

Ge L, Jiang S, Swennen R et al (2010) Chemical environment of cold seep carbonate formation on the northern continental slope of South China sea: evidence from trace and rare earth element geochemistry. Mar Geol 277(1–4):21–30

Ge L, Jiang S, Yang T et al (2011) Glycerol ether biomarkers and the carbon isotopic compositions in a cold seep carbonate chimney from the Shenhu area, northern South China Sea. Chin Sci Bull 56(16):1700–1707

Greinert J, Bohrmann G, Suess E (2001) Gas hydrate associated carbonates and methane–venting at Hydrate Ridge: classification, distribution and origin of authigenic lithologies. In: Paull CK, Dillon WP (eds) Natural gas hydrates: occurrence, distribution, and detection. American Geophysical Union, Washington, DC, USA, pp 91–113

Guan H, Feng D, Wu N et al (2016a) Methane seepage intensities traced by biomarker patterns in authigenic carbonates from the South China sea. Org Geochem 91:109–119

Guan H, Zhang M, Mao S et al (2016b) Methane seepage in the Shenhu area of the northern South China sea: constraints from carbonate chimneys. Geo-Mar Lett 36(3):175–186

Guan H, Liu L, Hu Y et al (2022) Rising bottom–water temperatures induced methane release during the middle Holocene in the Okinawa Trough. East China Sea. Chem Geol 590:120707

Guan H, Sun Y, Mao S et al (2014) Molecular and stable carbon isotopic compositions of hopanoids in seep carbonates from the South China Sea continental slope. J Asian Earth Sci 92:254–261

Guan H, Sun Y, Zhu X et al (2013) Factors controlling the types of microbial consortia in cold–seep environments: a molecular and isotopic investigation of authigenic carbonates from the South China Sea. Chem Geol 354:55–64

Guan H, Xu L, Wang Q et al (2019) Lipid Biomarkers and Their Stable Carbon Isotopes in Ancient Seep Carbonates from SW Taiwan. China. Acta Geol Sin-Engl Ed 93(1):167–174

Guan H, Birgel D, Peckmann J et al (2018) Lipid biomarker patterns of authigenic carbonates reveal fluid composition and seepage intensity at Haima cold seeps, South China Sea. J Asian Earth Sci 168:163–172

Haas A, Peckmann J, Elvert M et al (2010) Patterns of carbonate authigenesis at the Kouilou pockmarks on the Congo deep–sea fan. Mar Geol 268(1–4):129–136

Han X, Suess E, Huang Y et al (2008) Jiulong methane reef: Microbial mediation of seep carbonates in the South China Sea. Mar Geol 249(3–4):243–256

Himmler T, Birgel D, Bayon G et al (2015) Formation of seep carbonates along the Makran convergent margin, northern Arabian Sea and a molecular and isotopic approach to constrain the carbon isotopic composition of parent methane. Chem Geol 415:102–117

Hinrichs K–U, Boetius A (2002) The anaerobic oxidation of methane: new insights in microbial ecology and biogeochemistry. In: Wefer G, Billett D, Hebbeln D, Jørgensen BB, Schlüter M, van Weering, T.C.E. (Eds)Ocean Margin Systems. Springer, pp 457–477

Hinrichs K-U, Summons RE, Orphan V et al (2000) Molecular and isotopic analysis of anaerobic methane–oxidizing communities in marine sediments. Org Geochem 31(12):1685–1701

Jaekel U, Musat N, Adam B et al (2013) Anaerobic degradation of propane and butane by sulfate–reducing bacteria enriched from marine hydrocarbon cold seeps. ISME J 7(5):885–895

Jin M, Feng D, Huang KJ et al (2022) Magnesium Isotopes in Pore Water of Active Methane Seeps of the South China Sea. Front Mar Sci 9:858860

Joye SB (2020) The geology and biogeochemistry of hydrocarbon seeps. Annu Rev Earth Planet Sci 48:205–231

Joye SB, Boetius A, Orcutt BN et al (2004) The anaerobic oxidation of methane and sulfate reduction in sediments from Gulf of Mexico cold seeps. Chem Geol 205(3–4):219–238

Kallmeyer J, Boetius A (2004) Effects of temperature and pressure on sulfate reduction and anaerobic oxidation of methane in hydrothermal sediments of Guaymas Basin. Appl Environ Microbiol 70(2):1231–1233

Kleindienst S, Herbst F-A, Stagars M et al (2014) Diverse sulfatereducing bacteria of the Desulfosarcina/Desulfococcus clade are the key alkane degraders at marine seeps. ISME J 8(10):2029–2044

Kleindienst S, Knittel K (2020) Anaerobic hydrocarbon–degrading sulfate–reducing bacteria at marine gas and oil seeps. In: Teske A, Carvalho V (eds) Marine Hydrocarbon Seeps: Microbiology and Biogeochemistry of a Global Marine Habitat, Springer Oceanography. Springer International Publishing, Cham, pp 21–41

Kleindienst S, Ramette A, Amann R et al (2012) Distribution and in situ abundance of sulfate–reducing bacteria in diverse marine hydrocarbon seep sediments. Environ Microbiol 14(10):2689–2710

Kniemeyer O, Musat F, Sievert SM et al (2007) Anaerobic oxidation of short–chain hydrocarbons by marine sulphate–reducing bacteria. Nature 449(7164):898–901

Knittel K, Boetius A, Lemke A et al (2003) Activity, distribution, and diversity of sulfate reducers and other bacteria in sediments above gas hydrate (Cascadia margin, Oregon). Geomicrobiol J 20(4):269–294

Knittel K, Loekann T, Boetius A et al (2005) Diversity and distribution of methanotrophic archaea at cold seeps. Appl Environ Microbiol 71(1):467–479

Koga Y, Nishihara M, Morii H et al (1993) Ether polar lipids of methanogenic bacteria: Structures, comparative aspects, and biosyntheses. Microbiol Rev 57(1):164–182

Krake N, Birgel D, Smrzka D et al (2022) Molecular and isotopic signatures of oil–driven bacterial sulfate reduction at seeps in the southern Gulf of Mexico. Chem Geol 595:120797

Lai H, Fang Y, Kuang Z et al (2021) Geochemistry, origin and accumulation of natural gas hydrates in the Qiongdongnan Basin, South China Sea: Implications from site GMGS5–W08. Mar Pet Geol 123:104774

Lai H, Qiu H, Liang J et al (2022a) Geochemical Characteristics and Gas–to–Gas Correlation of Two Leakage–type Gas Hydrate Accumulations in the Western Qiongdongnan Basin, South China Sea. Acta Geol Sin – Engl Ed 96(2):680–690

Lai H, Qiu H, Kuang Z et al (2022b) Integrated signatures of secondary microbial gas within gas hydrate reservoirs: A case study in the Shenhu area, northern South China Sea. Mar Pet Geol 136:105486

Li Y, Fang YX, Zhou QZ et al (2021) Geochemical insights into contribution of petroleum hydro-carbons to the formation of hydrates in the Taixinan Basin, the South China Sea. Geosci Front 12(6):100974

Liang Q, Xiao X, Zhao J et al (2022) Geochemistry and sources of hydrate-bound gas in the shenhu area, northern south China sea: Insights from drilling and gas hydrate production tests. J Pet Sci Eng 208:109459

Liang Q, Hu Y, Feng D et al (2017) Authigenic carbonates from newly discovered active cold seeps on the northwestern slope of the South China Sea: Constraints on fluid sources, formation environments, and seepage dynamics. Deep-Sea Research Part I-Oceanogr Res Pap 124:31–41

Liu C, Meng Q, He X et al (2015) Characterization of natural gas hydrate recovered from Pearl River Mouth basin in South China Sea. Mar Pet Geol 61:14–21

Liu C, Morita S, Liao YH et al (2008) High resolution seismic images of the Formosa ridge off southwestern Taiwan where "hydrothermal" chemosynthetic community is present at a cold seep site. In: Proceedings of the 6th International Conference on Gas Hydrates (ICGH 2008), Vancouver, British Columbia, July, pp 6–10. doi: https://doi.org/10.14288/1.0041106

Losekann T, Knittel K, Nadalig T et al (2007) Diversity and abundance of aerobic and anaerobic methane oxidizers at the Haakon Mosby mud volcano. Barents Sea. Appl Environ Microbiol 73(10):3348–3362

Luff R, Wallmann K (2003) Fluid flow, methane fluxes, carbonate precipitation and biogeochem-ical turnover in gas hydrate–bearing sediments at Hydrate Ridge, Cascadia Margin: numerical modeling and mass balances. Geochim Cosmochim Acta 67(18):3403–3421

Milkov AV, Etiope G (2018) Revised genetic diagrams for natural gases based on a global dataset of >20,000 samples. Org Geochem 125:109–120

Naehr TH, Birgel D, Bohrmann G et al (2009) Biogeochemical controls on authigenic carbonate formation at the Chapopote asphalt volcano. Bay of Campeche. Chem Geol 266(3–4):390–402

Natalicchio M, Peckmann J, Birgel D et al (2015) Seep deposits from northern Istria, Croatia: a frst glimpse into the Eocene seep fauna of the Tethys region. Geol Mag 152(3):444–459

Nauhaus K, Treude T, Boetius A et al (2005) Environmental regulation of the anaerobic oxidation of methane: a comparison of ANME–I and ANME–II communities. Environ Microbiol 7(1):98–106

Niemann H, Elvert M (2008) Diagnostic lipid biomarker and stable carbon isotope signatures of microbial communities mediating the anaerobic oxidation of methane with sulphate. Org Geochem 39(12):1668–1677

Niu M, Fan X, Zhuang G et al (2017) Methane–metabolizing microbial communities in sediments of the Haima cold seep area, northwest slope of the South China Sea. FEMS Microbiol Ecol 93(9): fix101

Oba M, Sakata S, Tsunogai U (2006) Polar and neutral isopranyl glycerol ether lipids as biomarkers of archaea in near–surface sediments from the Nankai Trough. Org Geochem 37(12):1643–1654

Orphan VJ, House CH, Hinrichs K-U et al (2001) Methane–consuming archaea revealed by directly coupled isotopic and phylogenetic analysis. Science 293(5529):484–487

Peckmann J, Birgel D, Kiel S (2009) Molecular fossils reveal fluid composition and flow intensity at a Cretaceous seep. Geology 37(9):847–850

Peckmann J, Thiel V, Reitner J et al (2004) A microbial mat of a large sulfur bacterium preserved in a Miocene methane–seep limestone. Geomicrobiol J 21(4):247–255

Peters KE, Moldowan JM (1993) The Biomarker Guide: Interpreting Molecular Fossils in Petroleum and Ancient Sediments. Prentice Hall, Englewood Cliffs, NJ. ISBN 0-13-086752-7

Petro C, Jochum LM, Schreiber L et al (2019) Single–cell amplifed genomes of two uncultivated members of the deltaproteobacterial SEEP–SRB1 clade, isolated from marine sediment. Mar Genom 46:66–69

Reeburgh WS (2007) Oceanic methane biogeochemistry. Chem Rev 107(2):486–513

Rossel PE, Elvert M, Ramette A et al (2011) Factors controlling the distribution of anaerobic methanotrophic communities in marine environments: evidence from intact polar membrane lipids. Geochim Cosmochim Acta 75(1):164–184

Rossel PE, Lipp JS, Fredricks HF et al (2008) Intact polar lipids of anaerobic methanotrophic archaea and associated bacteria. Org Geochem 39(8):992–999

Rush D, Osborne KA, Birgel D et al (2016) The bacteriohopanepolyol inventory of novel aerobic methane oxidizing bacteria reveals new biomarker signatures of aerobic methanotrophy in marine systems. PLoS ONE 11(11):e0165635

Sassen R, Roberts HH, Aharon P et al (1993) Chemosynthetic bacterial mats at cold hydrocarbon seeps, Gulf of Mexico continental slope. Org Geochem 20(1):77–89

Sinninghe Damsté JS, Rijpstra WIC, Schouten S et al (2004) The occurrence of hopanoids in planctomycetes: implications for the sedimentary biomarker record. Org Geochem 35(5):561–566

Suess E (2005) RV SONNE cruise report SO 177, Sino–German cooperative project, South China Sea Continental Margin: Geological methane budget and environmental effects of methane emissions and gas hydrates. IFM–GEOMAR Reports. doi: https://doi.org/10.3289/ifm-geomar_rep_4_2005

Sprott GD (1992) Structures of archaebacterial membrane lipids. J Bioenerg Biomembr 24(6):555–566

Sprott GD, Dicaire CJ, Choquet CG et al (1993) Hydroxydiether lipid structures in Methanosarcina spp. and Methanococcus voltae. Appl Environ Microbiol 59 (3):912–914

Sprott GD, Ekiel I, Dicaire C (1990) Novel, acid–labile, hydroxydiether lipid cores in methanogenic bacteria. J Bioll Chem 265(23):13735–13740

Stadnitskaia A, Bouloubassi I, Elvert M et al (2008a) Extended hydroxyarchaeol, a novel lipid biomarker for anaerobic methanotrophy in cold seepage habitats. Org Geochem 39(8):1007–1014

Stadnitskaia A, Nadezhkin D, Abbas B et al (2008b) Carbonate formation by anaerobic oxidation of methane: evidence from lipid biomarker and fossil 16S rDNA. Geochim Cosmochim Acta 72(7):1824–1836

Talbot HM, Watson DF, Murrell JC et al (2001) Analysis of intact bacteriohopanepolyols from methanotrophic bacteria by reversed–phase highperformance liquid chromatography–atmospheric pressure chemical ionization mass spectrometry. J Chromatogr A 921(2):175–185

Thiel V (2018) Methane carbon cycling in the past: Insights from hydrocarbon and lipid biomarkers. In: Wilkes H (ed) Hydrocarbons, oils and lipids: diversity, origin, chemistry and fate. Switzerland, Springer International Publishing, Cham, pp 1–30

Thiel V, Peckmann J, Richnow HH et al (2001a) Molecular signals for anaerobic methane oxidation in Black Sea seep carbonates and a microbial mat. Mar Chem 73(2):97–112

Thiel V, Peckmann J, Schmale O et al (2001b) A new straight–chain hydrocarbon biomarker associated with anaerobic methane cycling. Org Geochem 32(8):1019–1023

Wallmann K, Pinero E, Burwicz E et al (2012) The Global Inventory of Methane Hydrate in Marine Sediments: A Theoretical Approach. Energies 5(7):2449–2498

Wang X, Guan H, Qiu JW et al (2022) Macro–ecology of cold seeps in the South China Sea. Geosystems and Geoenvironment 1(3):100081

Wegener G, Krukenberg V, Riedel D et al (2015) Intercellular wiring enables electron transfer between methanotrophic archaea and bacteria. Nature 526(7574):587–590

Wegener G, Niemann H, Elvert M et al (2008) Assimilation of methane and inorganic carbon by microbial communities mediating the anaerobic oxidation of methane. Environ Microbiol 10(9):2287–2298

Wei J, Liang J, Lu J et al (2019) Characteristics and dynamics of gas hydrate systems in the northwestern South China Sea-Results of the fifth gas hydrate drilling expedition. Mar Pet Geol 110:287–298

Wei J, Wu T, Zhu L et al (2021) Mixed gas sources induced co–existence of sI and sII gas hydrates in the Qiongdongnan Basin. South China Sea. Mar Pet Geol 128:105024

Wu N, Xu C, Li A et al (2022) Oceanic carbon cycle in a symbiotic zone between hydrothermal vents and cold seeps in the Okinawa Trough. Geosystems and Geoenvironment 1(3):100059

Wu N, Zhang H, Yang S et al (2011) Gas hydrate system of Shenhu area, northern South China Sea: Geochemical results. Journal of Geological Research 2011:370298

Xu LF, Guan HX, Liu L et al (2022a) Determining the double–bond positions of monounsaturated compounds in the alcohol fraction in seep carbonate. J Chromatogr A 1672:463009

Xu L, Guan H, Su Z et al (2022b) Diagenetic fate of glycerol ethers revealed by two novel isoprenoid hydroxyphytanyl glycerol monoethers and non–isoprenoid alkyl glycerol ethers. Org Geochem 163:104344

Yang K, Chu F, Zhu Z et al (2018) Formation of methane–derived carbonates during the last glacial period on the northern slope of the South China Sea. J Asian Earth Sci 168:173–185

Ye J, Wei J, Liang J et al (2019) Complex gas hydrate system in a gas chimney, South China Sea. Mar Pet Geol 104:29–39

Yu X, Han X, Li H et al (2008) Biomarkers and carbon isotope composition of anaerobic oxidation of methane in sediments and carbonates of northeastern part of Dongsha, South China Sea. Acta Oceanol Sin 30:77–84

Zhang CL, Li YL, Ye Q et al (2003) Carbon isotope signatures of fatty acids in Geobacter metallireducens and Shewanella algae. Chem Geol 195(1–4):17–28

Zhang W, Liang J, Wei J et al (2020) Geological and geophysical features of and controls on occurrence and accumulation of gas hydrates in the frst offshore gas–hydrate production test region in the Shenhu area, Northern South China Sea. Mar Pet Geol 114:104191

Zhuang C, Chen F, Cheng S et al (2016) Light carbon isotope events of foraminifera attributed to methane release from gas hydrates on the continental slope, northeastern South China Sea. Sci. China-Earth Sci 59(10):1981–1995

Zhuang G, Xu L, Liang Q et al (2019) Biogeochemistry, microbial activity, and diversity in surface and subsurface deep–sea sediments of South China Sea. Limnol Oceanogr 64(5):2252–2270

Chapter 12
Timing of Seep Activities and Potential Driving Forces

Dong Feng

Abstract One of the foremost topics in seep research is the timing of seep activities and their potential driving forces. In the South China Sea, seep activities are primarily driven by gas hydrate dissociation–destabilization of gas hydrate leads to a release of methane. As decreases in pressure and increases in temperature promote gas hydrate dissociation, ocean warming and sea level lowstands are proposed to cause the dissociation of gas hydrate deposits and consequently induce methane seepage at the seafloor. Cross-slope investigations suggest that the bottom water pressure–temperature conditions appear to have different impacts on seeps at different water depths. It is possible that seepage in the upper continental slope is more sensitive to sea level changes than that in the middle and lower continental slopes, which are more sensitive to bottom water temperature. Scientific drilling and the application of a range of geochemical and geophysical analytical approaches are proposed to advance our understanding of the temporal evolution of seep systems in the South China Sea.

12.1 Introduction

Understanding methane seepage dynamics in the past and present and the associated drivers and trigger mechanisms can be key for predicting methane emission scenarios in future oceans impacted by climate change (Boetius and Wenzhöfer 2013; Ruppel and Kessler 2017; Yao et al. 2020). Cold seeps emit materials into the oceans and leave characteristic footprints close to the seafloor, for example, extensive buildups of methane-derived authigenic carbonate minerals and a mass occurrence of seep-specific bivalves (Suess 2020). These footprints provide an excellent archive of past seepage and associated environmental parameters (Aharon et al. 1997;

D. Feng (✉)
College of Marine Sciences, Shanghai Ocean University, Shanghai 201306, China
e-mail: dfeng@shou.edu.cn

Laboratory for Marine Mineral Resources, Qingdao National Laboratory for Marine Science and Technology, Qingdao 266061, China

D. Chen and D. Feng (eds.), *South China Sea Seeps*,
https://doi.org/10.1007/978-981-99-1494-4_12

Bohrmann et al. 1998; Teichert et al. 2003; Bayon et al. 2009a, b, 2013, 2015; Feng et al. 2010; Chen et al. 2019; Himmler et al. 2019). Constraints on the parameters that control seepage dynamics can be obtained by dating seep carbonates and bivalve shells.

The South China Sea (SCS) is a natural laboratory for studying the development and evolution of methane seepage systems (Tong et al. 2013; Han et al. 2014; Feng and Chen 2015; Liang et al. 2017; Yang et al. 2018; Chen et al. 2019; Wei et al. 2020; Deng et al. 2021; Kunath et al. 2022; Wang et al. 2022), particularly the common trends across multiple geographic areas and the connection between gas hydrate dissociation and methane seepage dynamics (Feng et al. 2018). This chapter summarizes the materials and methods that have been used to constrain the timing of seepage in the SCS (Fig. 12.1) and explores how dating results can be exploited to study melt seepage activities under glacial-interglacial climate changes.

12.2 Materials and Methods

Methane-derived authigenic carbonate represents an appealing archive of seepage dynamics since carbonate deposits are present throughout geological records (e.g., Hovland et al. 1987; Peckmann and Thiel 2004). Seep carbonates exhibit high fidelity for several other tracers, including the isotopic composition of carbon and oxygen, which can be used to constrain the source of methane-rich fluids and their formation environments and possible connections with gas hydrate dissociation (e.g., Bohrmann et al. 1998; Aloisi et al. 2000; Suess 2020). The systematics of carbonate carbon and oxygen isotope systems are discussed in Chap. 10.

Earlier studies mainly focused on seabed seep carbonates. However, recent studies have started to investigate seep carbonates from hydrate-bearing drill cores, and these can be used to determine past gas hydrate dynamics throughout the Late Quaternary (Fig. 12.2; Tong et al. 2013; Han et al. 2014; Chen et al. 2019; Wei et al. 2020; Deng et al. 2021). While seafloor samples are easy to collect and seep carbonates are dated from the seabed, sampling bias cannot be excluded. Authigenic seep carbonates from drill cores represent a continuous record of gas hydrate dynamics that can help address the sampling bias issue (Fig. 12.3).

Uranium-thorium (U-Th) dating of carbonate offers a unique chronometer to assess the timing and duration of methane seepage on relatively long timescales up to approximately 500,000 years (Himmler et al. 2019; Wang et al. 2022). Most of the data accumulated in the SCS were obtained from applying U-Th techniques to seep carbonates through both solution-based multicollector inductively coupled-plasma mass spectrometry (MC-ICP-MS; Tong et al. 2013; Han et al. 2014; Yang et al. 2018; Chen et al. 2019; Wei et al. 2020; Deng et al. 2021; Wang et al. 2022) and the recently developed in situ laser ablation MC-ICP–MS methods (Wang et al. 2022). The in situ U-series geochronological method has the potential to date seep carbonates efficiently and reliably. Researchers have directly determined the initial $[^{230}Th/^{232}Th]$ ratios of seep carbonates and yielded a value of 0.7 ± 0.1 (2 SD,

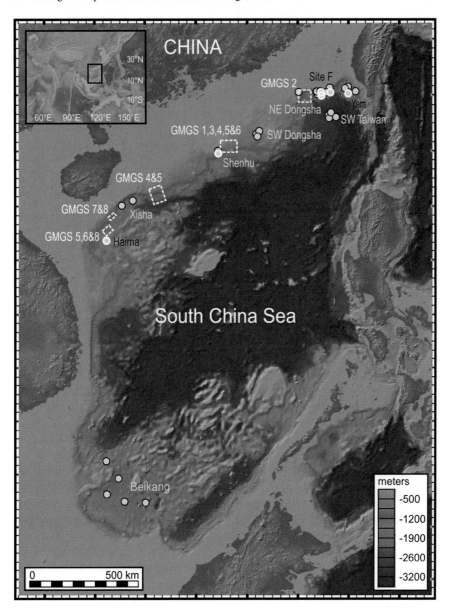

Fig. 12.1 Bathymetric image of the South China Sea showing locations of all hydrocarbon seeps that have been confirmed during the last two decades (yellow dots) (after Feng et al. 2018; Tseng et al. 2023). The locations of carbonate samples with dating results are illustrated (yellow dots with white circles). The white boxes are locations of gas hydrate drilling expeditions (GMGS 1–6) (after Wei et al. 2020)

Fig. 12.2 Typical morphologies of seep carbonate samples for dating. **a** A cut and polished surface of a carbonate sample from the seafloor at aite F. The sample was obtained during the *ROPOS* dive 2045 in 2018. The sample includes abundant cemented shell debris (arrows). The presence of shell fragments of chemosynthetic bivalves in the carbonates suggests that carbonate precipitation occurred close to the seafloor. **b** A cut and polished surface of a carbonate sample from the drilling site GMGS2-08F (at 61.8 m below the seafloor) in the dongsha area (from Chen et al. 2019). A microcrystalline matrix (yellow circle) usually represents initial seepage-related precipitation, whereas cavity-filling cement (red circle) typically reflects a later stage (Feng et al. 2010)

n = 12), which should be applied to future seep carbonate U-Th dating (Wang et al. 2022).

Microcrystalline matrix usually represents initial seepage-related precipitation. However, precise dating of these matrix is difficult because of the high initial ^{230}Th content due to variable amounts of detrital material and organic matter. In contrast, diagenetic cements in seep carbonates are often pure aragonites that have small uncertainties in their U-Th ages because they contain relatively less contamination. These diagenetic cements tend to reflect the last phase of carbonate precipitation (Fig. 12.2; Feng et al. 2010). In situ laser ablation has the potential to select appropriate regions to date within complex structures of seep carbonates and can obtain measurements efficiently (Table 12.1; Wang et al. 2022). Nevertheless, efforts directed toward applying U-Th dating to seep-specific bivalves proved to be infeasible because fossil mollusks often exhibit uranium contents that are many orders of magnitude greater than those in living shells, indicating uranium uptake after death and hence U-Th open-system behavior (Ayling et al. 2017). Due to the incorporation of methane-derived fossil carbon upon carbonate precipitation (Paull et al. 1989; Aharon et al. 1997), the utility of radiocarbon dating is limited to shell materials that are cemented in carbonates (Fig. 12.2).

Dating of authigenic carbonates preserved in sedimentary records can provide a chronology of past methane release events. This is usually achieved by the direct dating of seep carbonates using the uranium-series and by radiocarbon dating of shell materials cemented in seep carbonates. On the other hand, the timing of seeps can also be determined through dating of the host sediments in ancient sedimentary environments where authigenic carbonate is not present. However, one must assume that a shallow subsurface precipitation environment is present close to the sediment-water interface under extremely high methane flux conditions (e.g., Yao et al. 2020).

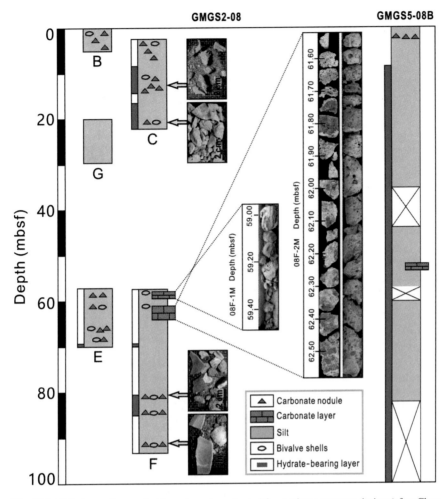

Fig. 12.3 Lithology of cores that have been investigated for methane seepage timing (after Chen et al. 2019; Wei et al. 2020). The carbonate thickness is exaggerated for greater visibility

Table 12.1 Comparison of methods for U-Th dating of seep carbonates

Method	Column chemistry	Isotope spike	^{238}U analyzed	$^{230}Th/^{238}U$ precision	Time/batch	Throughput/batch
α-counting	Needed	Needed	>1 μg	A few %	A week	A few
TIMS	Needed	Needed	A few ng	<1%	A week	Typically 10
MC-ICPMS	Needed	Needed	A few ng	<0.5%	A week	Typically 10
LA-MC-ICPMS	None	None	<1 ng	A few %	A day	Typically 50

TIMS, thermal ionization mass spectrometry; LA-MC-ICPMS, laser ablation-multicollector inductively coupled plasma-mass spectrometry. Table courtesy of Tianyu Chen at Nanjing University

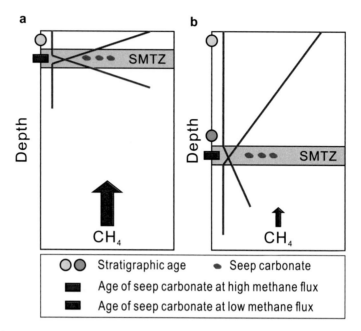

Fig. 12.4 Image illustrating the relationship between the stratigraphic age and age of seep carbonates at high (left, **a**) and low methane fluxes (right, **b**). Under extremely high methane flux conditions, the age of precipitated seep carbonates (blue rectangle) is close to that of the stratigraphic chronosequence (yellow dot in the left image). Under extremely low methane flux conditions, the age of precipitated seep carbonates (red rectangle) is close to that of the stratigraphic chronosequence of the subsurface sediments (yellow dot in the right image) but much younger than that of the stratigraphic chronosequence of the neighboring sediments (green dot)

The depth of the SMTZ may be independent from the stratigraphic chronosequence (Fig. 12.4). Dating the surrounding sediment is not a method for dating authigenic carbonates but a method to establish their maximum age (Yao et al. 2020; Guan et al. 2022). Such a method has been used to reveal methane seep dynamics since the last glacial maximum in the Shenhu area of the SCS (Zhang et al. 2022a, b).

12.3 Methane Seepage and Hydrate Dissociation

From passive margins of the SCS, Tong et al. (2013), Han et al. (2014), and Yang et al. (2018) were the first to report chronology constraints on the timing of seepages across the continental slope of the northern SCS. These studies not only confirmed that the formation of seep carbonates in the northern SCS is inextricably linked to gas hydrates but also provided evidence for extensive seep activity during sea level lowstands, hence suggesting that hydrostatic pressure was the most important factor controlling gas hydrate stability. Recently, Wang et al. (2022) investigated the

U-Th geochemistry and geochronology of 31 cold-seep carbonates recovered from the northern SCS by laser ablation in situ mass spectrometric analysis. The new age data suggest that cold seeps were likely continuously active since at least ~72 ka in the northern SCS, but different sites have contrasting durations (Fig. 12.5). The contrasting age spectra of seep carbonates from the upper continental slope of the northern SCS indicate that bottom water pressure–temperature conditions might yield contrasting stability conditions for methane hydrates at different depths in the same geological setting. Accurate information on the bottom water pressure–temperature history would thus be necessary to reliably quantify past marine cold seepage flux from the continental margins. While seep carbonate age data provide first-order constraints on past methane leakage activities, more geochemical proxy work is needed to reconstruct the magnitude of methane release in the geological past.

In a recent U-Th investigation of drilled seep carbonates from the Dongsha area, Chen et al. (2019) suggested that increased bottom-water temperature during Marine Isotope Stage (MIS) 5e caused methane hydrate destabilization at the SCS margin. This finding provided the first direct evidence that methane seepage possibly intensified during full sea level highstands. Additional investigation of deeply buried seep carbonates is required to assess whether this hypothesis holds true for earlier interglacial periods and to further test the relationships that link both methane hydrate stability and seepage intensity to Late Quaternary climate change. The assumption of intensified methane seepage was further confirmed by Deng et al. (2021) using the same sets of samples. Wei et al. (2020) analyzed three seep carbonate samples (3, 52.1 and 53.6 mbsf) recovered from site W08B in the Qiongdongnan Basin, China, and suggested that the formation of the carbonates was induced by the dissociation of gas hydrates. The dissociations took place at 12.2 ka BP and 131–136 ka BP.

12.4 Timescales of Methane Seepage

Knowledge of the timescales of gas hydrate dissociation and subsequent methane release is critical for understanding the impact of marine gas hydrates on the ocean-atmosphere system (Crémière et al. 2016; Ruppel and Kessler 2017; Ruppel and Waite 2020). However, assessments of the timescales over which gas hydrate systems respond to the processes that drive changes in pressure and temperature in the subsurface are limited. Site F and the Haima active methane seeps have been extensively investigated to study seep-impacted sediments, pore fluids, authigenic carbonates, and seep-dwelling faunas. The use of these materials or a combination thereof will ultimately allow us to constrain seepage intensities at different time scales (Feng et al. 2018). Each of these materials has its own validity in revealing the characteristics and mechanisms of seepage (Feng et al. 2018). For example, the geochemical data obtained from the solid fraction of sediments and from authigenic carbonates provide time-averaged information on biogeochemical processes on a timescale of

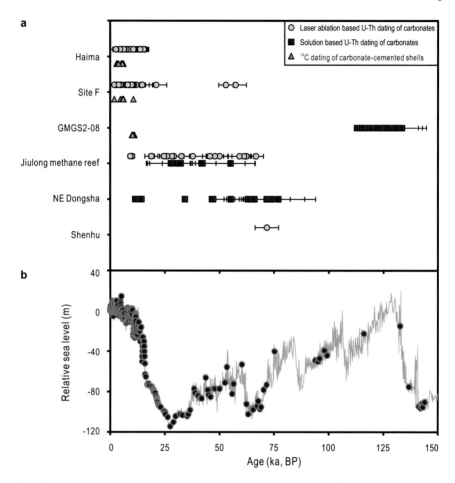

Fig. 12.5 a The seepage duration of cold seeps in the northern South China Sea includes four seafloor cold seep sites (Shenhu, Jiulong methane reef, Haima and Site F) and sediment core GMGS2-08 (Chen et al. 2019; Deng et al. 2021). The yellow circles represent the U-Th ages obtained by in situ U-Th dating by Wang et al. (2022); previously reported data with high $[^{232}Th/^{238}U]$ or $\delta^{234}U_{ini}$ values that deviate from the seawater value are not included (Wang et al. 2022). The red rectangles represent the U-Th age data of the seep cements from the Jiulong methane reef (Tong et al. 2013; Han et al. 2014; Wang et al. 2022) and GMGS2-08 cores (Chen et al. 2019; Deng et al. 2021). The green triangles represent ^{14}C dating on shells cemented in the cold seep carbonate samples from Site F (Feng and Chen 2015; Wang et al. 2022), Haima (Liang et al. 2017; Wang et al. 2022) and GMGS2-08 cores (Deng et al. 2021). **b** Previously reported timing of seep carbonate formation worldwide versus global sea-level changes (Chen et al. 2019 and references therein). Sea level curve from Rohling et al. (2009)

years to centuries (Kiel et al. 2014; Oppo et al. 2020). Sediment pore waters and seep-dwelling fauna, on the other hand, provide information on much shorter timescales, spanning from days to months (cf., Valentine et al. 2005). A relevant case study was conducted by Luo et al. (2015) to estimate the time of pockmark formation in the SW Xisha Uplift of the SCS using reaction-transport modeling. They found that the pockmarks formed at least 39 kyr B.P. and that the termination of fluid seepage may be ascribed to gas hydrate stabilization or to complete depletion after the sea levels reached a relative highstand. The proposed method could be a significant tool for illuminating the temporal evolution of fluid seepage in pockmarks, especially if reliable age control is lacking, because most pockmarks on continental margins are presently dormant. Appropriate modeling tools are critical for establishing links between these multiproxy approaches to improve our understanding of seep systems.

12.5 Summary and Perspectives

This chapter summarizes findings from the first decade of seep geochronology in the SCS, which have revealed a number of global, regional, local, and geological processes that can render different seep dynamics with distinct timings. Since the seep activities of the SCS depend largely on the dissociation of locally abundant gas hydrates, age determination of seeps can be used to glean unique insights into the dynamics of gas hydrate systems, especially those located in the subsurface of marine sediments (Fig. 12.6). Despite recent progress, a number of key uncertainties remain in determining the timing of seeps, and these need to be resolved.

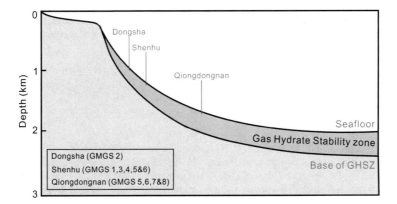

Fig. 12.6 Schematic diagram of the gas hydrate stability zone (GHSZ) extension in marine sediments of the South China Sea. The eight Guangzhou Marine Geological Survey (GMGS) gas hydrate drilling expeditions in the Dongsha, Shenhu, and Qiongdongnan areas are illustrated

Attaining a better understanding of the evolution of seepage through time and its links to gas hydrate reservoirs, seep ecosystems, and the amount of hydrocarbon escaping seabeds remains challenging (Skarke et al. 2014). Recent drilling campaigns and coring programs have contributed greatly to our understanding of gas hydrate dynamics in the SCS and in marine environments worldwide. The time is now ripe to start a dedicated program that targets these drilled cores and uses synergistic modeling–analytical approaches to shed new light on the timing of periods of enhanced seepage and its relationship to the dynamics affecting hydrate reservoirs.

The records of seepage indicators are usually incomplete; for example, many bivalves and carbonate nodules either on the seabed or in the drilling cores have not been dated or included in discussions (Fig. 12.4). In addition, sample-based case studies are mainly focus on longer timescales. For the short term, determining how much methane is being released to the oceans on regional to basin-wide scales remains a central challenge. Observing seafloor methane seeps in various environments is therefore critical.

Hydrocarbon seeps are uniquely prone to perturbation resulting from global change (e.g., Kennett et al. 2000; Ketzer et al. 2020; Kim and Zhang 2022; Weldeab et al. 2022). Focused paleoceanographic studies should also constrain bottom-water temperature changes on the continental slope of the SCS during glacial-interglacial cycles. In addition to continued sample analyses, more sophisticated numerical simulations are required to demonstrate the assumed gas hydrate dissociation effects before researchers hypothesize that seeps could have played a significant role as a climate feedback mechanism during the known large temperature and sea level changes of the Quaternary glacial-interglacial cycles (Li et al. 2023). The next decade of seep chronology is poised to offer many novel insights into the impact of seepage in response to typical tectonic, sedimentary, or climate triggers.

Acknowledgments Maoyu Wang, Xudong Wang, and Zihan Zheng are acknowledged for preparation of Figs. 12.5 and 12.6. Yu Hu and Maoyu Wang are thanked for constructive comments, which have greatly improved the quality of the chapter.

References

Aharon P, Schwarcz HP, Roberts HH (1997) Radiometric dating of submarine hydrocarbon seeps in the gulf of mexico. Geol Soc Am Bull 109(5):568–579

Aloisi G, Pierre C, Rouchy JM et al (2000) Methane-related authigenic carbonates of eastern mediterranean sea mud volcanoes and their possible relation to gas hydrate destabilisation. Earth Planet Sci Lett 184(1):321–338

Ayling BF, Eggins S, McCulloch MT et al (2017) Uranium uptake history, open-system behaviour and uranium-series ages of fossil tridacna gigas from huon peninsula, papua new guinea. Geochim Cosmochim Acta 213:475–501

Bayon G, Dupre' S, Ponzevera E et al (2013) Formation of carbonate chimneys in the mediterranean sea linked to deepwater oxygen depletion. Nat Geosci 6:755–760

Bayon G, Henderson GM, Bohn M (2009) U-Th stratigraphy of a cold seep carbonate crust. Chem Geol 260(1–2):47–56

Bayon G, Henderson GM, Etoubleau J et al (2015) U-Th isotope constraints on gas hydrate and pockmark dynamics at the Niger delta margin. Mar Geol 370:87–98

Bayon G, Loncke L, Dupré S et al (2009) Multidisciplinary investigation of fluid seepage on an unstable margin: the case of the central nile deep sea fan. Mar Geol 261(1–4):92–104

Boetius A, Wenzhöfer F (2013) Seafloor oxygen consumption fuelled by methane from cold seeps. Nat Geosci 6(9):725–734

Bohrmann G, Greinert J, Suess E et al (1998) Authigenic carbonates from the cascadia subduction zone and their relation to gas hydrate stability. Geology 26(7):647–650

Chen F, Wang X, Li N et al (2019) Gas hydrate dissociation during sea-level highstand inferred from U/Th dating of seep carbonate from the south china sea. Geophys Res Lett 46(23):13928–13938

Crémière A, Lepland A, Chand S et al (2016) Timescales of methane seepage on the norwegian margin following collapse of the scandinavian ice sheet. Nat Commun 7:11509

Deng Y, Chen F, Guo Q et al (2021) Possible links between methane seepages and glacial-interglacial transitions in the South China Sea. Geophys Res Lett 48(8): e2020GL091429

Feng D, Chen D (2015) Authigenic carbonates from an active cold seep of the northern south china sea: new insights into fluid sources and past seepage activity. Deep-Sea Res Part II-Top Stud Oceanogr 122:74–83

Feng D, Roberts HH, Cheng H et al (2010) U/Th dating of coldseep carbonates: an initial comparison. Deep-Sea Res Part II-Top Stud Oceanogr 57(21–23):2055–2060

Feng D, Qiu JW, Hu Y et al (2018) Cold seep systems in the south china sea: an overview. J Asian Earth Sci 168:3–16

Guan HX, Liu L, Hu Y et al (2022) Rising bottom-water temperatures induced methane release during the middle holocene in the okinawa trough, East China Sea. Chem Geol 590:120707

Han X, Suess E, Liebetrau V et al (2014) Past methane release events and environmental conditions at the upper continental slope of the south china sea: constraints by seep carbonates. Int J Earth Sci 103:1873–1887

Himmler T, Sahy D, Martma T et al (2019) A 160,000-year-old history of tectonically controlled methane seepage in the Arctic. Sci Adv 5(8):eaaw1450

Hovland M, Talbot MR, Qvale H et al (1987) Methane-related carbonate cements in pockmarks of the north sea. J Sediment Res 57(5):881–892

Kennett JP, Cannariato KG, Hendy IL et al (2000) Carbon isotopic evidence for methane hydrate instability during quaternary interstadials. Science 288(5463):128–133

Ketzer M, Praeg D, Rodrigues LF et al (2020) Gas hydrate dissociation linked to contemporary ocean warming in the southern hemisphere. Nat Commun 11(1):3788

Kiel S, Hansen C, Nitzsche KN et al (2014) Using 87Sr/86Sr ratios to date fossil methane seep deposits: methodological requirements and an example from the great valley group, California. J Geol 122(4):353–366

Kim B, Zhang YG (2022) Methane hydrate dissociation across the oligocene-miocene boundary. Nat Geosci 15(3):203–209

Kunath P, Crutchley G, Chi WC et al (2022) Episodic venting of a submarine gas seep on geological time scales: formosa ridge, northern south china sea. J Geophys Res: Solid Earth 127(9):e2022JB024668

Liang Q, Hu Y, Feng D et al (2017) Authigenic carbonates from newly discovered active cold seeps on the northwestern slope of the south china sea: constraints on fluid sources, formation environments, and seepage dynamics. Deep-Sea Res Part I-Oceanogr Res Pap 124:31–41

Luo M, Dale AW, Wallmann K et al (2015) Estimating the time of pockmark formation in the SW xisha uplift (South China Sea) using reaction-transport modeling. Mar Geol 364:21–31

Oppo D, De Siena L, Kemp DB (2020) A record of seafloor methane seepage across the last 150 million years. Sci Rep 10(1):2562

Paull CK, Martens CS, Chanton JP et al (1989) Old carbon in living organisms and young CaCO3 cements from abyssal brine seeps. Nature 342:166–168

Peckmann J, Thiel V (2004) Carbon cycling at ancient methane-seeps. Chem Geol 205(3–4):443–467

Rohling EJ, Grant K, Bolshaw M et al (2009) Antarctic temperature and global sea level closely coupled over the past five glacial cycles. Nat Geosci 2(7):500–504

Ruppel CD, Waite WF (2020) Timescales and processes of methane hydrate formation and break-down, with application to geologic systems. J Geophys Res-Solid Earth 125(8):e2018JB016459

Ruppel CD, Kessler JD (2017) The interaction of climate change and methane hydrates. Rev Geophys 55(1):126–168

Skarke A, Ruppel C, Kodis M et al (2014) Widespread methane leakage from the sea floor on the northern US Atlantic margin. Nat Geosci 7(9):657–661

Suess E (2020) Marine cold seeps: background and recent advances. In: Wilkes (ed) Hydrocarbons, Oils and Lipids: Diversity, Origin, Chemistry and Fate. Springer International Publishing, Cham, pp 747–767

Teichert BMA, Eisenhauer A, Bohrmann G et al (2003) U/Th systematics and ages of authigenic carbonates from hydrate ridge, cascadia margin: recorders of fluid flow variations. Geochim Cosmochim Acta 67(20):3845–3857

Tong H, Feng D, Cheng H et al (2013) Authigenic carbonates from seeps on the northern continental slope of the South China Sea: New insights into fluid sources and geochronology. Mar Pet Geol 43:260–271

Tseng Y, Römer M, Lin S et al (2023) Yam seep at four-way closure ridge–a prominent active gas seep system at the accretionary wedge SW offshore Taiwan. Int J Earth Sci. https://doi.org/10.1007/s00531-022-02280-4

Valentine DL, Kastner M, Wardlaw GD et al (2005) Biogeochemical investigations of marine methane seeps, hydrate ridge, Oregon. J Geophys Res 110(G2):G02005

Wang M, Chen T, Feng D et al (2022) Uranium-thorium isotope systematics of cold-seep carbonate and their constraints on geological methane leakage activities. Geochim Cosmochim Acta 320:105–121

Wei J, Wu T, Zhang W et al (2020) Deeply buried authigenic carbonates in the qiongdongnan basin, south china sea: implications for ancient cold seep activities. Minerals 10(12):1135

Weldeab S, Schneider RR, Yu J et al (2022) Evidence for massive methane hydrate destabilization during the penultimate interglacial warming. Proc Natl Acad Sci U S A 119(35):e2201871119

Yang K, Chu F, Zhu Z et al (2018) Formation of methane-derived carbonates during the last glacial period on the northern slope of the South China Sea. J Asian Earth Sci 168:173–185

Yao H, Niemann H, Panieri G (2020) Multi-proxy approach to unravel methane emission history of an Arctic cold seep. Quat Sci Rev 244:106490

Zhang Q, Wu D, Jin G et al (2022a) Novel use of unique minerals to reveal an intensified methane seep during the last glacial period in the South China Sea. Mar Geol 452:106901

Zhang Q, Wu D, Jin G et al (2022b) Methane seep in the shenhu area of the South China sea using geochemical and mineralogical features. Mar Pet Geol 144:105829

Li N, Wang X, Feng J et al. (2023) Intermediate water warming caused methane hydrate instability in South China Sea during past interglacials: GSA Bulletin. https://doi.org/10.1130/B36859.1

Chapter 13
Cold Seepage in the Southern South China Sea

Niu Li and Junxi Feng

Abstract Extensive submarine cold seep areas, i.e., the Beikang Basin and the Nansha Trough, were discovered on the southern continental slope of the South China Sea. Bottom-simulating reflections are widespread in these areas and show a close relationship to the cold seep system. High-resolution 2-D seismic data and multibeam bathymetry data have confirmed the existence of deep-routed conduits—mud volcanoes, diapirs, and gas chimneys. The geochemical characteristics of seep carbonates and headspace gas indicate that the fluid was mainly sourced from biogenic gas, with contributions from deep-rooted thermogenic gases. Additionally, negative pore water chloride anomalies and positive $\delta^{18}O$ values ($3.7‰ < \delta^{18}O < 5.0‰$) of the seep carbonates provided indicators of hydrate water addition during carbonate precipitation. The negative $\delta^{13}C$ excursion of planktonic foraminifera from the Nansha Trough indicated two methane release events, which occurred approximately 29–32 ka and 38–42 ka before present, and the driving mechanism for methane seepage in this area is possibly related to overpressure from the large sediment accumulation that occurred during sea level lowstands.

N. Li (✉)
Key Laboratory of Ocean and Marginal Sea Geology, South China Sea Institute of Oceanology, Innovation Academy of South China Sea Ecology and Environmental Engineering, Chinese Academy of Sciences, Guangzhou 510301, China
e-mail: liniu@scsio.ac.cn

J. Feng
MLR Key Laboratory of Marine Mineral Resources, Guangzhou Marine Geological Survey, Guangzhou 510075, China
e-mail: fengjx123@163.com

National Engineering Research Center of Gas Hydrate Exploration and Development, Guangzhou 510075, China

© The Author(s) 2023
D. Chen and D. Feng (eds.), *South China Sea Seeps*,
https://doi.org/10.1007/978-981-99-1494-4_13

13.1 Introduction

Cold seepage systems are widely developed in the South China Sea (SCS), and more than 40 seep sites have been found along the northern slope of the SCS (Feng et al. 2018a). Various seep-related samples, including seep-derived authigenic carbonates, seep-impacted sediments, gas samples, and dead and living seep fauna, have been collected over the past decade. Ongoing studies provide information about the geomorphological characteristics and associated subsurface migration pathways of fluids, about biogeochemical processes, fluid origins and methane seepage histories, and about the macroecology of chemosynthesis-based ecosystems in the SCS (Feng et al. 2018a).

To date, using high-resolution 2-D seismic data, bottom-simulating reflections (BSRs) indicating the presence of gas hydrate and free gas had been identified in the offshore south Vietnam (Lee and Watkins 1998), the northwestern (NW) Sabah/Borneo (Behein et al. 2003; Gee et al. 2007; Laird and Morley 2011; Paganoni et al. 2016), the Nansha Trough (Berner and Faber 1990; Chen et al. 2007), the Nanweixi Basin (Deng et al. 2004), and the Beikang Basin (He et al. 2018; Huang et al. 2022). Various fluid seepage conduits, including mud volcanoes, gas chimneys, pockmark, diapirs, fluid-escape pipes, and paleo-uplift associated faults had also been identified by using 2D and 3D seismic analyses and multibeam bathymetry in some areas of the southern SCS (Traynor and Sladen 1997; Lee and Watkins 1998; Gee et al. 2007; Laird and Morley 2011; Paganoni et al. 2018; Yan et al. 2020; Zhang et al. 2020; Huang et al. 2022). Methane-rich fluid plumes and fluid seeps were found in the headwall region of a giant landslide offshore NW Borneo, indicating the presence of active fluid venting, which may provide a mechanism for weakening and triggering slope failure (Rehder and Suess 2001; Gee et al. 2007). The sediment pore water chemical composition in the Nansha Trough and Beikang Basin has been used to calculate the depth of the sulfate methane transition zone (SMTZ) and estimate the rate of anaerobic methane oxidation, revealing the occurrence of shallow hydrocarbon gas (Berner and Faber 1990; Feng et al. 2018b, 2021; Huang et al. 2022). Authigenic carbonates derived from oil/gas seepage were discovered on the seafloor in the offshore southern Vietnam (Traynor and Sladen 1997; Wetzel 2013), offshore NW Borneo (Warren et al. 2010) and the Beikang Basin (Huang et al. 2022; Zhang et al. 2023). Fluid geochemical compositions and their sources are acquired according to the headspace gas and seep carbonates (Huang et al. 2022). Sediment geochemistry indicated two methane release events in the southern SCS (Li et al. 2018, 2021). The history of seep activity in the Nansha Trough has also been reconstructed by using the stable carbon isotope characteristics of planktonic foraminifera (Zhou et al. 2020). In recent years, new discoveries and studies on seep activities in the southern SCS have been performed by Chinese teams in two areas with seep activities, namely the Beikang Basin and the Nansha Trough (Fig. 13.1).

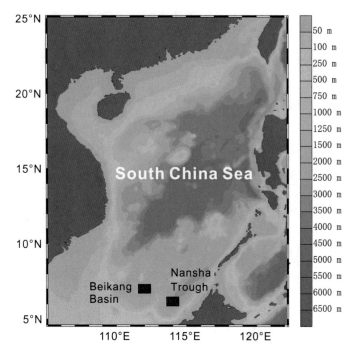

Fig. 13.1 Location of the two seep sites in the southern South China Sea. The map was generated with ocean data view

13.2 Geological Setting

The southern SCS is located in a region where three tectonic stresses (tensile, shear, and compression) are superimposed (Hutchison 2004). The study area includes the Beikang Basin and the Nansha Trough (Fig. 13.1). The Beikang Basin, which has a water depth of 1200 to 2000 m, is a Cenozoic fracture basin with sedimentary thickness ranging from 1000 to 12,000 m, and many fault terraces, mud diapirs, and fold-thrust belts have developed (Fig. 13.2; Liu et al. 2011). The tectonic activity, block migration, and South China Sea expansion influence basin development and sedimentary evolution, which provide good geological conditions for oil and gas accumulation. At present, several oil and gas fields have been found in the Beikang Basin (Liu 2005). In addition, seafloor BSRs have been identified in this area, indicating the presence of gas hydrates in unconsolidated sediments (He et al. 2018). Simultaneously, observed anomalies of mercury and natural aluminum in sediments are considered to be related to cold seepage (Chen et al. 2010; Liu et al. 2011).

The Nansha Trough is the largest trough developed along the SW–NE rift in the southern SCS, and its sedimentary thickness reaches 2000 m. The tectonic activity since the Early Miocene induced the formation of hydrocarbon gases in source rocks,

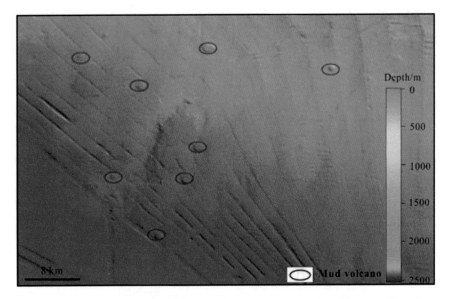

Fig. 13.2 Multibeam image showing the locations and topography of mud volcanoes in Beikang Basin (from Huang et al. 2022). Reprinted from Marine and Petroleum Geology, 139, Huang et al. (2022) Geological, geophysical, and geochemical characteristics of deep-routed fluid seepage and its indication of gas hydrate occurrence in the Beikang Basin, southern South China Sea, 105,610, Copyright (2022), with permission from Elsevier

and the gases migrated through permeable migration channels, eventually accumulating abundant oil and gas resources (Ingram et al. 2004). In addition, BSR was found in this area, suggesting gas hydrate accumulation (Berner and Faber 1990).

13.3 Geochemical Constraints on Fluid Sources and Biogeochemical Processes

The depth of the SMTZ is an effective indicator of methane flux (Borowski et al. 1996). According to piston cores from the Beikang Basin, marine pore water depth profiles of methane and sulfate indicate that the SMTZ is located at approximately 3.3–6.6 m (Fig. 13.3; Feng et al. 2018b, 2021; Huang et al. 2022). The methane oxidation rate calculated by numerical simulation ranged from 27.5 mmol m^{-2} yr^{-1} to 43.1 mmol m^{-2} yr^{-1}, and almost all the methane sourced from subsurface sediments was depleted within the SMTZ (Feng et al. 2018b), which indicated diffusive methane seepage.

Seep carbonate is formed as a result of increased pore water alkalinity due to the sulfate-driven anaerobic oxidation of methane (AOM), and the mineralogy and

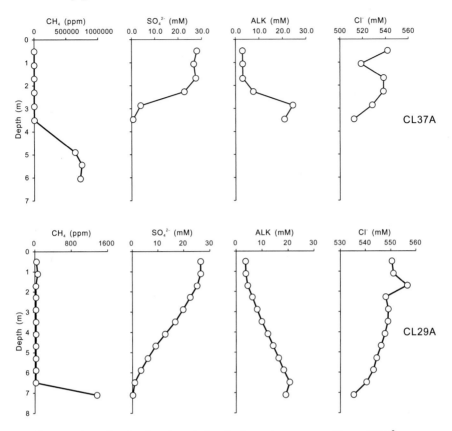

Fig. 13.3 Depth profiles showing the variations in the methane content, Cl⁻ and SO₄²⁻ concentrations, and alkalinity (ALK) of the headspace gas and pore water from the Beikang Basin (Modified from Huang et al. 2022). Reprinted from Marine and Petroleum Geology, 139, Huang et al. (2022) Geological, geophysical, and geochemical characteristics of deep-routed fluid seepage and its indication of gas hydrate occurrence in the Beikang Basin, southern South China Sea, 105,610, Copyright (2022), with permission from Elsevier

stable isotopic signature of seep carbonates can provide information on fluid sources and formation conditions (Peckmann and Thiel 2004). The negative $\delta^{13}C$ values ($-58.7‰$ to $-50.8‰$) of seep carbonates from the piston core of the Beikang Basin suggest that the dissolved inorganic carbon is mainly derived from methane oxidation and possibly from biogenic methane (Huang et al. 2022; Zhang et al. 2023). This conclusion is also supported by the headspace gas data. The plot of $\delta^{13}C_1$ versus C_1/C_{2+} suggested that the methane mostly originated from biogenic gas, with some deep-sourced thermogenic gas (Fig. 13.4; Huang et al. 2022). The oxygen isotopic composition of seep carbonate is dependent on the fluid source and temperature of carbonate formation (Bohrmann et al. 1998). The positive $\delta^{18}O$ values (3.7 to 5.0‰) of the seep carbonates from the Beikang Basin show disequilibrium with present-day bottom water temperatures, which suggests an influence of ^{18}O-enriched fluids

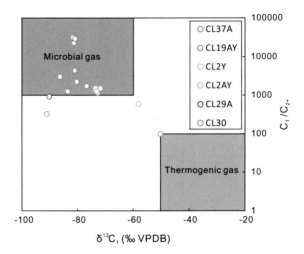

Fig. 13.4 Plot of $\delta^{13}C_1$ versus C_1/C_{2+} for the headspace gas from piston cores in the Beikang Basin (Modified from Huang et al. 2022). Reprinted from Marine and Petroleum Geology, 139, Huang et al. (2022) Geological, geophysical, and geochemical characteristics of deep-routed fluid seepage and its indication of gas hydrate occurrence in the Beikang Basin, southern South China Sea, 105,610, Copyright (2022), with permission from Elsevier

on carbonate formation (Huang et al. 2022; Zhang et al. 2023). Given the negative pore water chloride anomalies in the same core (Fig. 13.3), it is likely that these ^{18}O-enriched fluids mostly originated from gas hydrate dissociation (Huang et al. 2022).

13.4 Origin and History of Methane Seepage

AOM increases pore-fluid alkalinity and favors the precipitation of authigenic seep carbonates with low $\delta^{13}C$ values. Therefore, markedly low $\delta^{13}C$ values of total inorganic carbon (TIC) in bulk sediments could serve as significant indicators of sulfate-driven AOM (Peckmann and Thiel 2004). In addition, intensive and long-term sulfate-driven AOM would result in a dissolved sulfate limitation; thus, iron sulfide minerals are typically more abundant within or near the SMTZ than in other zones and are enriched in ^{34}S (Jørgensen et al. 2004).

Based on the decreased TOC/TS ratios, the positive sulfur isotopic value of chromium reducible sulfur (CRS), and the negative carbon isotopic value of TIC in the sediments, two methane release events were identified both in the Beikang Basin (Fig. 13.5) and the Nansha Trough (Fig. 13.6). Numerous studies have shown that foraminifera are one of the best carriers for recording seep activity. The negative excursion of the carbon isotopes of foraminiferal shells can be used to identify seep activity in geological history (Kennett et al. 2000). In a study of the sediments

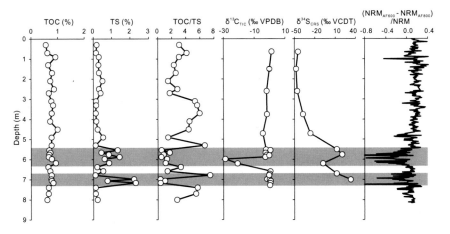

Fig. 13.5 Depth profiles of elemental, isotopic, and magnetism data in core 58S from the Beikang Basin (Modified from Li et al. 2018). The shaded areas indicate the sections influenced by the anaerobic oxidation of methane as characterized by high TS/TOC ratios. Reprinted from Journal of Asian Earth Sciences, 168, Li et al. (2018) Paleo-cold seep activity in the southern South China Sea: Evidence from the geochemical and geophysical records of sediments, 106–111, Copyright (2018), with permission from Elsevier

in the Nansha Trough in the southern SCS, it was found that the carbon isotopes of planktonic foraminifera in the sediments indicated two methane release events, roughly from 29–32 ka and 38–42 ka (Fig. 13.6). Since the seep carbonates also developed in a layer containing negative carbon isotopes of foraminifera, the lowest carbon isotope value reached $-27.4‰$. Therefore, in foraminiferal tests, the negative carbon isotopes of foraminiferal shells mainly result from the cement consisting of authigenic carbonates (Zhou et al. 2020).

Multiple driving mechanisms have been proposed to explain methane emissions from cold seepage, including bottom water warming, mass wasting processes, ice sheet dynamics, seismic activity, and low sea levels (Ruppel and Kessler 2017). For the Nansha Trough, from 29–32 ka before present (BP) and 38–42 ka BP, the site has been inside the methane hydrate stability field. Gas hydrates can reduce sediment permeability and cause overpressure build-up at the base of the gas hydrate stability zone. The resultant hydrofracturing forms fluid seepage conduits for overpressured fluids to migrate upward (Elger et al. 2018). Therefore, the driving mechanism for methane seepage in this area is possibly not related to gas hydrate dissociation due to thermodynamic instability outside the gas hydrate stability zone and is more likely due to the release of overpressured pore fluids due to sediment loading.

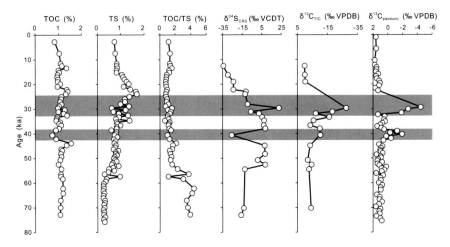

Fig. 13.6 Depth profiles of elemental and isotopic data in core 2PC from the Nansha Trough (Modified from Gao et al. 2019; Zhou et al. 2020; Li et al. 2021). The shaded areas indicate the two carbon isotopic excursions of planktonic foraminifera. Reprinted from Ore Geology Reviews, 129, Li et al. (2021) Persistent oxygen depletion of bottom waters caused by methane seepage: Evidence from the South China Sea, 103,949, Copyright (2021), with permission from Elsevier

13.5 Summary

The development of methane seepage in the Nansha Trough and Beikang Basin in the southern South China Sea is due to the abundant oil and gas resources and extensive deep-routed fluid seepage conduits. The fluid is mainly sourced from biogenic gas, with contributions from deep-sourced thermogenic gas and water released by hydrate dissociation. Sediment pore water data in the southern SCS show that cold seeps may be active, but no definitive evidence has been put forth. Their presence could be confirmed in the future by submersible diving field investigations. Two methane release events that occurred at approximately 29–32 and 38–42 ka BP were identified in the Nansha Trough. It should be emphasized here that the age of a foraminifer is only a rough constraint for the timing of cold seep activities. The dating of seep carbonates in the future could allow the estimation of the chronology of seepage. Compared with the northern SCS, seeps in the southern SCS have been investigated less. Seafloor observations, sampling, and further work are therefore critical.

Acknowledgements We are grateful to the Guangzhou Marine Geological Survey (GMGS) for the geochemical data. Sponsorship is by the National Natural Science Foundation of China (41976061, 42202174), the Major Program of Guangdong Basic and Applied Research (Grant: 2019B030302004), and the Guangdong Basic and Applied Basic Research Foundation (2019A1515110306, 2022A1515011822).

References

Behain D, Fertig J, Meyer H et al (2003) Properties of a gay hydrate province on a subduction-collision related margin off Sabah, NW Borneo (POPSCOMS). In EGS-AGU-EUG Joint Assembly (p. 10008)

Berner U, Faber E (1990) Hydrocarbon gases in surface sediments of the South China Sea. In: Ocean C (ed) Marine Geology and Geophysics of the South China Sea, Jin X L. Press, Beijing, China, pp 199–211

Bohrmann G, Greinert J, Suess E et al (1998) Authigenic carbonates from the cascadia subduction zone and their relation to gas hydrate stability. Geology 26(7):647–650

Borowski WS, Paull CK, Ussler W III (1996) Marine pore water sulfate profiles indicate in situ methane flux from underlying gas hydrate. Geology 24(7):655–658

Chen Z, Yan W, Huang C et al (2007) Geological settings and indicators of potential gas hydrates in the nansha trough area. South China Sea. Front Earth Sci 14(6):299–308 (in Chinese with English abstract)

Chen Z, Huang CY, Wu BH et al (2010) Discovery of native aluminum and its possible origin from prospective gas hydrate areas in the South China Sea. Sci China-Earth Sci 53(3):335–344

Deng H, Yan P, Liu H (2004) Seismic characteristics of gas hydrate in the Nansha waters. Mar Geol Q Geol 24(4):89–94 (in Chinese with English abstract)

Elger J, Berndt C, Ruepke L et al (2018) Submarine slope failures due to pipe structure formation. Nat Commun 9:715

Feng D, Qiu JW, Hu Y et al (2018a) Cold seep systems in the South China Sea: an overview. J Asian Earth Sci 168:3–16

Feng JX, Yang SX, Liang JQ et al (2018b) Methane seepage inferred from the porewater geochemistry of shallow sediments in the beikang basin of the southern South China Sea. J Asian Earth Sci 168:77–86

Feng JX, Li N, Liang JQ et al (2021) Using multi-proxy approach to constrain temporal variations of methane flux in methane-rich sediments of the southern South China Sea. Mar Pet Geol 132:105152

Gao HH, Yang XQ, Zhang JP et al (2019) Paleomagnetic records since−80 ka from the southern South China Sea. Chinese J Geophys-Chinese Ed 62(12):4750–4765 (in Chinese with English abstract)

Gee MJR, Uy HS, Warren J et al (2007) The Brunei slide: a giant submarine landslide on the north west borneo margin revealed by 3D seismic data. Mar Geol 246(1):9–23

He Y, Kuang Z, Xu M (2018) Seismic reflection characteristics and triggering mechanism of mass transport deposits of quaternary in Beikang Basin. Geol Sci Technol Inf 37(4):258–268 (in Chinese with English abstract)

Huang W, Meng MM, Zhang W et al (2022) Geological, geophysical, and geochemical characteristics of deep-routed fluid seepage and its indication of gas hydrate occurrence in the beikang basin, Southern South China Sea. Mar Pet Geol 139:105610

Hutchison CS (2004) Marginal basin evolution: the southern South China Sea. Mar Pet Geol 21(9):1129–1148

Ingram GM, Chisholm TJ, Grant CJ et al (2004) Deepwater north west borneo: hydrocarbon accumulation in an active fold and thrust belt. Mar Pet Geol 21(7):879–887

Jørgensen BB, Böttcher ME, Lüschen H et al (2004) Anaerobic methane oxidation and a deep H2S sink generate isotopically heavy sulfides in black sea sediments. Geochim Cosmochim Acta 68(9):2095–2118

Kennett JP, Cannariato KG, Hendy IL et al (2000) Carbon isotopic evidence for methane hydrate instability during quaternary interstadials. Science 288(5463):128–133

Laird AP, Morley CK (2011) Development of gas hydrates in a deep-water anticline based on attribute analysis from three-dimensional seismic data. Geosphere 7:240–259

Lee GH, Watkins JS (1998) Seismic sequence stratigraphy and hydrocarbon potential of the phu khanh basin, offshore central Vietnam, South China Sea. AAPG Bulletin 82(9):1711–1735

Li N, Yang XQ, Peng J et al (2018) Paleo-cold seep activity in the southern South China Sea: evidence from the geochemical and geophysical records of sediments. J Asian Earth Sci 168:106–111

Li N, Yang XQ, Peckmann J et al (2021) Persistent oxygen depletion of bottom waters caused by methane seepage: evidence from the South China Sea. Ore Geol Rev 129:103949

Liu HL, Yao YJ, Deng H (2011) Geological and geophysical conditions for potential natural gas hydrate resources in southern South China Sea waters. J Earth Sci 22(6):718–725

Liu ZH (2005) Distribution of sedimentary basins and petroleum potential in southern South China Sea. Geotect Metall 29(3):410–417 (in Chinese with English Abstract)

Paganoni M, Cartwright JA, Foschi M et al (2016) Structure II gas hydrates found below the bottom-simulating reflector. Geophys Res Lett 43:5696–5706

Paganoni M, Cartwright JA, Foschi M et al (2018) Relationship between fluid-escape pipes and hydrate distribution in offshore Sabah (NW Borneo). Mar Geol 395:82–103

Peckmann J, Thiel V (2004) Carbon cycling at ancient methane-seeps. Chem Geol 205(3–4):443–467

Rehder G, Suess E (2001) Methane and pCO2 in the kuroshio and the South China Sea during maximum summer surface temperatures. Mar Chem 75(1–2):89–108

Ruppel CD, Kessler JD (2017) The interaction of climate change and methane hydrates. Rev Geophys 55(1):126–168

Traynor JJ, Sladen C (1997) Seepage in vietnam—onshore and offshore examples. Mar Pet Geol 14(4):345–362

Warren JK, Cheung A, Cartwright I (2010) Organic geochemical, isotopic, and seismic indicators of fluid flow in pressurized growth anticlines and mud volcanoes in modern deep-water slope and rise sediments of offshore brunei darussalam: implications for hydrocarbon exploration in other mud-and salt-diapir provinces. In: Wood L (ed.), Shale Tectonics. AAPG Memoir Vol. 93, pp. 163–196

Wetzel A (2013) Formation of methane-related authigenic carbonates within the bioturbated zone—an example from the upwelling area off Vietnam. Palaeogeogr Palaeocl 386:23–33

Yan W, Zhang G, Zhang L et al (2020) Focused fluid flow systems discovered from seismic data at the southern margin of the South China Sea. Interpretation 8(3):555–567

Zhang K, Guan Y, Song H et al (2020) A preliminary study on morphology and genesis of giant and mega pockmarks near andu seamount, nansha region (South China Sea). Mar Geophys Res 41:1–12

Zhang W, Chen C, Su P et al (2023) Formation and implication of cold-seep carbonates in the southern South China Sea. J Asian Earth Sci 241:105485

Zhou Y, Di PF, Li N et al (2020) Unique authigenic mineral assemblages and planktonic foraminifera reveal dynamic cold seepage in the southern South China Sea. Minerals 10(3):1–13

Chapter 14
In Situ Detection and Seafloor Observation of the Site F Cold Seep

Xin Zhang, Zhendong Luan, and Zengfeng Du

Abstract The in situ detection and seafloor observation of the Site F cold seep began after its discovery. Research on deep−sea cold seep systems often begins with descriptions of topography and geomorphology. The earliest platform for topographic and geomorphologic exploration was the scientific expedition vessel. With the development of underwater vehicles, autonomous underwater vehicles (AUVs) and remote operated vehicles (ROVs) have become platforms for geophysical exploration of the seafloor. Thus, the spatial resolution of exploration has also been enhanced to the centimeter level. At the same time, sampling and in situ detection technology have gradually become the main research methods for cold seep systems. Based on the obtained samples and in situ data, research on the geochemistry and bioecology of cold seep systems has been carried out. Many technologies have been developed and may be used to promote the limit of detection of spectral−based methods to broaden the application range. Long−term detection for in situ experiments with specific scientific targets under natural cold seep environments is another trend for detection and observation in cold seep areas.

X. Zhang (✉) · Z. Luan · Z. Du
Key Lab of Marine Geology and Environment and Center of Deep Sea Research, Institute of Oceanology, Center for Ocean Mega-Science, Chinese Academy of Sciences, Qingdao 266071, PR China
e-mail: xzhang@qdio.ac.cn

Z. Luan
e-mail: luan@qdio.ac.cn

Z. Du
e-mail: duzengfeng@qdio.ac.cn

X. Zhang
Laboratory for Marine Geology, Pilot Laboratory for Marine Science and Technology (Qingdao), Qingdao 266061, China

University of Chinese Academy of Sciences, Beijing 101408, China

D. Chen and D. Feng (eds.), *South China Sea Seeps*,
https://doi.org/10.1007/978-981-99-1494-4_14

14.1 Introduction

Cold seeps have been discovered since the 1980s (Paull et al. 1984) and have become a research focus because they play a significant role in the geochemical cycle as a mass and energy exchange path between the lithosphere, biosphere, and hydrosphere. Investigating the cold seep system is significantly beneficial to monitoring the marine environment and the element cycle in the deep ocean. Recovering and transporting fluid samples back to the laboratory for geochemical analysis has been a commonly used method for studying cold seep systems in recent decades (Du et al. 2020). However, the compositions in cold seep fluids, which are sensitive to ambient conditions (such as P, T, pH, and Eh), may decompose during the sample acquisition and recovery process (Lilley et al. 2003). In situ detection has become a growing trend for studying cold seep systems. Here, we briefly review the application of the in situ detection and seafloor observation of the Site F cold seep of the South China Sea.

14.2 Fine Topographic and Geomorphologic Detection

The Site F active cold seep is located southwest of Taiwan (Fig. 14.1). It is part of the passive continental margin and the leading edge of the southwestern Taiwan deformation zone (Liu et al. 2010). The ridge line is NW–SE−trending, the water depth at the top of the ridge is approximately 1120 m, the longitudinal length along the ridge line is approximately 30 km, and the widest transverse part of the vertical ridge line is approximately 5 km. The topographic contours on the ridge are closed, and the slopes of the two wings are steep, with a maximum of more than 30° (Berndt et al. 2014). Typical seabed features of cold seep systems, such as large−scale exposed authigenic carbonate rocks, reduced sediments, chemosynthetic communities, and gas flares, have also been found (Lin et al. 2007).

According to the submarine topographic map formed by shipborne multibeam bathymetric data and the topographic profile close to the parallel ridge, Formosa Ridge is surrounded by submarine canyons on the N, W and E sides, making it a nearly independent positive topography. Formosa Ridge has two ridges, the north and the south ridges, with an interval of approximately 2 km. The water depths of the north and south ridges are approximately 1100 m and 1125 m, respectively. The seafloor topography tilts from north to south. On the topographic profile close to the vertical ridge line, the ridges and canyons are arranged parallel and alternately in the direction of NNW–SSE, and the combination forms a "W" topographic feature. The height differences between the top of the ridge and the bottom of the surrounding canyon are 300−850 m, from NW to SE, and the height difference tends to increase at first and then decrease. On the same section of the vertical ridge line, the height difference in the southwest is greater than that in the northeast, and the maximum height difference appears between the top of the south ridge and the bottom of the southwest canyon, which is approximately 850 m. The width of the ridge is basically

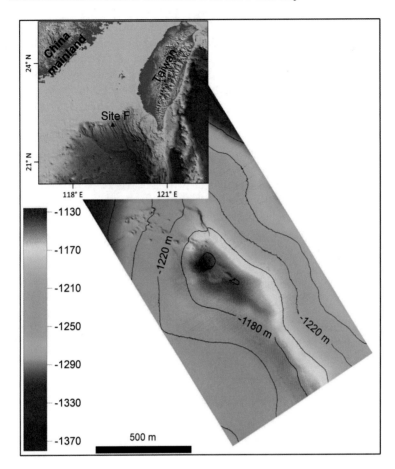

Fig. 14.1 The location of the Site F active cold seep

the same to the north of the top of the ridge in the south, i.e., approximately 5 km (Fig. 14.1), while in the south, it tends to become larger and then smaller, and the widest point is approximately 7 km. The slopes of the two wings of the ridge are similar and steep, the overall slope angle is more than 10°, and the slope of the southwest wing is larger than that of the northeast wing. Along the ridge from NW to SE, the slope of the two wings increases gradually, and the slope of the southwest wing reaches 34° in the 2D section. Sporadic secondary canyons are present on both wings of the ridge and are mainly concentrated on the west side of the north ridge top and the east side of the south ridge top. The secondary canyon converges into the submarine canyon at the foot of the two slopes, and there is no obvious submarine fan accumulation at the mouth of the canyon. Researchers speculate that the sediment transported by these secondary canyons is carried away by the turbidity current or undercurrent flowing through the main submarine canyon.

Wang et al. (2021) systematically identified and analyzed the seabed features of the cold seep area of Site F through the massive high−definition imaging data obtained by ROV geophysical data and a 3D laser scanning system. The results show that the seabed features of this cold seep area mainly include 6 types: cold seep vents with gas flares, carbonate rocks of different shapes, dense benthic chemosynthetic communities, mussel beds with low population densities, shells and/or shell debris areas, and reduced sediments. All the abovementioned seabed features are distributed in an approximately zoned pattern with gas plumes and benthic communities in the innermost part of Site F and reduced sediments in the outermost part (Wang et al. 2021).

A comprehensive analysis reveals that two main sets of dominant fluid trans-port channel systems have developed in the authigenic carbonate mound at Site F, including the large−scale fluid channel under the flourishing biological area in the south and the transport system composed of multiple dominant channels in the luxu-riant biological region in the north (as shown in Fig. 14.2). This discovery reflects the interaction between the cold seep fluid transport channel and the sediment type: the location of the dominant cold seep fluid transport channel and the intensity of fluid activity control the biological distribution characteristics on the authigenic carbonate mound; the development of chemosynthetic communities reacts to the activity of seepages; and authigenic carbonate rocks can converge longitudinally and channel laterally to cold seep fluids.

14.2.1 In Situ Detection and Findings

Since the discovery of cold seep systems on the seafloor in the 1980s, scientists have gradually realized that cold seep systems connect the oceanic lithosphere, marine hydrosphere and benthic biosphere. These systems are transport pathways for deep−sea matter and energy. The phenomenon of deep−sea seepage is of great signif-icance to the reaction mechanism of biogeochemical processes and the interaction of different deep−sea layers.

The components of the fluids erupting from the cold seep were studied by sampling and then transported to the laboratory for analysis. However, changes in the temper-ature, pressure and redox conditions during the sampling process led to the escape or chemical reaction of the components in the fluid, resulting in the failure of the laboratory analysis results to obtain the real in situ concentration of the components in the fluid.

With the development of Raman spectroscopy, it has been gradually used for in situ detection in the deep ocean. The Monterey Bay Aquarium Research Institute (MBARI) claimed the lead in applying laser Raman spectroscopy to deep sea in situ detection and developed the first international deep ocean Raman in situ spectrom-eter (DORISS), which has been updated to DORISS II (Brewer et al. 2004; Zhang et al. 2012). The Ocean University of China (OUC) has developed a deep ocean Raman spectrometer with two excitation lasers (Du et al. 2015). The Institute of

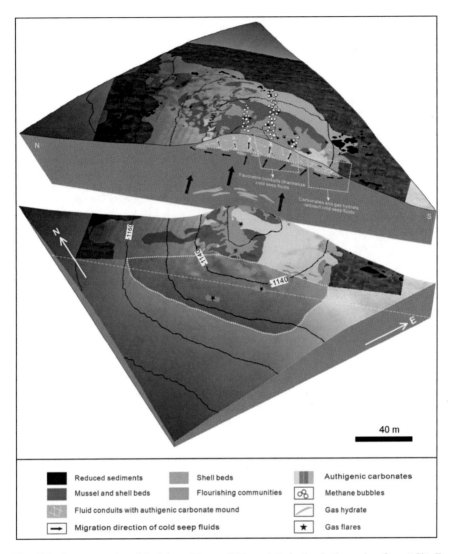

Fig. 14.2 A conceptual model of the cold seep fluid conduits in the shallow subsurface at Site F based on the distribution patterns of the seabed features

Oceanology, Chinese Academy of Sciences (IOCAS) has developed a deep ocean Raman spectrometer with serial insertion probes (Zhang et al. 2017b), and the Dalian Institute of Chemical Physics, Chinese Academy of Sciences (DICP) has developed a deep ocean Raman spectrometer with a ultraviolet laser. The Technical University of Berlin (TUB) has developed Raman spectrometers for field applications based on SERS technology (Schmidt et al. 2004). Depending on the system performance, the

Raman spectrometers mentioned above have been used to detect different kinds of targets from shallow water to the deep sea.

14.2.2 In Situ Detection of Fluids in the Cold Seep Area

For most cold seep systems, the cold seep fluid is mainly composed of seawater and CH_4 and may include liquefied fine−grained sediments and other components. There are two main sources of CH_4 components in the fluid: biogenic gas and thermal degradation gas (Stakes et al. 1999; Judd and Hovland 2009). The former is the gas formed by the degradation of organic matter in shallow sediments with the participation of microorganisms, while the latter is the gas formed by thermal degradation of kerogen in deep strata under the action of microorganisms. Of course, the methane component in the fluid may also be a mixed product of the two genetic gases. For these two genetic gases, their CH_4 content and carbon isotope characteristics are obviously different, so they can be used as an important index to assess the source of methane. The study of cold seep systems often begins with the analysis of various indicative components in fluids.

The RiP was used for in situ detection of the cold seep fluids at Site F in the South China Sea. By detecting the spectra of sediment pore water at different depths (as shown in Fig. 14.3), the concentration profiles of methane, sulfate and hydrogen sulfide in sediment pore water were obtained. A total of 22 profiles were acquired during past scientific expeditions. The results show that the sulfate−methane interface is shallow in the reduced sediment area, with sufficient methane supply, while the sulfate−methane interface is deeper in the common sediment area.

The cold seep fluids erupted from the active vent at Site F, and the fluids in the chemosynthetic communities near the active cold seep vent were also detected in situ using the RiP system (Du et al. 2018a). The in situ Raman spectra revealed that the presence of CH_4 and C_3H_8 in the fluids erupted from the cold seep vent, consistent with the results of gas chromatography analysis. Finding C_3H_8 provides a new possible explanation for the carbon source of this cold seep area. The in situ Raman spectra of the fluids at different depths of the chemosynthetic communities show that the concentration of SO_4^{2-} decreases with increasing depth, whereas the concentrations of CH_4 and S_8 increase in fluids in the chemosynthetic communities but without H_2S (Fig. 14.4). These results indicate that CH_4 has been oxidized by SO_4^{2-} and that S_8 is formed. This process usually occurs in reduced sediments as anaerobic oxidation of methane. Overall, the findings provide new insight into the geochemical analysis of cold seep vent fluids and in situ proof of the oxidation of methane in chemosynthetic communities near cold seeps.

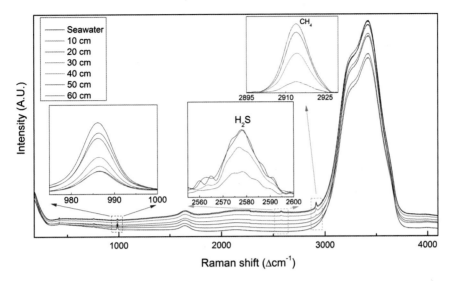

Fig. 14.3 The in situ Raman spectra of the sediment pore water at different depths

14.2.3   In Situ Detection of the Solid Targets

14.2.3.1   Detection of Authigenic Carbonates

Raman spectroscopy shows great advantages in the in situ detection of deep−sea fluids, gases and gas hydrates and authigenic carbonates because the formation and evolution of authigenic carbonates are closely related to cold seep fluids. In situ detection of authigenic carbonate rocks in cold seep areas is helpful to accurately understand the formation environment of authigenic carbonate rocks and the inter- action between fluids and rocks. The application of laser Raman spectroscopy to the in situ detection of deep−sea rocks can reveal preliminary information, such as the rock mineral composition and relative content, which makes our sampling work directional and purposeful. Previous studies on Raman spectra of carbonate minerals have shown that carbonate minerals are very sensitive to Raman scattering (Wang et al. 1995). White (2009) effectively distinguished the difference between calcite and aragonite in a deep−sea extreme environment (White 2009) based on Raman spectroscopy. The Raman spectra of biogenic carbonate minerals, such as calcite, dolomite and aragonite, have been systematically studied, laying the foundation for qualitative identification (Herman et al. 1987; Urmos et al. 1991; Buzgar and Apopei 2009). Raman spectroscopy is also used to study the mechanism of carbonate forma- tion (Bonales et al. 2013). The Raman shift of carbonate can indicate the difference of the magnesium content in magnesium calcite (Borromeo et al. 2017). The above studies show that Raman spectroscopy can effectively reflect the carbonate minerals in authigenic carbonate rocks, which lays a foundation for deep−sea in situ explo- ration of authigenic carbonate rocks. The Raman shift of the main carbonate peak of

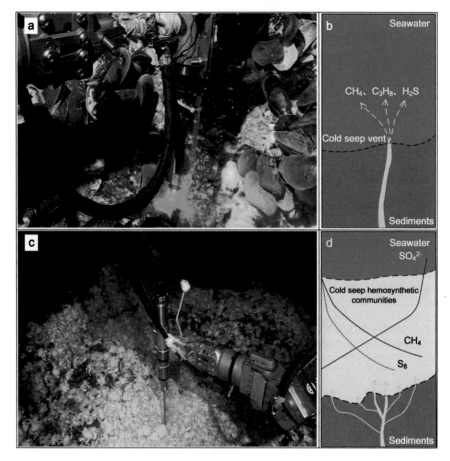

Fig. 14.4 The in situ Raman quantitative detection and results of cold seep vents and fluids in chemosynthetic communities near an active cold seep vent in the South China Sea

Site F is $1084-1087$ cm^{-1}, which is three wavenumbers less than that of Site P (f the Pear River Mouth basin) (Fig. 14.5). The results show that there is a significant component difference between them, indicating that the difference in the element composition affects the Raman shift.

In situ Raman detection of the authigenic carbonates and fluids at different locations with different fauna has been accomplished simultaneously. The in situ Raman quantitative analyses of the fluids in the dense chemosynthetic communities (total of 30 cm) show that the concentrations of Cl$^-$ and SO$_4^{2-}$ decreased while CH$_4$ increased with increasing depth (Fig. 14.6). On the other hand, the minerals at the different positions with different faunal densities also show significant differences. The in situ Raman spectra of the authigenic carbonates show that the full width at half maximum (FWHM) of the Raman peak of aragonite increased from the fauna−rich

Fig. 14.5 The in situ Raman spectra of the authigenic carbonate acquired at the Formosa Ridge (Site F) and the eastern region of the Pear River Mouth basin (Site P)

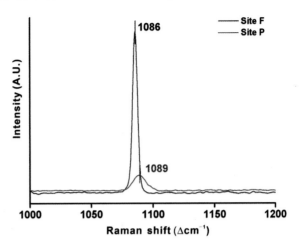

area to the desert edge area, suggesting that the degree of destruction of the aragonite increased due to the effects of the surrounding fluids.

The changes in the components suggested that complex biochemical processes occurred in the communities (Du et al. 2018b). Xi et al. proposed that one of the reactions may be the sulfate−driven anaerobic oxidation of methane (Xi et al. 2020). The reactions also contribute to the formation of authigenic carbonates and the regular aragonite structure in the shallow seafloor (Luff and Wallmann 2003):

$$CH_4 + SO_4^{2-} \rightarrow HCO_3^- + HS^- + H_2O \tag{14.1}$$

$$2HCO_3^- + Ca^{2+} \rightarrow CaCO_3 + CO_2 + H_2O \tag{14.2}$$

Furthermore, compared with the reducing sediments in the seafloor, the high fluid perturbance in the chemosynthetic communities increased the dissolved oxygen content in the communities, which contributed to the oxidation of methane:

$$CH_4 + 2O_2 \rightarrow CO_2 + 2H_2O \tag{14.3}$$

With the upward diffusion of the methane−rich fluid in the communities, the methane content showed a decreasing trend due to reactions (14.1) and (14.3). As an important reactant, SO_4^{2-} is efficiently reduced in reaction (14.1) due to the addition of upward methane. Moreover, as the product of the above three reactions, the addition of H_2O efficiently lowered the salinity of the fluid in the chemosynthetic communities, especially at the bottom of the communities. The fluid with low Cl^- and SO_4^{2-} can inhibit the corrosion of authigenic carbonates to a great extent under long−term erosion (Xi et al. 2020).

Based on common features of the fluids and authigenic carbonates covered by chemosynthetic communities with different densities and the high−resolution videos

Fig. 14.6 The in situ Raman quantitative analyses of the fluids from different locations. **a** Quantitative analysis of SO_4^{2-} concentration in the fluid at different locations. The SO_4^{2-} concentrations in seawater at different locations ranged from 28.5 to 30 mM. **b** Quantitative analysis of CH_4 concentration in the fluid at different locations

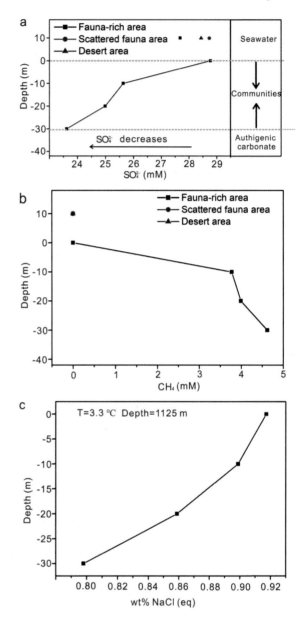

of the cold seep system, Xi et al. (2020) proposed a detailed model of the system at Site F to study the evolution of a seepage system by the differences in the spatial distribution characteristics of the authigenic carbonate and overlying chemosynthetic communities.

In addition, combined with the advantages of quantitative analysis of Raman spectroscopy, the compositional changes of fluids and the Raman spectral characteristics of authigenic carbonate minerals in the corresponding fluids are quantitatively analyzed. This analysis not only provides effective data and a new understanding for the simultaneous analysis of the interaction between cold seep fluid and authigenic carbonate rocks but also expands the scope of laser Raman spectroscopy in deep−sea in situ detection.

14.2.3.2 Detection of Gas Hydrates

The properties of gas hydrate have been studied in the laboratory by simulations in past decades. The structures of the methane hydrate are transformed with changes in the pressure. The methane gas hydrate in natural fluid inclusions has been used to identify the formation conditions in fluid inclusions. With the development of submarine vehicles and in situ detection technologies, the deep sea has become a natural testing ground for studying the physical and chemical properties of natural gas hydrate (NGH). Raman spectroscopy has been increasingly applied for the in situ detection and analysis of natural gas hydrate (NGH). Zhang et al. (2017a) conducted the first in situ detection of gas hydrates exposed on the seafloor at Site F in the South China Sea using the Raman insertion probe for gas hydrate (Fig. 14.7). The results reveal that the hydrates are primarily Structure I hydrates but contain significant amounts of C_3H_8 and H_2S, indicating that they may have originated from thermogenic sources. Free gas can exist in hydrate fabric during very rapid hydrate formation, and a new growth model for the early stages of crystalline gas hydrate formation was proposed based on the in situ data. Du et al. (2018b) conducted in situ detection of synthetic gas hydrate (SGH) that formed quickly by cold seep fluids erupting from active cold seeps at Formosa Ridge in the South China Sea. Authigenic carbonate debris, which is found in the in situ Raman spectra of SGH samples for the first time, may be one factor that can promote the formation of synthetic gas hydrate. The gas−water interface also contributes to the rapid formation of SGH samples. Time series Raman spectra of the four SGH samples demonstrate that methane large−to−small cage occupancy ratios of the hydrates vary from 1.01 to 1.39, and the ratios of methane large−to−small gas hydrates increase from 0.53 to 1.55. The results indicate that gas hydrate is continuously formed by the occluded gaseous CH_4 after quick formation. The Raman spectra of the SGH sample, which was surprisingly found and recovered after 410 days on the seafloor, reveal that a methane−rich environment is a necessary condition for the stable existence of synthetic gas hydrate (Fig. 14.8).

14.2.3.3 Detection of Biological Samples

With the aid of micro−techniques, Raman spectroscopy has been used for the in situ detection of different bacteria, which is the method of direct detection in the natural environment via an environment simulator without sample preparation (Kirschner

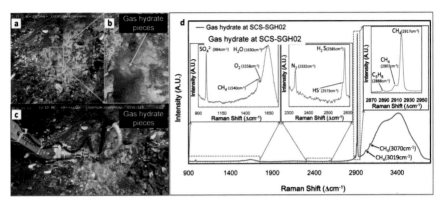

Fig. 14.7 The in situ optical and Raman observations of the gas hydrates exposed on the seafloor at the SCS−SGH02 site. **a** The exposed hydrate inside the channel of a cold seep vent surrounded by lush chemosynthetic communities. **b** A close−up view of the exposed hydrate attached to the inner wall of the channel. **c** The release of small pieces of gas hydrate from the vent during the Raman observations performed by the RiP−Gh probe, which is held by the arm of the ROV *Faxian*. **d** In situ Raman spectrum of the natural gas hydrate from 900 to 3,800 cm^{-1} at the SCS−SGH02 site

et al. 2001). The distribution pattern of bacterial aggregates over a relatively large area can also be obtained by confocal Raman microscopy (CRM) because this technology has been established. With the advantages of high lateral and axial resolution, CRM has been regarded as a powerful tool to acquire noninvasive and in situ information on microbial communities (Pätzold et al. 2006). The microbial distribution of nitrifiers and anammox bacteria are scanned and mapped in situ based on the resonance Raman effect of cytochrome c (Cyt c), which may be a promising indicator for the metabolic condition of the bacteria (Pätzold et al. 2006). CRM is used to confirm the zero−valent sulfur compounds (S^0) formed during anaerobic oxidation of methane (AOM) through a new pathway for dissimilatory sulfate reduction by methanotrophic archaea (Pasteris and Wopenka 2004; Milucka et al. 2012). CRM is also deployed to classify mussel foot (Liang et al. 2020) and mussel shells from cold seep areas (Cui et al. 2020). The application of Raman spectroscopy for biological samples provides new insight for real−time investigation of the life process and strategies of adaptation to an extreme environment of deep−sea macrofaunals. The symbiotic mechanism between metazoans and chemoautotrophs is an essential factor in maintaining the stability and sustainability of life activities in deep−sea ecosystems, and it has become an important issue in deep−sea research. However, there is no in situ detection method for deep−sea extracellular metabolism. Because of the influence of the Raman detection limit, it is difficult to detect deep−sea microbial metabolites or intermediates at low concentrations. Surface−enhanced Raman scattering (SERS) has been applied to enhance the Raman detection limit. A SERS substrate of coccinella septempunctata−like silver nanospheres coated with silver nanoisland structures (Ag NI@Ag NSs) was developed by a simple annealing process

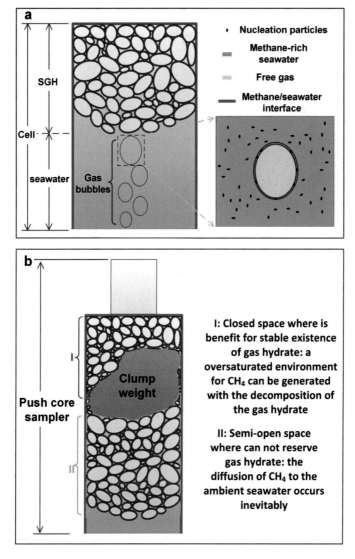

Fig. 14.8 **a** The proposed model for the quick formation of SGH; **b** a schematic diagram of the preliminary model for the stable existence of gas hydrate

(Fig. 14.9), and the detection of a low concentration of the archaeal biomarker PETA was realized under a simulated deep−sea pressure of 1100 m (11 MPa) (Wang et al. 2022).

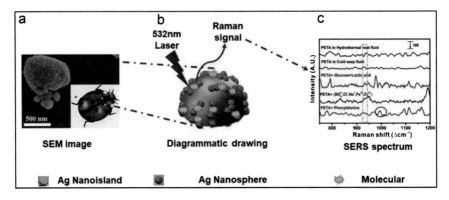

Fig. 14.9 **a** SEM diagram of the SERS base (the morphology was similar to the pattern of coccinella septempunctata in the illustration); **b** enhancement principle of the SERS base; **c** Raman spectra obtained from SERS detection in complex environments

14.3   In Situ Experiment and Long−Term Detection Using Mulit−RiPs

Fluid seepages in deep sea cold seep areas link the lithosphere, the hydrosphere and the seafloor biosphere. The strong inhomogeneity of cold seep systems in terms of temporal and spatial scales requires in situ, long−term and multitarget detection. Many sensors, including deep ocean Raman spectrometers, have been developed and used for in situ detection in cold seep areas. However, the limitation of detecting a single target presents a large challenge to obtaining a more comprehensive understanding of cold seep systems.

A novel long−term ocean observation platform (LOOP) and multichannel Raman insertion probes (Multi−RiPs) were developed and deployed at Site F of the SCS during 2020, 2021 and 2022 (Fig. 14.10). Using the deep−sea cold seep area as a natural test site, the designed in situ experiment was conducted continuously for a long time (Du et al. 2022). A new steerable mode for launching and recovery with the aid of the research vessel and submarine vehicles was used during the deployments of the LOOP at Site F in the SCS. The LOOP can be operated in an online real−time control mode, allowing landing site selection and adjustment of observation parameters during the launching process, with a subsequent switch to an offline stand−alone operation mode for long−term, continuous observation (Du et al. 2023). Thus, in situ experiments can be conducted with the aid of ROV and detected continuously in situ by Multi−RiPs.

A calibration switching module (CSM) is used to connect the light path between the laser, the Raman insertion probe and the spectrometer (Fig. 14.11). Thus, Raman probes can share core components (the laser, spectrometer, etc.) that are sequentially controlled by the CSM and detect the specified targets in situ according to the preprogrammed schedule. Combined with the developed series of Raman insertion

Fig. 14.10 The Mulit−RiPs was deployed by LOOP in three consecutive years (**a**: 2020; **b**: 2021; **c**: 2022) at Site F for the long−term detection of deep−sea in situ experiments

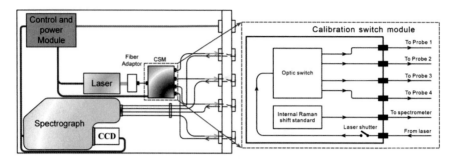

Fig. 14.11 The layout diagram of key optical components and the optical switching module of the Multi−RiPs (modified from Du et al. 2022)

probes, the Multi−RiPs can simultaneously detect multiple targets with different phase states, such as fluid, solid and gas, in deep−sea cold seep systems.

By analyzing the continuous long−term spectra obtained by two Raman probes placed near the cold seep vents at different depths in the chemosynthetic communities, it was found that the most significant difference at different depths in the chemosynthetic communities was the difference in elemental sulfur: elemental sulfur was not detected at the middle depth in the chemosynthetic communities, but more elemental sulfur was found at the bottom of the chemosynthetic communities.

The continuous long−term Raman spectra obtained from the in situ synthesized gas hydrate show that after the quick formation of gas hydrate, enough gaseous methane with a closed honeycomb structure causes the hydrate formation process to continue dominating the dynamic process of the hydrate system. At the same time, it is also confirmed that the stable existence of natural gas hydrate requires a methane−rich environment in addition to the stable pressure and temperature field of natural gas hydrate.

Gas flows are probably transient events. At certain times, gas flows through the sediments and eventually into the bottom water. Along the way, some gas forms hydrates along fractures and in weakening horizontal zones within the sediments. While the gas flow is episodic, the hydrates act as dynamic "capacitors" from which methane is constantly diffusing into the direction of the methane−depleted bottom water, thereby creating stable chemical gradients for microbes carrying out anaerobic oxidation of methane and, subsequently, the occurrence of chemosynthetic organisms. By placing a Raman cold−seep eruption fluid collection device attached above the cold−seep vent, natural gas hydrate, or free−form methane gas, can be captured by the collection device, and when the cold seep is not erupting, the seawater is captured by the device. The time series Raman spectra confirmed that the cold seeps are intermittently erupting, which is consistent with previous studies. The results also indicate that the gas hydrate at shallow sediment depths and exposed on the seafloor acting as a capacitor dominate the biogeochemical processes in the cold seep area.

14.4   Summary and Perspectives

The substance brought by seepages from the deep earth has a complicated composition, and the concentrations of the targets in the fluid plume even a few centimeters away from the cold seep vent decrease sharply. They have become the two barriers that limit laser spectroscopy from having broader applications in the in situ detection of seepages in the deep ocean. One limitation is the low detection sensitivity of laser spectroscopy to the targets. To improve the detection sensitivity of conventional Raman spectroscopy, Raman signal enhancement methods based on multiple reflection, liquid core optical fiber, resonance enhancement and surface enhancement have been proposed and used.

The other limitation of laser spectroscopy is that one kind of spectroscopy may be sensitive to certain types of targets (such as molecules for laser Raman spectroscopy (LRS) and atoms or elements for laser−induced breakdown spectroscopy (LIBS)). Therefore, sufficient and effective information cannot be obtained by a single measurement. To solve the problem of insufficient detection ability, combining different kinds of in situ detection technologies to achieve effective detection for different types of targets (such as acid ions and metal cations) has been proposed in past decades.

To further expand the application field of laser spectroscopy for in situ detection, reliable LRS and LIBS signal enhancement methods suitable for deep sea conditions and quantitative analysis methods with high accuracy for underwater LIBS need to be developed. Underwater multispectral in situ integrated detection systems have become a tendency for in situ detection.

Another trend for more applications of in situ laser spectroscopy is the long−time and real−time detection deployed in submarine observation networks or underwater observatory landers for seafloor ecosystems and geological systems. These two directions are also the main development trends of laser spectroscopy in the field of deep−sea in situ detection.

Acknowledgments Sincere thanks to the crews and scientists on board R/V Kexue and the pilots of the ROV for their full support and invaluable help during work at sea. This work was supported by the National Natural Science Foundation of China (92058206, 52001303, 41822604), the Strategic Priority Research Program, CAS (XDB42040302, XDA22050102), the Senior User Project of RV *KEXUE* (KEXUE2019GZ06), Key Deployment Project of Centre for Ocean Mega−Research of Science, CAS (COMS2020J03), the Young Taishan Scholars Program (tsqn201909158).

References

Berndt C, Crutchley G, Klaucke I et al (2014) Geological controls on the gas hydrate system of formosa ridge, south china sea. In: OCEANS 2014−TAIPEI. IEEE pp 1–4
Bonales LJ, Muñoz−Iglesias V, Santamaría−Pérez D et al (2013) Quantitative raman spectroscopy as a tool to study the kinetics and formation mechanism of carbonates. Spectroc Acta Pt A−Molec Biomolec Spectr 116:26–30

Borromeo L, Zimmermann U, Andò S et al (2017) Raman spectroscopy as a tool for magnesium estimation in Mg−calcite. J Raman Spectrosc 48(7):983–992

Brewer PG, Malby G, Pasteris JD et al (2004) Development of a laser raman spectrometer for deep−ocean science. Deep−Sea Res Part I−Oceanogr Res Pap 51(5):739–753

Buzgar N, Apopei AI (2009) The raman study of certain carbonates. Geol Tomul L 2:97–112

Cui NN, Du ZF, Zhang X et al (2020) The application of confocal Raman spectroscopy in mussels shell. Spectrosc Spectr Anal 40(3):750–754

Du Z, Li Y, Chen J et al (2015) Feasibility investigation on deep ocean compact autonomous raman spectrometer developed for in−situ detection of acid radical ions. Chin J Oceanol Limnol 33(2):545–550

Du Z, Zhang X, Luan Z et al (2018a) In situ Raman quantitative detection of the cold seep vents and fluids in the chemosynthetic communities in the South China Sea. Geochem Geophys Geosyst 19(7):2049–2061

Du Z, Zhang X, Xi S et al (2018b) In situ raman spectroscopy study of synthetic gas hydrate formed by cold seep flow in the south china sea. J Asian Earth Sci 168:197–206

Du Z, Zhang X, Xue B et al (2020) The applications of the in situ laser spectroscopy to the deep−sea cold seep and hydrothermal vent system. Solid Earth Sci 5(3):153–168

Du Z, Xi S, Luan Z et al (2022) Development and deployment of lander−based multi−channel raman spectroscopy for in−situ long−term experiments in extreme deep−sea environment. Deep−Sea Res Part I−Oceanogr Res Pap 190:103890

Du Z, Zhang X, Lian C et al (2023) The development and applications of a controllable lander for in-situ long-term observation of deep sea chemosynthetic communities. Deep-Sea Res Part I-Oceanogr Res Pap 193:103960

Herman RG, Bogdan CE et al (1987) Discrimination among carbonate minerals by raman spectroscopy using the laser microprobe. Appl Spectrosc 41(3):437–440

Judd AG, Hovland M (2009) Seabed fluid flow: the impact on geology. Cambridge University Press, Biology and the Marine Environment

Kirschner C, Maquelin K, Pina P et al (2001) Classification and identification of enterococci: a comparative phenotypic, genotypic, and vibrational spectroscopic study. J Clin Microbiol 39(5):1763–1770

Liang ZW, Du ZF, Li CL et al (2020) Confocal raman micro−spectroscopy analysis of mussel foot. Spectrosc Spectr Anal 40(3):755–759

Lilley MD, Butterfield DA, Lupton JE et al (2003) Magmatic events can produce rapid changes in hydrothermal vent chemistry. Nature 422(6934):878–881

Lin S, Lim Y, Liu C−S et al (2007) Formosa ridge, a cold seep with densely populated chemosynthetic community in the passive margin, southwest of taiwan. Geochim Cosmochim Acta 71(15):A582

Liu C, Hsu H, Morita S et al (2010) Seismic imaging of a cold seep site offshore southwestern Taiwan. In AGU Fall Meeting Abstracts, Vol. 2010, pp. OS44A−06

Luff R, Wallmann K (2003) Fluid flow, methane fluxes, carbonate precipitation and biogeochemical turnover in gas hydrate−bearing sediments at hydrate ridge, cascadia margin: numerical modeling and mass balances. Geochim Cosmochim Acta 67(18):3403–3421

Milucka J, Ferdelman TG, Polerecky L et al (2012) Zero−valent sulphur is a key intermediate in marine methane oxidation. Nature 491(7425):541–546

Pasteris J, Wopenka B, Freeman JJ et al (2004) Raman spectroscopy in the deep ocean: successes and challenges. Appl Spectrosc 58(7):195A-208A

Pätzold R, Keuntje M, Anders−Von Ahlften A (2006) A new approach to non−destructive analysis of biofilms by confocal raman microscopy. Anal Bioanal Chem 386(2):286–292

Paull CK, Hecker B, Commeau R et al (1984) Biological communities at the florida escarpment resemble hydrothermal vent taxa. Science 226(4677):965–967

Schmidt H, Bich Ha N, Pfannkuche J et al (2004) Detection of PAHs in seawater using surface−enhanced raman scattering (SERS). Mar Pollut Bull 49(3):229–234

Stakes DS, Orange D, Paduan JB et al (1999) Cold−seeps and authigenic carbonate formation in monterey bay, California. Mar Geol 159(1–4):93–109

Urmos J, Sharma SK, Mackenzie FT (1991) Characterization of some biogenic carbonates with raman spectroscopy. Am Miner 76(3–4):641–646

Wang A, Jolliff BL, Haskin LA (1995) Raman spectroscopy as a method for mineral identification on lunar robotic exploration missions. J Geophys Res−Planets 100(E10):21189–21199

Wang B, Du Z, Luan Z et al (2021) Seabed features associated with cold seep activity at the formosa ridge, south china sea: integrated application of high−resolution acoustic data and photomosaic images. Deep−Sea Res Part I−Oceanogr Res Pap 177:103622

Wang SY, Xi SC, Pan RH et al (2022) One−step method to prepare coccinellaseptempunctate−like silver nanoparticles for high sensitivity SERS detection. Surf Interfaces 35:102440

White SN (2009) Laser raman spectroscopy as a technique for identification of seafloor hydrothermal and cold seep minerals. Chem Geol 259(3–4):240–252

Xi S, Zhang X, Luan Z et al (2020) Biogeochemical implications of chemosynthetic communities on the evolution of authigenic carbonates. Deep−Sea Res Part I−Oceanogr Res Pap 162:103305

Zhang X, Kirkwood WJ, Walz PM et al (2012) A review of advances in deep−ocean raman spectroscopy. Appl Spectrosc 66(3):237–249

Zhang X, Du Z, Luan Z et al (2017) In situ raman detection of gas hydrates exposed on the seafloor of the South China Sea. Geochem Geophys Geosyst 18(10):3700–3713

Zhang X, Du Z, Zheng R et al (2017) Development of a new deep−sea hybrid raman insertion probe and its application to the geochemistry of hydrothermal vent and cold seep fluids. Deep−Sea Res Part I−Oceanogr Res Pap 123:1–12

Printed in the United States
by Baker & Taylor Publisher Services